Ramsar Wetlands of the North American West Coast and Central Pacific

This is the first comprehensive international atlas featuring all ecological services provided by Ramsar wetlands, with complete views of all Ramsar sites, through remote sensing and mapping. Written by an international expert on wetlands and remote sensing, this atlas is for a broad audience and compiles much-needed information on how the Ramsar wetlands are of significant value to the planet and society and can and should be managed in such a way that supports planetary sustainability. Focused on the 72 designated Ramsar sites along the western coasts of Alaska, Canada, California, Mexico, and the Central Pacific islands, each wetland is articulately documented with respect to its specific ecological functions and services.

FEATURES

- Provides a comprehensive assessment of the key biophysical and societal elements of each Ramsar-designated wetland along the North American West Coast and Central Pacific
- Brings all designated Ramsar wetlands to the reader in one visually appealing compendium using geospatial technology
- Aids in highlighting the importance of and options for wetland conservation and restoration worldwide
- Explains the important role that wetlands play in environmental sustainability, directly supporting the global sustainable development goals of the United Nations
- Introduces the contributions of the Ramsar Convention on Wetlands to global conservation and restoration

This atlas is intended for wetland managers and policymakers involved in the Ramsar Convention activities and for wetland ecologists and other allied environmental scientists and practitioners, such as hydrologists, microbiologists, and botanists. It is also a valuable resource for researchers, faculty, and graduate students affiliated with programs such as wetland ecology, wetland management, environmental studies, environmental management, and survey of wetlands.

Ramsar Wetlands of the North American West Coast and Central Pacific
An Atlas

Ricardo D. Lopez

CRC Press
Taylor & Francis Group
Boca Raton London New York

CRC Press is an imprint of the
Taylor & Francis Group, an **informa** business

First edition published 2024
by CRC Press
2385 NW Executive Center Drive, Suite 320, Boca Raton FL 33431

and by CRC Press
4 Park Square, Milton Park, Abingdon, Oxon, OX14 4RN

CRC Press is an imprint of Taylor & Francis Group, LLC

© 2024 Taylor & Francis Group, LLC

ISBN: 978-0-367-85798-1 (hbk)
ISBN: 978-0-367-49489-6 (pbk)
ISBN: 978-1-003-04674-5 (ebk)

DOI: 10.1201/9781003046394

Typeset in Times
by codeMantra

Dedication

No Mud, No Lotus.

- Thich Nhat Hanh (1926–2022)

Contents

Author xi

Chapter 1 Introduction to Ramsar 1

Background and Importance of "Ramsar Wetlands"... 1
Summary of Ramsar Wetlands of Coastal Western North America and the
Central Pacific 3
Organizational Structure of the Ramsar Convention.. 3
Partners of the Ramsar Convention 4
How the Ramsar Convention Impacts Our World 5
References 5

Chapter 2 The United States and Canada 6

Izembek Lagoon National Wildlife Refuge, Alaska, USA: Overview 6
Site Details 6
Fraser River Delta, British Columbia, Canada: Overview.................................... 13
Site Details 13
Laguna De Santa Rosa Wetland Complex, California, USA: Overview 16
Site Details 19
Tomales Bay, California, USA: Overview 20
Site Details 20
Bolinas Lagoon, California, USA: Overview 23
Site Details 26
San Francisco Bay/Estuary, California, USA: Overview 26
Site Details 29
Grassland Ecological Area, California, USA: Overview.. 30
Site Details 31
Elkhorn Slough, California, USA: Overview 33
Site Details 34
Tijuana River National Estuarine Research Reserve, California,
USA: Overview 37
Site Details 37
Kawainui and Hāmākua Marsh Complex, Hawai'i, USA: Overview 41
Site Details 41
Palmyra Atoll National Wildlife Refuge, Equatorial Islands, USA: Overview........... 44
Site Details 44
References 48

Chapter 3 Mexico: States of Baja California and Baja California Sur 50

Humedales Del Delta Del Río Colorado, Baja California and Sonora: Overview 51
Site Details 51
Estero De Punta Banda, Baja California: Overview 54
Site Details 54
Bahía De San Quintín, Baja California: Overview.. 57
Site Details 60
Corredor Costero La Asamblea-San Francisquito, Baja California: Overview........... 61
Site Details 62

Isla Rasa, Baja California: Overview 65
Site Details 65
Laguna Ojo De Liebre, Baja California Sur: Overview 69
Site Details 69
Laguna San Ignacio, Baja California Sur: Overview 72
Site Details 72
Humedal La Sierra De Guadalupe, Baja California Sur: Overview 75
Site Details 77
Humedal Los Comondú, Baja California Sur: Overview 80
Site Details 80
Oasis Sierra De La Giganta, Baja California Sur: Overview 83
Site Details 86
Parque Nacional Bahía De Loreto, Baja California Sur: Overview 87
Site Details 87
Oasis De La Sierra El Pilar, Baja California Sur, Mexico: Overview 91
Site Details 91
Balandra, Baja California Sur, Mexico: Overview 94
Site Details 94
Humedales Mogote-Ensenada La Paz, Baja California Sur, Mexico: Overview 98
Site Details 98
Parque Nacional Cabo Pulmo, Baja California Sur: Overview 101
Site Details 103
Sistema Ripario De La Cuenca and Estero De San José Del Cabo, Baja
California Sur: Overview 105
Site Details 105
References 109

Chapter 4   Mexico: States of Sonora and Sinaloa 111
Sistema De Humedales Remanentes Del Delta Del Río Colorado, Baja
California and Sonora: Overview 111
Site Details 111
Humedales De Bahía Adair, Sonora: Overview 116
Site Details 116
Humedales De Bahía San Jorge, Sonora: Overview 120
Site Details 120
Canal Del Infiernillo and Esteros Del Territorio Comcaác, Sonora: Overview 123
Site Details 126
Humedales De La Laguna La Cruz, Sonora: Overview 127
Site Details 130
Isla San Pedro Mártir, Sonora: Overview 131
Site Details 133
Estero El Soldado, Sonora: Overview 135
Site Details 135
Complejo Lagunar Bahía Guásimas-Estero Lobos, Sonora: Overview 139
Site Details 139
Humedales De Yavaros-Moroncarit, Sonora: Overview 143
Site Details 143
Lagunas De Santa María-Topolobampo-Ohuira, Sinaloa: Overview 147
Site Details 147
Sistema Lagunar San Ignacio-Navachiste-Macapule, Sinaloa: Overview 150
Site Details 153

Laguna Playa Colorada-Santa María La Reforma, Sinaloa: Overview 154
Site Details 157
Ensenada De Pabellones, Sinaloa: Overview... 158
Site Details 160
Sistema Lagunar Ceuta, Sinaloa: Overview... 162
Site Details 162
Playa Tortuguera El Verde Camacho, Sinaloa: Overview............................ 165
Site Details 165
Laguna Huizache-Caimanero, Sinaloa: Overview....................................... 168
Site Details 168
References 172

Chapter 5   Mexico: States of Nayarit and Jalisco 177

Marismas Nacionales, Sinaloa and Nayarit: Overview................................ 177
Site Details 177
Parque Nacional Isla Isabel, Nayarit: Overview 182
Site Details 182
La Tovara, Nayarit: Overview 186
Site Details 186
Islas Marietas, Nayarit: Overview... 189
Site Details 191
Sistema Lagunar Estuarino Agua Dulce-El Ermitaño, Jalisco: Overview 193
Site Details 193
Estero El Chorro, Jalisco: Overview 196
Site Details 196
Estero Majahuas, Jalisco: Overview 200
Site Details 200
Laguna Xola-Paramán, Jalisco: Overview...203
Site Details 203
Laguna Chalacatepec, Jalisco: Overview..206
Site Details 209
Reserva De La Biosfera Chamela-Cuixmala, Jalisco: Overview............................ 210
Site Details 212
Estero La Manzanilla, Jalisco: Overview 215
Site Details 215
Laguna Barra De Navidad, Jalisco: Overview... 218
Site Details 220
References 222

Chapter 6   Mexico: States of Colima, Michoacan, and Guerrero..224

Laguna De Cuyutlán Vasos III Y IV, Colima: Overview 224
Site Details 224
Reserva De La Biosfera Archipiélago De Revillagigedo: Overview 229
Site Details 230
Santuario Playa Boca De Apiza-El Chupadero-El
Tecuanillo, Colima: Overview 233
Site Details 233
Playa De Colola, Michoacan: Overview 237
Site Details 237
Playa De Maruata, Michoacan: Overview 240
Site Details 243

Playón Mexiquillo, Michoacan: Overview 244
Site Details 244
Laguna Costera El Caimán, Michoacan: Overview 248
Site Details 248
Playa Tortuguera Tierra Colorada, Guerrero: Overview 251
Site Details 251
Playa Tortuguera Cahuitán, Guerrero: Overview 255
Site Details 258
References 260

Chapter 7   Mexico: States of Oaxaca and Chiapas 262

Lagunas De Chacahua, Oaxaca: Overview 262
Site Details 262
Cuencas and Corales De La Zona Costera De Huatulco, Oaxaca: Overview 267
Site Details 269
Playa Barra De La Cruz, Oaxaca: Overview 271
Site Details 271
Sistema Estuarino Puerto Arista, Chiapas: Overview 275
Site Details 275
Sistema Estuarino Boca Del Cielo, Chiapas: Overview 278
Site Details 278
Reserva De La Biosfera La Encrucijada, Chiapas: Overview 282
Site Details 282
Zona Sujeta A Conservación Ecológica Cabildo-Amatal, Chiapas: Overview 285
Site Details 287
Zona Sujeta A Conservación Ecológica El Gancho-Murillo,
Chiapas: Overview 289
Site Details 291
References 293

Spanish to English Translations/Definitions for Terms Used 295

Annotated Acronyms Used 296

Appendix A (Taxa Cited) 299

Index 339

Author

Ricardo D. Lopez, PhD, has provided leadership on diverse environmental science issues over the last four decades. During his tenure in both academia and public service with the U.S. Environmental Protection Agency, the U.S. Forest Service, and the U.S. Fish and Wildlife Service, Ric has inspired scientific research teams and natural resource managers across a variety of geographic areas, from Asia and Indo-Pacific Regions to the Americas, focusing on shared learning and public service among the diverse communities of the world. A natural thread woven throughout his life and career and writings is the importance of sustaining all of the planet's ecosystems and the people who rely upon them. Among the first areas of research Ric focused on was the wetlands of the Laurentian Great Lakes, and he is proud to have contributed to that particular "community of science and conservation" during the past several decades, a time when wetlands have been recognized for their critical role in maintaining life on our changing planet.

Most recently, Ric has built upon his years of community-based applied science and conservation as CRC Press author and editor of *Societal Dimensions of Environmental Science: Global Case Studies of Collaboration and Transformation* and as CRC Press series editor of *Environmental and Societal Dimensions of Sustainable Development Goals*. This book is his first in a series of compilations and focal assessments of the approximately 2,400 Ramsar Wetlands of International Importance as designated by 172 Ramsar Contracting Parties (i.e., nations) of the world.

Ric earned a BS in Ecology, Behavior, and Evolution at the University of California, San Diego, and a Master's and PhD in Environmental Sciences at the Ohio State University, with an emphasis on landscape ecology and wetland ecology. He is currently Administrator of the Pacific Islands Area of the U.S. Fish and Wildlife Service, conserving and serving 'āina (Hawaiian language meaning *the land, earth*) among the 22 National Wildlife Refuges and 4 Marine National Monuments of the Pacific.

1 Introduction to Ramsar

BACKGROUND AND IMPORTANCE OF "RAMSAR WETLANDS"

The "Ramsar Convention on Wetlands of International Importance especially as Waterfowl Habitat," commonly referred to in the more concise form of the "Ramsar Convention," was signed on February 2, 1971 in Ramsar, Iran, and was made effective on December 21, 1975. The Ramsar Convention is an international treaty that provides for the conservation and sustainable uses of wetlands around the world, and thus has been even further simplified by some as the "Convention on Wetlands." The signatories for the Ramsar Convention are the Director General of the United Nations Educational, Scientific, and Cultural Organization (UNESCO) and 172 parties (the "Contracting Parties"), each of which represents one of the 195 countries on earth, which are customarily referred to as the Ramsar Convention Member States. The primary purpose of the Ramsar Convention's mission is "the conservation and wise use of all wetlands through local and national actions and international cooperation, as a contribution towards achieving sustainable development throughout the world."

Every 3 years the representatives of the Contracting Parties meet at what is referred to as the "Conference of the Contracting Parties" (COP), which is the policy-making organ of the Convention that adopts all decisions (i.e., resolutions and recommendations) to administer the work of the Convention, and to improve the way in which the Parties implement objectives.

There are currently over 2,400 Wetlands of International Importance covering more than 2.5 million km^2 of the earth's surface. The country (i.e., COP Member) with the greatest number of Ramsar Sites is the United Kingdom, with a total of 175 sites, and the country with the greatest areal coverage of listed sites is Bolivia, with over 140,000 km^2 of designated wetlands. Currently there are 23 Transboundary Ramsar Sites, which cross national borders, as well as 20 Ramsar Regional Initiatives that cover a variety of areas in the Mediterranean Region, Asia, Africa, and South America.

The global wetland outlook for our planet is bleak, and if the anticipated trajectory of degradation and loss of these critical ecosystems does not dramatically improve, soon the lives of future generations will be dramatically altered. This clarion call for demonstrable leadership at local to global scales to avert the dire environmental outcomes that result from wetland degradation and loss clearly demonstrates the focal strategy of designating precious wetland acreages as the practical and actionable conservation method of the Ramsar Convention does. The Convention's clarion call is thus articulated by outlining the key issues that wetland scientists, managers, and decision makers need to better understand in order to better conserve and restore these extremely economically valuable ecosystems, such as their functional role in regulating and mediating global climate and hydrologic processes. Among the most alarming declarations of the above warnings by the members of the Ramsar Convention, despite decades of concern about global atmospheric carbon budget surpluses, wetlands (one of two main carbon sequestering ecosystems of the world) are disappearing at three times the rate of another critical ecosystem type, upland forests. Ramsar Convention Secretary General, Martha Rojas Urrego, described the findings in the Ramsar Convention's 2018 report as a "wake-up call," not only concerning the steep loss rate for the world's wetlands but also concerning the concomitant losses of critical ecological services they provide. Speaking about our global wetlands, Urrego also invoked a bold new United Nations' initiative, remarking that, "without [wetlands], the Global Agenda on Sustainable Development will not be achieved," referring specifically to the United Nations Sustainable Development Goals, which simultaneously build economic growth that serves a host of societal needs, whilst tackling climate change and promoting environmental protection (https://www.un.org/sustainabledevelopment/development-agenda/).

DOI: 10.1201/9781003046394-1

Given the urgency to address global wetland conservation and restoration, this book series was created to celebrate the multitude of Ramsar wetlands across the globe, region by region, and in a manner that supports the Ramsar Convention's strong tradition of action-oriented approaches to global wetland conservation and restoration. Namely, in the face of the worldwide history of devastating wetland destruction and functional degradation of associated watersheds, this book series is a practical teaching tool to provide easy access and further interest in the wetlands of the world by explaining to a variety of readers and audiences the implications for loss and ecological impacts of any further steps in this destructive and degenerative trajectory. To this end, in this volume, we succinctly outline the key biophysical and sociological elements of the rich and productive wetlands of Coastal Western North America and the Central Pacific. This book provides initial background on the Ramsar Convention for context, moving quickly to a detailed description of each of the 72 designated Ramsar sites within the focal geography of this volume.

In 1983, Frayer et al. outlined for the first time the status of wetlands in the continental United States of America, painting a stark picture that was quite telling simply by the numbers: "total acreage of wetlands and deepwater habitats, in the 48 conterminous United States, in the 1950s was 179.5 million acres; in the 1970s it was 171.9 million acres, a net loss of 7.6 million acres [in just 20 years]; average annual net loss for this 20-year period was about 380 thousand acres." For comparison, this loss of 7.6 million acres of wetlands is roughly the size of the US State of Maryland, which one might say is not an extremely large area compared to the remaining US states; however considering that this is merely the loss of the rather specialized ecosystem type "wetland," and the losses occurred over just a relatively short 20-year period, it is a substantial and rapid loss of ecosystem acreage. Contributing causal factors for this rapid and substantial acreage loss of wetlands in the United States was the simultaneous and tumultuous period of urban and agricultural development in the United States during this same 20-year period. However, these wetland losses were just a drop in the bucket compared to the overall losses and destruction of wetlands in prior years within the United States, which destroyed what has gradually become better understood to be one of humanity's precious and truly unique natural and cultural resources. Among the notable acreage losses in the United States is the estimated 50% of wetlands converted within the State of California, having been mostly transformed into agricultural land and associated human developments between 1860 and 1900 (Dahl and Allord 1996). Similarly, the States of Nevada, Hawai'i, and Alaska all participated in the willful destruction of wetlands during the 19th century and throughout much of the 20th century (Mitsch and Gosselink 2015).

Additional threats to deepwater habitats, also known as reefs, are recognized among a number of impacts from changes in global climate change, specifically from increased carbon dioxide in the atmosphere and sea temperatures, which ultimately are causing "coral bleaching" and the associated effects on other sea life that depend on these rare and vital reef ecosystems. As scientists study these atrophying coral ecosystems, rough estimates are that 20% of the planet's coral reefs have been lost or severely damaged in this way, which is approximately twice the rate of worldwide tropical rainforest losses observed during the same period. These recent trends are a microcosm of the global trends for all wetland losses in the last two centuries, including the nearly 50% loss of earth's wetlands since 1900. It is within the context of such wetland loss trends, and the scientific and societal understanding of, and re-connections with, these wetlands of the world, that the formation of the "Ramsar Convention on Wetlands of International Importance especially as Waterfowl Habitat" occurred in 1971.

This book series recognizes the work of the Ramsar Convention and provides an in-depth journey to each of the Ramsar Wetlands of International Importance in Coastal Western North America and the Central Pacific, as currently designated. Accordingly, this book also focuses on the full breadth of each subject wetland, including its history, physical and human geography, relevant weather and atmospheric conditions, and its watershed, floral, faunal, and geological characteristics. Each of the chapters in this book emphasizes the interactions of these biophysical, societal, and other characteristics through a thorough yet concise exploration of the ecology and history of each wetland site.

The main intent of this book is to provide the reader with a full experience in each of the presented wetlands, which are emblematic of their type and landscape conditions, achieving a full understanding of the variety and uniqueness of Ramsar Wetlands in Coastal Western North America and the Central

Pacific. The inciteful distillations of the vast amount of information and perspectives available for each of the sites are provided in this succinct and useful compendium of all 72 wetlands, which total 4,162,474 ha (41,625 km²), which for reference is approximately the area of the country of Switzerland.

SUMMARY OF RAMSAR WETLANDS OF COASTAL WESTERN NORTH AMERICA AND THE CENTRAL PACIFIC

The three countries that comprise North America, i.e., the Dominion of Canada (Canada), the United States of America (The United States), and the United Mexican States (Mexico) are the broad focus of this volume, including those areas in the Central Pacific that fall within the political sphere of North America (i.e., the United States). Consequently, and due to the vast expanse covered in this book, there is a wide diversity of these wetlands, and their associated socio-biophysical surroundings discussed in the following chapters. Each chapter of this book begins with an introduction to a geographic region, followed by a series of site descriptions for each of the wetlands in that region.

Among the three countries of North America, Canada was first to become a member of the Conference of the Contracting Parties (COP; May 15, 1981), with the United States following in those footsteps 5 years later (December 18, 1986), along with Mexico becoming a COP Member to the Ramsar Convention on November 4, 1986. There are currently 37 wetland sites in Canada designated as Wetlands of International Importance, totaling 13,086,767 ha, nationwide; 41 wetland sites in the United States designated as Wetlands of International Importance, totaling 1,884,551 ha nationwide; and 144 sites in Mexico designated as Wetlands of International Importance, totaling 8,721,911 ha nationwide. Of the 37 Ramsar Wetlands in Canada, one of them is in Coastal Western Canada, i.e., British Columbia; of the 41 Ramsar Wetlands in the United States, 10 of them are in the Coastal Western United States and the Central Pacific, i.e., Equatorial Islands, Hawai'i, Alaska, and California; and of the 144 Ramsar Wetlands in Mexico, 61 of them are in the Coastal Western Mexico.

The full complement of 72 Ramsar Wetlands in Coastal Western North America and the Central Pacific in this volume allows for both a deep dive into the details of sites, while providing the flavor of the amazingly broad range of sites and surrounding biophysical environmental conditions that exist within, from submersed reef environments, to humid tropical conditions, to temperate wetlands within both freshwater and saline environments. The Ramsar Wetlands of Coastal Western North America and the Central Pacific also span the diverse communities and cultures of the region, from those of Pacific Islanders and Alaskans, to the broad ranging communities and cultures found in of the State of California and in Western Mexico. Each of these amazing ecosystems and places are detailed, by region, and then from north to south with that region.

ORGANIZATIONAL STRUCTURE OF THE RAMSAR CONVENTION

The Ramsar Convention comprises several key organizational elements and functioning partners, which interact in a specific manner to improve efficient negotiations, dissemination of information, and conservation actions. These key elements are:

- The Conference of the Contracting Parties (COP)
- The Scientific and Technical Review Panel (STRP)
- The Standing Committee
- The Secretariat
- International Organization Partners (IOPs)
- Other organizational and public partners

The Conference of the Contracting Parties (COP) is the Convention's main governing body that represents all of the current governments that have ratified the treaty, and thus participate in Convention activities. The COP is the ultimate authority body that reviews all activity progress under the Convention, and they identify any new priorities for the Convention so that work plans for member nations can be well founded and coordinated. The COP has the power to make amendments to the existing Convention, design and create expert advisory entities, review the progress reports submitted by member nations, and collaborate in a variety of ways with the Convention's other associated international organizations via agreements.

Established in 1993, the Scientific and Technical Review Panel (STRP) provides scientific and technical guidance to the Conference of the Parties, the Standing Committee, and the Ramsar Secretariat. The STRP's Work Plan for each triennium is built around the priority tasks approved by the Standing Committee, and is based on the requests of the COP by means of its Strategic Plan and COP Resolutions and Recommendations. The Standing Committee is the intersessional executive body, which represents the COP between its triennial meetings, within the framework of the decisions made by the COP. The Contracting Parties that are members of the Standing Committee are elected by each meeting of the COP to serve for the coming 3-year period. The Secretariat carries out day-to-day coordination of the Convention's activities, and is based at the headquarters of the International Union for the Conservation of Nature in Gland, Switzerland. The implementation of the Ramsar Convention is a partnership between the Convention Secretariat and the Contracting Parties, and Standing Committee, with the advice of the subsidiary expert body, the STRP, and with the support of the International Organization Partners (IOPs).

PARTNERS OF THE RAMSAR CONVENTION

The Ramsar Convention is allied with a plethora of external partners, which is one of several powerful aspects of this global organization. Among their many partners, there are six major organizations that support the bulk of these leveraged activities within the Ramsar Convention organization, as follows:

- BirdLife International (BI, https://www.birdlife.org/)
- International Union for Conservation of Nature (IUCN, https://www.iucn.org/)
- International Water Management Institute (IWMI, http://www.iwmi.cgiar.org/)
- Wetlands International (WI, https://www.wetlands.org/)
- Wildfowl and Wetlands Trust (WWT, https://www.wwt.org.uk/)
- World Wildlife Fund (WWF, https://www.worldwildlife.org/)

These six major partner organizations comprise the Ramsar Convention's International Organization Partners (IOPs). The fundamental roles of the IOPs are the support of the Convention's work by providing expert technical advice, helping implement field studies, and providing financial support. The IOPs also participate regularly as observers in all meetings of the COP, and they also function as full members of the Scientific and Technical Review Panel (STRP).

In addition to the IOP, the Ramsar Convention also collaborates amongst a vast network of other partnerships, including:

- Biodiversity-related conventions, such as:
  - The Convention on Biological Diversity (CBD, https://www.cbd.int/);
  - The Convention on International Trade in Endangered Species (CITES, www.cites.org/);
  - The Convention on Migratory Species (CMS, https://www.cms.int/);
  - The Convention to Combat Desertification (UNCCD, https://www.unccd.int/);
  - The World Heritage Convention (WHC, http://whc.unesco.org/)
- Project funding organizations, such as:
  - Global environmental funders;
  - Multilateral development banks; and
  - Bilateral donors

- United Nations Agencies, such as:
  - The United Nations Development Program (UNDP, https://www.undp.org/content/undp/en/home/);
  - The United Nations Economic Commission for Europe (UNECE, http://www.unece.org/info/ece-homepage.html);
  - The United Nations Educational, Scientific, and Cultural Organization (UNESCO, https://en.unesco.org/); and
  - The United Nations Environmental Program (UNEP, https://www.unenvironment.org/)
- Non-governmental organizations, such as:
  - Conservation International (CI, https://www.conservation.org/Pages/default.aspx);
  - The International Association for Impact Assessment (IAIA, https://www.iaia.org/);
  - The Nature Conservancy (TNC, https://www.nature.org/en-us/); and
  - The Society of Wetland Scientists (SWS, https://www.sws.org/)
- Private sector organizations and consortia, such as:
  - The Danone Group (https://www.danone.com/) including the Evian brand; and
  - The Star Alliance airline network (i.e., 27 international airline carriers, https://www.staralliance.com/en/).

HOW THE RAMSAR CONVENTION IMPACTS OUR WORLD

Since its inception in 1971 the Ramsar Convention has brought needed focus to the world's wetlands. Although the importance of wetlands is well recognized and increasingly understood, it is necessary to understand that in 1971 the status of wetlands as functioning ecosystems around the world was poor, and wetland ecology, management, and conservation were not well recognized by the public. For context, in 1971 the discipline of wetland ecology was not among the possible college degrees that a biologist or ecologist could obtain. This is not to say that there were no wetland ecologists amongst the ranks of biologists, ecologists, and land managers, but the formal recognition of wetland ecology as a scientific discipline unto itself was not founded in the era of the initiation of the Ramsar Convention. Today, wetland ecology programs within academia, government, and private industry are commonplace and the discipline of wetland ecology is currently flourishing in the United States and abroad. Jobs for wetland ecologists, wetland managers, and other wetland specialists abound relative to the 1970s and 1980s. It is clear that Ramsar had, and continues to have, a global influence on the consciousness of the world's inhabitants about wetlands, such as their initiation and continued celebration of "World Wetlands Day," first celebrated in 1997. Subsequently, every February 2 has been commemorated as the foundation of the Ramsar Convention on Wetlands, on February 2, 1971. World Wetlands Day is intended to establish and increase global awareness of wetlands and their value to humanity, and to our planet as a whole. The special celebration of World Wetland Day has succeeded in increasing awareness of the uniqueness and importance of wetlands. For example, since World Wetlands Day was first celebrated in 1997 it has grown extraordinarily, with hundreds of large events occurring across the world each year (https://www.worldwetlandsday.org/).

REFERENCES

Dahl, T.E. and G.J. Allord. 1996. Technical aspects of wetlands: History of wetlands in the conterminous United States (pp. 19–26). In: *USGS (United States Geological Survey), National Water Summary on Wetland Resources*. USGS Water-Supply Paper 2425. Washington, DC.

Frayer, W.E., T.J. Monahan, D.C. Bowden, and F.A. Graybill. 1983. *Status and trends of wetlands and deepwater habitats in the conterminous United States, 1950s to 1970s*. Colorado State University, Fort Collins. 31pp.

Mitsch, W.J. and J.G. Gosselink. 2015. *Wetlands*. John Wiley and Sons, Hoboken. 743pp.

2 The United States and Canada

The Ramsar wetlands along the western coasts of the United States range widely in latitude from the colder northern regions of Alaska, to the balmy central Pacific, and with most of them along the temperate to arid areas of northern and southern California, respectively. Canada's western coastal wetlands support a wide variety of species and biophysical environments, among which is included its single Ramsar site in the Greater Vancouver area. These 11 Ramsar sites comprise the diverse and unique coastal wetland grouping that is described in this chapter (Figure 2.1):

* Izembek Lagoon National Wildlife Refuge, USA
* Fraser River Delta, Canada
* Laguna de Santa Rosa Wetland Complex, USA
* Tomales Bay, USA
* Bolinas Lagoon, USA
* San Francisco Bay/Estuary, USA
* Grassland Ecological Area, USA
* Elkhorn Slough, USA
* Tijuana River National Estuarine Reserve, USA
* Kawainui and Hāmākua Marsh Complex, USA
* Palmyra Atoll National Wildlife Refuge, USA

IZEMBEK LAGOON NATIONAL WILDLIFE REFUGE, ALASKA, USA: OVERVIEW

Izembek Lagoon National Wildlife Refuge (Figures 2.2 and 2.3) consists of a series of lagoons and surrounding marshes on the north coast of the Alaska Peninsula. Izembek Lagoon is the largest wetland of its type in Alaska and contains one of the world's largest beds of eelgrass, which is a primary food source for many geese and ducks. The site provides staging habitats for waterfowl that undertake transoceanic flights and nest or winter along the Pacific Rim. The site supports a wide variety of waterbird species, as well as a large number of sea otters (Enhydra lutris) and Steller sea lions (Eumetopias jubatus), which are endangered and near-threatened, respectively, according to the IUCN Red List (2007).

As the sole Ramsar wetland site in the State of Alaska, USA (designated December 18, 1986), and as the second (of 41 total) Ramsar wetland sites designated within the United States, this 168,433 ha complex of lagoons and adjacent marshes along the northern coast of the Alaska Peninsula is also a US Fish and Wildlife Service National Wildlife Refuge, established in 1980. The site has a large tidal embayment of open water, with mudflats and marshes, originally inhabited by the Unangan People (also referred to as the Aleuts by later explorers to the region). Izembek Lagoon National Wildlife Refuge is approximately 1000 km southwest of Anchorage, the most populous city in Alaska, with an approximate population of 288,000.

SITE DETAILS

The Izembek National Wildlife Refuge (NWR) is located near the extreme western portion of the Alaska Peninsula. The 168,433 ha NWR resides between the Bering Sea (Bristol Bay) and the Northern Pacific Ocean, and is all within a range of elevation from sea level up to 1,500 m. The Izembek NWR consists of a series of lagoons surrounded by marshes on the north side of the Alaska Peninsula, facing northwest into Bristol Bay and the Bering Sea. The site is about 77 km long, oriented from the

DOI: 10.1201/9781003046394-2

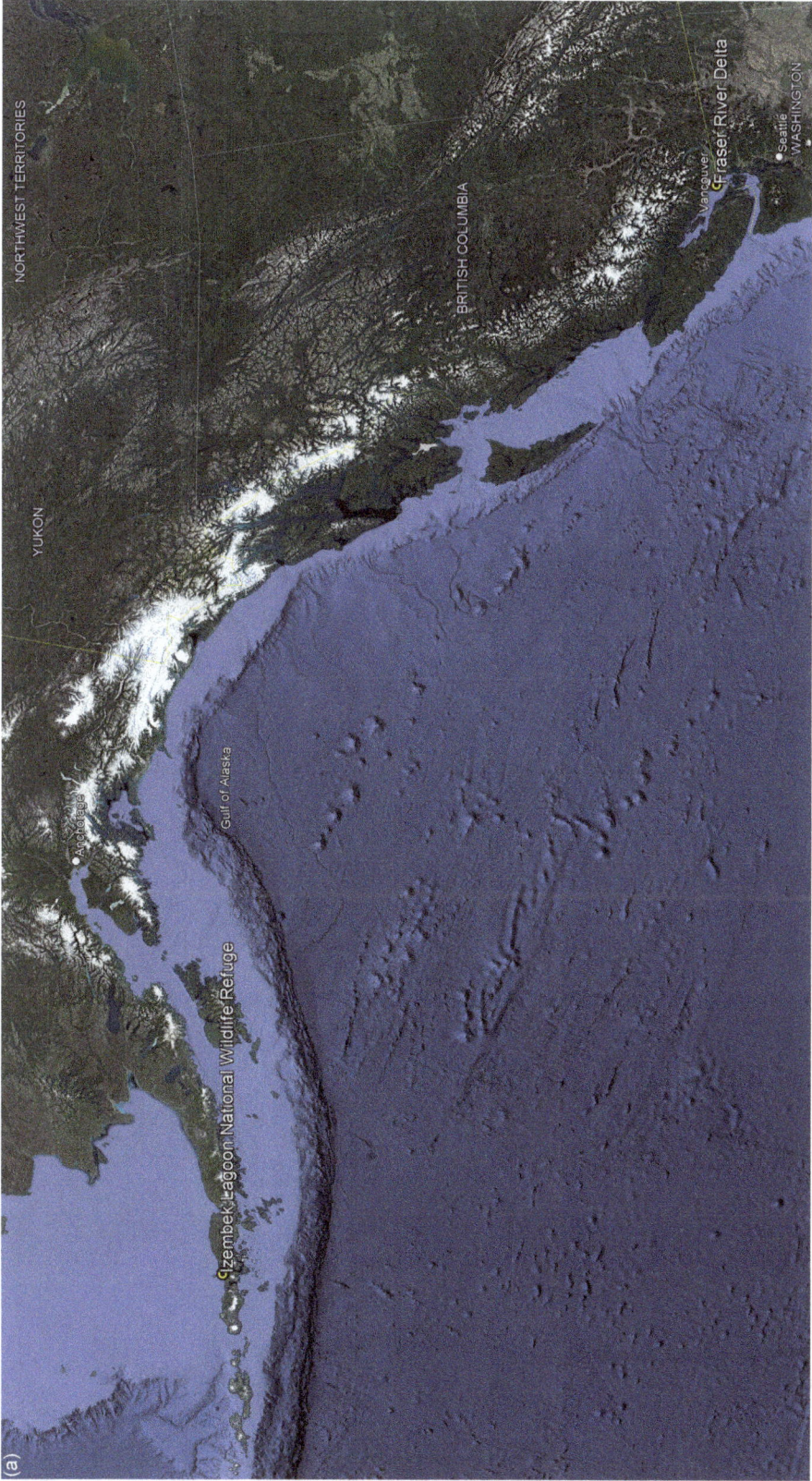

FIGURE 2.1 Ramsar wetlands of coastal western United States and Canada: (a) the state of Alaska [USA] and Canada; (b) Central Pacific; and (c) the state of California [USA].

(Continued)

FIGURE 2.1 (*Continued*) Ramsar wetlands of coastal western United States and Canada: (a) the state of Alaska [USA] and Canada; (b) Central Pacific; and (c) the state of California [USA].

FIGURE 2.1 (Continued) Ramsar wetlands of coastal western United States and Canada: (a) the state of Alaska [USA] and Canada; (b) Central Pacific; and (c) the state of California [USA].

FIGURE 2.2 Composite aerial image of Izembek Lagoon national wildlife refuge. Designated December 18, 1986; Area: 168,433 ha; Coordinates: 55°18′55″N 162°53′08″W; Ramsar site number: 349.

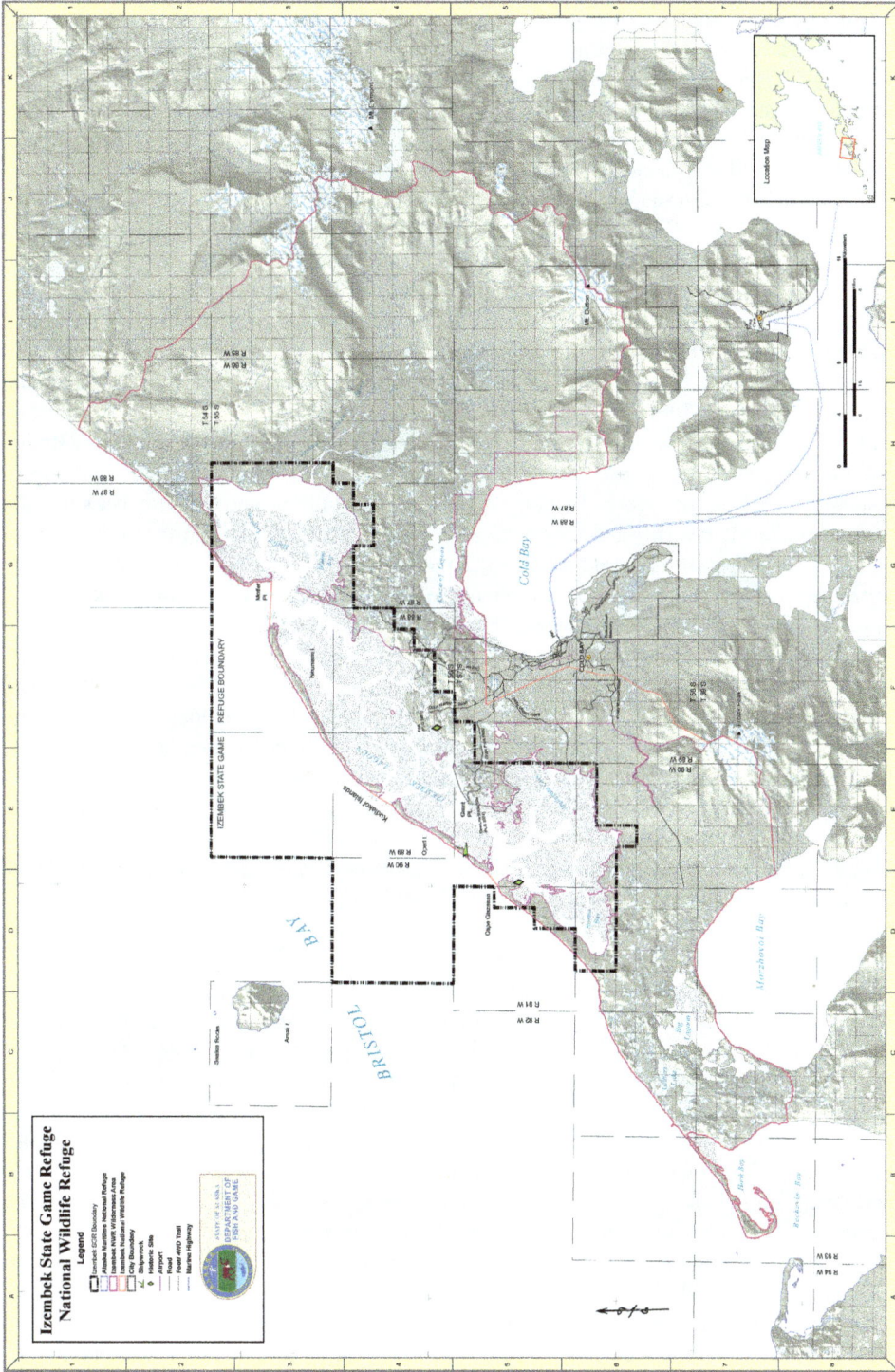

FIGURE 2.3 Map of Izembek Lagoon national wildlife refuge. Designated December 18, 1986; Area: 168,433 ha; Coordinates: 55°18'55"N 162°53'08"W; Ramsar site number: 349.

southwest edge to the northeast edge, with a width along that length that varies from 8 km to 40 km, containing approximately 130,000 ha of upland ecosystems and 38,450 ha of tidelands and coastal lagoons. Izembek Lagoon, the largest lagoon of its type in Alaska, is nearly 48 km long and ranges from 5 km to 10 km in width along this same southwest-northeast orientation.

As a result of these vast areas of uninterrupted wetlands, a whole host of species thrive at the site, including nearly 150 species of birds and 23 species of mammals as residents and/or migrants. The physical conditions at the site also attract and support large populations of shorebirds and water-fowl, especially black brant, emperor and Canada geese, dabbling ducks, and eiders during certain times of the year, as well as ptarmigan, brown bear, and caribou.

Izembek Lagoon is the largest and most unique lagoon of its type in Alaska. The major attractive force for the waterfowl that visit the site are the world's largest eelgrass beds that are flourishing there. Staging habitats at Izembek Lagoon provide the last opportunity for migrating birds to build lipid reserves prior to long over-water flights to wintering areas. The birds that stage at the Izembek NWR nest and/or winter along distant areas along the Pacific Rim including Russia, Canada, Japan, Australia, Mexico, and the U.S. Since the eelgrass is a primary food source for geese and ducks that use the area, Izembek Lagoon and its vast eelgrass beds are truly (and literally) of international importance to the migratory birds.

The Izembek area supports nearly the entire North American population of black brant (Branta bernicla nigricans) during spring and fall migrations; in some years substantial numbers (up to 30,000) may overwinter on the lagoon. In addition to brant, tens of thousands of Taverner's Canada geese (Branta canadensis taverni) and emperor geese (Chen canagica) migrate through the area. Emperor geese overwinter in moderate numbers. Tundra swans (Cygnus columbianus) are a key nesting waterfowl species on the site. Migrant populations of up to 300,000 dabbling ducks use this area during spring and fall migration periods. The most abundant ducks are pintail (Anas acuta) and mallard (A. platyrhynchos). Other ducks occurring in fairly large numbers include gadwall (A. strepera) and American widgeon (A. americana). Wintering populations of Steller's eider (Polysticta stelleri) may approach 100,000 birds on the Izembek refuge and adjacent lagoons, a major portion of the world population of this species. Other sea ducks wintering in abundance include oldsquaw (Clangula hyemalis), bufflehead (Bucephala albeola), greater scaup (Aythya marila), common gold-eneye (Bucephala clangula), common and king eider (Somateria mollissima and S. spectabilis), and black and white-winged scoter (Melanitta nigra and M. fusca).

In all seasons, but particularly during the fall, the site is host to enormous numbers of shore-birds, especially rock sandpiper (Calidris ptilocnemis), dunlin (C. alpina), and sanderling (C. alba). Wintering populations of Steller's eider (Polysticta stelleri) may approach 100,000 birds on Izembek NWR and adjacent lagoons, a major portion of the world population of this species. The area in the vicinity of Izembek NWR supports nearly the entire North American populations of black brant and emperor geese during spring and fall. Tens of thousands of Taverner's Canada geese migrate through the area and up to 300,000 dabbling ducks at a time use the area during spring and fall migrations. Additionally, the Izembek NWR also supports bald eagles (Haliaeetus leucocephalus) and peregrine falcons (Falco peregrinus pealei) as year-round residents. Important upland species using the refuge include the Alaskan brown bear (Ursus arctos), barren ground caribou (Rangifer tarandus), wolverine (Gulo gulo), mink (Mustela vison), willow ptarmigan (Lagopus lagopus), river otter (Lontra canadensis), and gray wolf (Canis lupus). Sand spits and barrier island beaches sur-rounding Izembek Lagoon are haulout sites for harbor seals (Phoca vitulina), with as many as 5,000 present at one time. An estimated 500 to 1,000 sea otters (Enhydra lutris) and 100 Steller sea lions (Eumetopias jubatus) frequent waters in the Lagoon and nearby offshore areas for feeding and rest-ing; both species are listed as endangered red status in the IUCN Red List.

Four species of Pacific salmon (Oncorhynchus spp.), i.e., sockeye, chinook, pink, and chum, rou-tinely enter the freshwater streams of the Izembek area from the Bering Sea and the Pacific Ocean to spawn. Dolly Varden (Salvelinus malma) and rainbow trout (Salmo gairdneri) also inhabit many of the streams and lakes. A minimum of 39 species of fish use the site as migratory, spawning, or nursery habitat.

The undulating glacial outwash/coastal plains in the area are dominated by a mixture of low ericaceous shrubs and graminoid tundra. Typical vegetation includes crowberry (Empetrum nigrum), mountain cranberry (Vaccinium vitis-idaea), bluejoint grass (Calamagrostis canadensis), cottongrass (Eriophorum scheuchzen), arctic willow and other willows (Salix spp.), reindeer mosses (Cladonia spp.), and several species of sphagnum moss. Wet meadows and marshes, dominated by beach ryegrass (Elymus arenaria) and sedges (Carex spp.), adjoin the lagoon.

FRASER RIVER DELTA, BRITISH COLUMBIA, CANADA: OVERVIEW

Fraser River Delta (Figures 2.4 and 2.5) is formed by six components (Burns Bog, Sturgeon Bank, South Arm Marshes, Boundary Bay, Serpentine, and the former 'Alaksen' Ramsar Site), all in the Metro Vancouver Region and part of the most important river-delta and estuary for fish and birds on the west coast of Canada. The complex provides an internationally critical migratory stopover area for the western sandpiper (Calidris mauri), one of the most common shorebirds in the western hemisphere. It provides feeding and roosting sites to about 250,000 migrating and wintering waterfowl and 1 million shorebirds, regularly. A number of provincially and federally listed fish species of concern can be found within the estuarine habitats, including Acipenser transmontanus, Acipenser medirostris, and Thaleichthys pacificus. The complexity of ecosystems found at the site, such as estuarine marsh, mudflats, floodplains, sloughs, and river channels provides critical feeding and rearing areas for anadromous salmon during their transition between river and marine stages of their life cycle. Some of the delta's subsites are used for low-impact recreation, but the site is mostly reserved for wildlife habitat conservation. The site was renamed and vastly extended in 2012 from 586 ha to 20,682 ha.

All sub-components of the Frasier River Delta site are in the relatively densely-populated Vancouver Metropolitan Region, located in the south-west corner of the province of British Columbia, Canada. The area is home to an estimated 2.5 million people within 24 municipalities, which is summarized as follows by sub-components:

Burns Bog: Located 17 km south-east of the center of the city of Vancouver, British Columbia (BC). It lies within the Corporation of Delta (a municipal government), BC, which has a population of approximately 100,000 people.

Sturgeon Bank: Located at the mouth of the Fraser River along the western edge of the City of Richmond, BC, 7 km from the City of Vancouver. It forms the western border of Sea, Lulu, and Iona islands. Richmond has a population of approximately 216,000.

South Arm Marshes: Located 17 km south of the center of the city of Vancouver, BC. The islands of South Arm Marshes are at the mouth of the Fraser River located in the City of Richmond.

Boundary Bay: Located 25 km south-east of the center of the City of Vancouver, BC. It encompasses intertidal and near-shore subtidal provincial land bordered to the east by the City of Surrey (approximate population 518,000) and to the south by the US-Canada border.

Serpentine: Located within the City of Surrey, BC, and is 30 km south-east of Vancouver's city center.

Alaksen: Located within the Corporation of Delta, BC, 17 km south of the center of the City of Vancouver.

SITE DETAILS

The site's productive estuarine habitats and nutrient rich waters provide the foundation for diverse and seasonally dense fish populations. In particular, Fraser River is the largest producer of salmon on the entire Pacific Coast of North America. Annually, millions of anadromous salmon migrate through the estuary upstream to spawn along numerous tributaries. Millions of young fish later descend to the estuary on their way out to oceanic habitats. The Sturgeon Bank, Alaksen, and South Arm Marshes

FIGURE 2.4 Composite aerial image of Fraser River Delta. Designated May 24, 1982; Area: 20,682 ha; Coordinates: 49°06′00″N 123°03′00″W; Ramsar site number: 243.

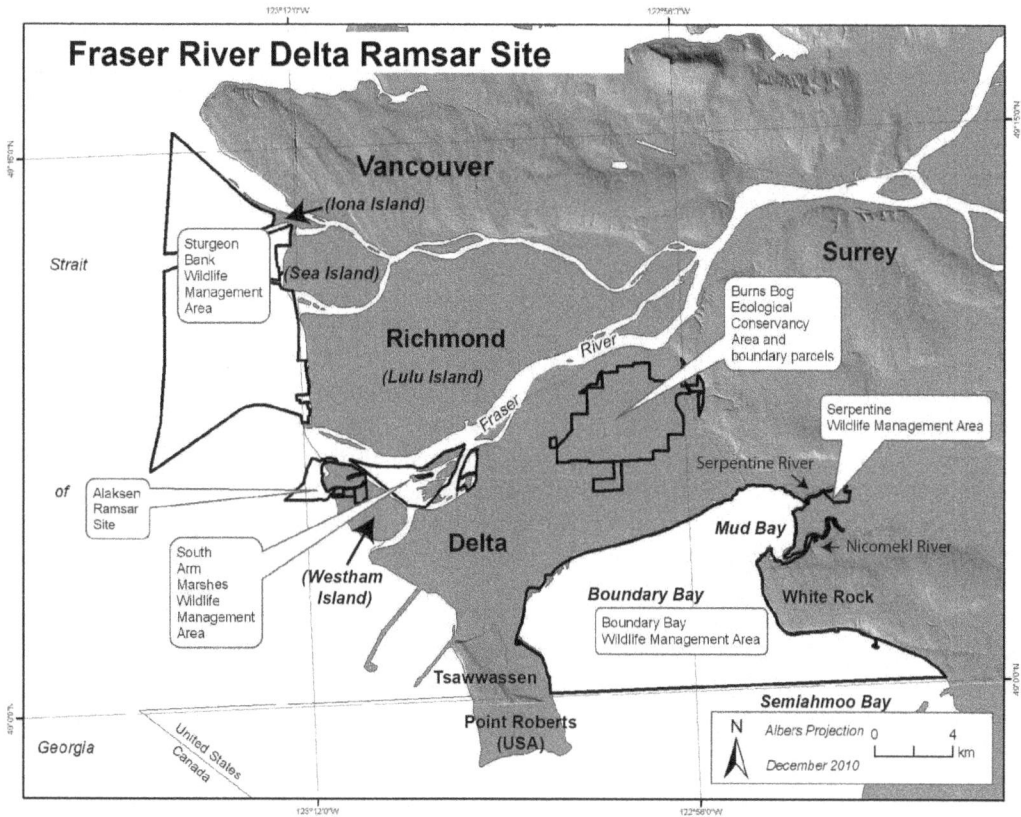

FIGURE 2.5 Map of Fraser River Delta. Designated May 24, 1982; Area: 20,682 ha; Coordinates: 49°06′00″N 123°03′00″W; Ramsar site number: 243.

sub-components form an almost contiguous band of protected habitats within the Fraser River Delta that is important to survival of these juveniles leaving the Fraser River. Estuarine marsh, mudflats, floodplains, sloughs, and river channels are all critical feeding and rearing areas for these fish during their transition between river and marine stages of their life cycle. One of the largest eelgrass meadows in Canada is found in Boundary Bay, which supports an abundance of small fish and shellfish.

The Fraser River Delta ecosystem is an extensive area of wetlands, which existed throughout the entirety of the Fraser River lowlands prior to the mid-1800s. Ecologically, the area was extremely productive and provided valuable habitat for large numbers of fish and wildlife. The area supported over a dozen First Nations groups and continues to provide seasonally abundant food and resources to them.

Draining and conversion of wetlands throughout the Fraser River lowlands for agriculture and urban development have resulted in an estimated loss of over 85% of historical wetlands. The effect of this has been to concentrate waterfowl and many other migratory birds in the estuary and remaining floodplain marshes. Today, the Fraser River estuary and tributary marshes are recognized as globally critical habitat for hundreds of thousands of migratory and over-wintering birds along the Pacific Flyway, mostly wintering and migrating shorebirds and waterfowl but also including high densities of wintering raptors as well as a variety of songbirds; for a full list of species that are known to occur on the Fraser River estuary see Butler and Campbell (1986). The site's productive estuarine habitats and nutrient rich waters provide the foundation for diverse and seasonally dense fish and wildlife populations and comprise the most important area of aquatic bird habitat in British Columbia.

The Fraser River is the largest single salmon-producing river system on the Pacific Coast of North America and supports healthy runs and some subpopulations that are of conservation concern. Whereas normal returns are typically less than 10 million, 34 million sockeye (Oncorhynchus nerka) returned to the Fraser River in 2010. Five species of chinook (Oncorhynchus tshawytscha), chum (Oncorhynchus keta), pink salmon (Oncorhynchus gorbuscha), and coho salmon (Oncorhynchus kisutch) as well as anadromous steelhead trout (Oncorhynchus mykiss) rely on the freshwater and brackish marshes at the mouth of the Fraser River, as a critical life history transition to a life at sea by juvenile salmonids.

Many of the Fraser River Delta wetlands are part of an area identified as Canada's most significant Important Bird Areas (National Audubon Society), and are considered a critical migration stopover for shorebirds and waterfowl. Fifty shorebird species have been recorded in the area. Significant concentrations of western sandpiper (Calidris mauri), dunlin (Calidris alpina), and black-bellied plovers (Pluvialis squatarola) occur annually. Large concentrations of American wigeons (Anas americana), northern pintails (Anas acuta), lesser snow geese (Chen caerulescens caerulescens), and trumpeter swans (Cygnus buccinator) gather on the delta annually during their wintering and winter migration periods. The site is also frequented by gray whales and killer whales.

A research strategy to support potential benefits to the ecosystem is in place to varying degrees among subunits, and Sturgeon Bank and Boundary Bay are frequently used by faculty and staff at the two local universities, University of British Columbia, and Simon Fraser University, as well as by federal and provincial biologists to conduct field research on the foraging, migratory, and behavioral ecology of waterfowl and wading birds. Bird Studies Canada's (2010) Coastal Waterbird Survey has been running continuously for decades to monitor waterbirds within the delta area. Pacific Wildlife Foundation has documented whales in Boundary Bay for decades. Since 1972, the Pacific and Yukon Canadian Wildlife Service office has been located on the Alaksen sub-component, from which on-going bird research and monitoring are conducted.

Recreation opportunities at the site include nature trails for birding, wildlife viewing, and experiencing coastal habitats. A public education and interpretation facility is located on the George C. Reifel Migratory Bird Sanctuary, operated by the British Columbia Waterfowl Society, a non-governmental organization. The sanctuary area includes trails, artificial ponds, and an observation tower.

LAGUNA DE SANTA ROSA WETLAND COMPLEX, CALIFORNIA, USA: OVERVIEW

The Laguna de Santa Rosa Wetland Complex (Figures 2.6 and 2.7) is composed of seasonal and perennial freshwater wetlands such as creeks, ponds, marshes, vernal pools, swales, floodplains, riparian forest, and grasslands located in the Laguna de Santa Rosa Watershed. The complex includes an array of public and privately owned units with a variety of conservation statuses that range from wildlife areas to mitigation banks. The site is considered a biological hotspot due to its various types of rare and unique wetlands, such as vernal pools with their associated rare and endemic plants, such as Sonoma sunshine (Blennosperma bakeri) and animal species such as the California tiger salamander (Ambystoma californiense). Besides its high biological value, the site provides flood control, scenic beauty, and recreation services to the majority of Sonoma County's human population. The Laguna de Santa Rosa Complex's main threats are associated with recent land use changes in the area, such as wetland drainage for farming and expansion of urban areas, pollution due to excessive use of fertilizers in the Santa Rosa Plain, and hydrological changes due to the construction of drainage and flood control channels. Currently, site managers are using a restoration and management plan, originally published in 2006, to implement the conservation goals in the Laguna de Santa Rosa Wetland Complex.

The site is located in Sonoma County, California, west of the City of Santa Rosa (approximate population 177,000), east of Sebastopol (approximate population 7,500), and 78 km north of the City of San Francisco (approximate population 815,000). The site is situated within the Laguna de Santa Rosa watershed, with the Laguna de Santa Rosa being the largest tributary to the Russian River, to the north of the site. The Russian River drains into the Pacific Ocean, approximately 50 km, due west of the site.

FIGURE 2.6 Composite aerial image of Laguna de Santa Rosa Wetland complex. Designated April 16, 2010; Area: 1576 ha; Coordinates: 38°24′N 122°47′W; Ramsar site number: 1,930.

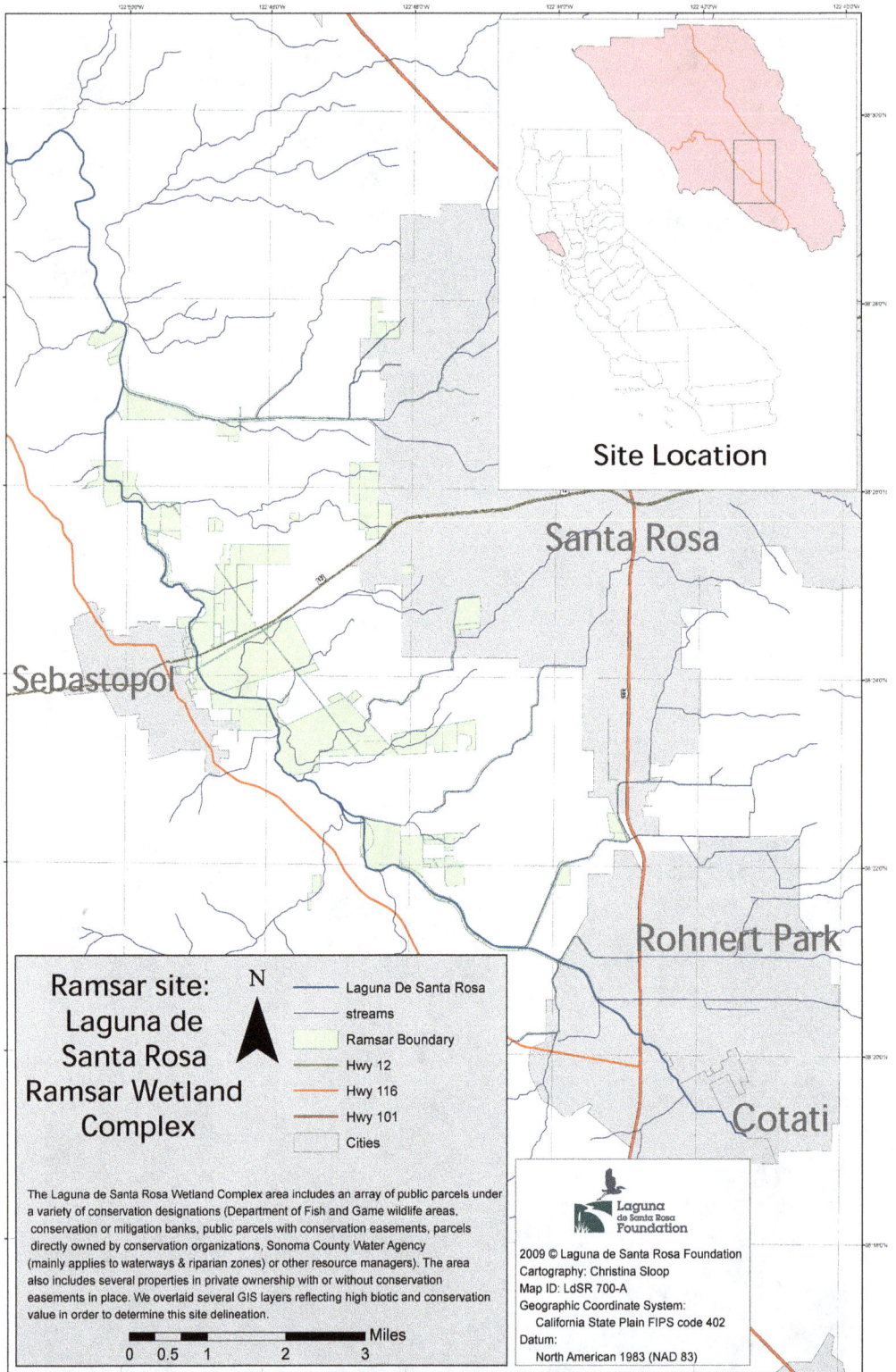

Site Location

Santa Rosa

Sebastopol

Rohnert Park

Cotati

Ramsar site:
Laguna de
Santa Rosa
Ramsar Wetland
Complex

N

——	Laguna De Santa Rosa
——	streams
▨	Ramsar Boundary
——	Hwy 12
——	Hwy 116
——	Hwy 101
▨	Cities

The Laguna de Santa Rosa Wetland Complex area includes an array of public parcels under a variety of conservation designations (Department of Fish and Game wildlife areas, conservation or mitigation banks, public parcels with conservation easements, parcels directly owned by conservation organizations, Sonoma County Water Agency (mainly applies to waterways & riparian zones) or other resource managers). The area also includes several properties in private ownership with or without conservation easements in place. We overlaid several GIS layers reflecting high biotic and conservation value in order to determine this site delineation.

Miles
0   0.5   1   2   3

Laguna
de Santa Rosa
Foundation

2009 © Laguna de Santa Rosa Foundation
Cartography: Christina Sloop
Map ID: LdSR 700-A
Geographic Coordinate System:
   California State Plain FIPS code 402
Datum:
   North American 1983 (NAD 83)

FIGURE 2.7   Map of Laguna de Santa Rosa Wetland complex. Designated April 16, 2010; Area: 1576 ha; Coordinates: 38°24′N 122°47′W; Ramsar site number: 1,930.

SITE DETAILS

The site is composed of seasonal and perennial freshwater wetlands containing perennial and seasonal creeks, ponds, marshes, vernal pools and swales, floodplains, riparian forests, oak woodlands, and grasslands. Vernal pools are temporary seasonal wetlands that once occurred throughout California grasslands, providing habitat for many rare endemic organisms. The Santa Rosa Plain, immediately east of the Laguna de Santa Rosa, contains the remaining 15% of this unique habitat type in the region, lost in recent decades due to changes in land use. The permanent waterways, marshes, and floodplain of the Laguna de Santa Rosa capture and slow the storm waters that drain the entire watershed each winter and provide year-round habitat for local and migrating wildlife, and the drainage likely functions as an important wildlife corridor and refuge for animals.

The Laguna de Santa Rosa Wetland Complex is habitat to several animals, plants, or plant communities considered rare, threatened, or endangered by national, state, and international legislation. The Santa Rosa Plain vernal pool complex has ecologically distinctive flora and fauna. The unique ecological community includes four plant species that are listed on both the U.S. Federal and the State of California's endangered species list. Three of these four plants are endemic and occur primarily on the Santa Rosa Plain of Sonoma County.

Listed endangered, threatened, or special concern plants native to the Laguna de Santa Rosa Wetland Complex include:

- Sonoma sunshine (Blennosperma bakeri), restricted to remaining vernal pool habitats within the Santa Rosa Plain and Sonoma Valley, Sonoma County. According to California Natural Diversity Database and the Laguna de Santa Rosa Foundation Vernal Pool Survey database, there are currently 23 extant occurrences, while approximately 30% of the historic occurrences have been eliminated or seriously damaged. Most of the remaining sites are threatened by urbanization, wastewater effluent irrigation, agricultural land conversion, and competition from non-native invasive species. Westward expansion of the City of Santa Rosa threatens at least half the remaining habitat (USFWS 1991);
- Sebastopol meadowfoam (Limnanthes vinculans) occurs in seasonally wet meadows, swales, and vernal pools in the Santa Rosa Plain, Sonoma County, and with one occurrence in Napa County, according to the California Natural Diversity Database and the Laguna de Santa Rosa Foundation Vernal Pool Survey database; there are 37 extant occurrences on the Santa Rosa Plain. Any activities that cause the destruction of the plants or hydrologic changes in their habitats represent the primary threats to the species, such as urbanization, industrial development, agricultural land conversion, off-highway vehicle use, horseback riding, trampling by grazing cattle, and road widening (USFWS 1991);
- Burke's goldfields (Lasthenia burkei). This vernal pool species is known only from southern portions of Lake and Mendocino Counties and from northeastern Sonoma County. Historically, 39 populations were known from the Santa Rosa Plain, two sites in Lake County, and one site in Mendocino County;
- California tiger salamander (Ambystoma californiense) is listed as vulnerable in the IUCN Red List, federally threatened, and as a California species of Concern. California tiger salamanders found in the Santa Rosa Plain in Sonoma County are geographically separated from other California tiger salamander populations;
- Coho salmon (Oncorhynchus kisutch), which is listed as federally endangered and state threatened, occurs in Mark West and Santa Rosa creeks;
- Steelhead trout (Oncorhynchus mykiss), which is federally threatened, occurs in Mark West, Santa Rosa, and Copeland creeks;
- California freshwater shrimp (Syncaris pacifica), which is federally endangered, occurs in the Blucher Creek, a tributary to the Laguna de Santa Rosa;

- Western pond turtle (Clemmys marmorata) is listed as vulnerable in the IUCN Red List (2007) and as a California species of concern-There are remnant populations reported in the wetland complex; and
- American bald eagle (Haliaeetus leucocephalu), a species of special concern.

The two major human settlements on the periphery of the site, Sebastopol and Santa Rosa, are both very bustling areas with local and visitor traffic; as you leave Santa Rosa and make your way westward toward the town of Sebastopol, you find yourself in a peaceful gap between the two towns where the small unincorporated area of Llano sits, currently containing very few structures and residents, and which lies just to the east of the main area of the wetland complex.

Archaeological data in Southern Sonoma County indicate that the land surrounding the Laguna de Santa Rosa was under the control of three Pomo sub-tribes, which together controlled about 90,650 ha. Current land use of the entire Laguna de Santa Rosa watershed consists of urban and rural areas that include approximately 2200 ha of vineyards, 1600 ha of pasture, 1100 ha of dairies, and 500 ha of mixed agriculture. Land-use within the 100-year floodplain, where the largest portion of the site is, includes hay production, dairies, pastures, vineyards, horses, cattle, poultry, and a wastewater treatment facility.

TOMALES BAY, CALIFORNIA, USA: OVERVIEW

Tomales Bay (Figures 2.8 and 2.9) is a marine-coastal wetland consisting of geomorphologically dynamic estuaries, eelgrass beds (Zostera marina), sand dune ecosystems, and restored emergent tidal marshes, which floods the northern 20 km of the San Andreas Fault-generated Olema Valley on the Central California coast. Approximately 90% of the bay's 28.5 km^2 area is subtidal with a much greater area of open water at low tide than most other Pacific coast estuaries, thus becoming suitable waterbird habitat through the tidal cycle. Because the 58,000 ha watershed is non-industrial and has a small human population density, the bay is relatively pristine. The site supports several endangered or threatened plant and animal species, and is an important waterbird migratory stopover site and over-wintering ground along the Pacific flyway; the site regularly hosts over 20,000 individuals in the winter months, most notably the surf scoter (Melanitta perspicillata), bufflehead (Bucephala albeola), and greater scaup (Aythya marila). In the past, the site has been affected by industrial and agricultural activities, which have since been terminated or mitigated. Local authorities and private and non-governmental organizations have conducted a number of watershed protection measures, and conservation and restoration projects over the past 40 years, in the area.

SITE DETAILS

The abundance and diversity of waterbirds supported by Tomales Bay, especially during the autumn, winter, and spring months, indicate its importance as a migratory stop-over site and over-wintering ground along the Pacific flyway. Various studies have documented its biogeographic importance as waterbird habitat (Shuford et al. 1989, Kelly and Tappen 1998, Kelly 2001) and its existence as part of a complex of estuarine habitats (Bolinas Lagoon, Bodega Bay, Abbott's Lagoon, Estero de Limantour, and Drakes Estero) that support a diverse array of avian species. The importance of the estuary for fish is also indicated by the diversity of species present, and its historical importance as a spawning ground for Pacific herring and coho salmon. Eelgrass beds at the site are critical foraging habitat for migratory black brant (Branta bernicla nigricans) during spring and fall and support an estimated 30% of the California population during winter (Kelly and Tappen 1998). Numerous marine invertebrates are also associated with this habitat. The tidal marsh and transitional vegetation provides important refuge for the threatened California black rail (Laterallus jamaicensis coturniculus) and other salt marsh dependent species during periods of extreme tidal inundation (Evens and Page 1996). Tomales Bay may provide the most important winter habitat for bufflehead

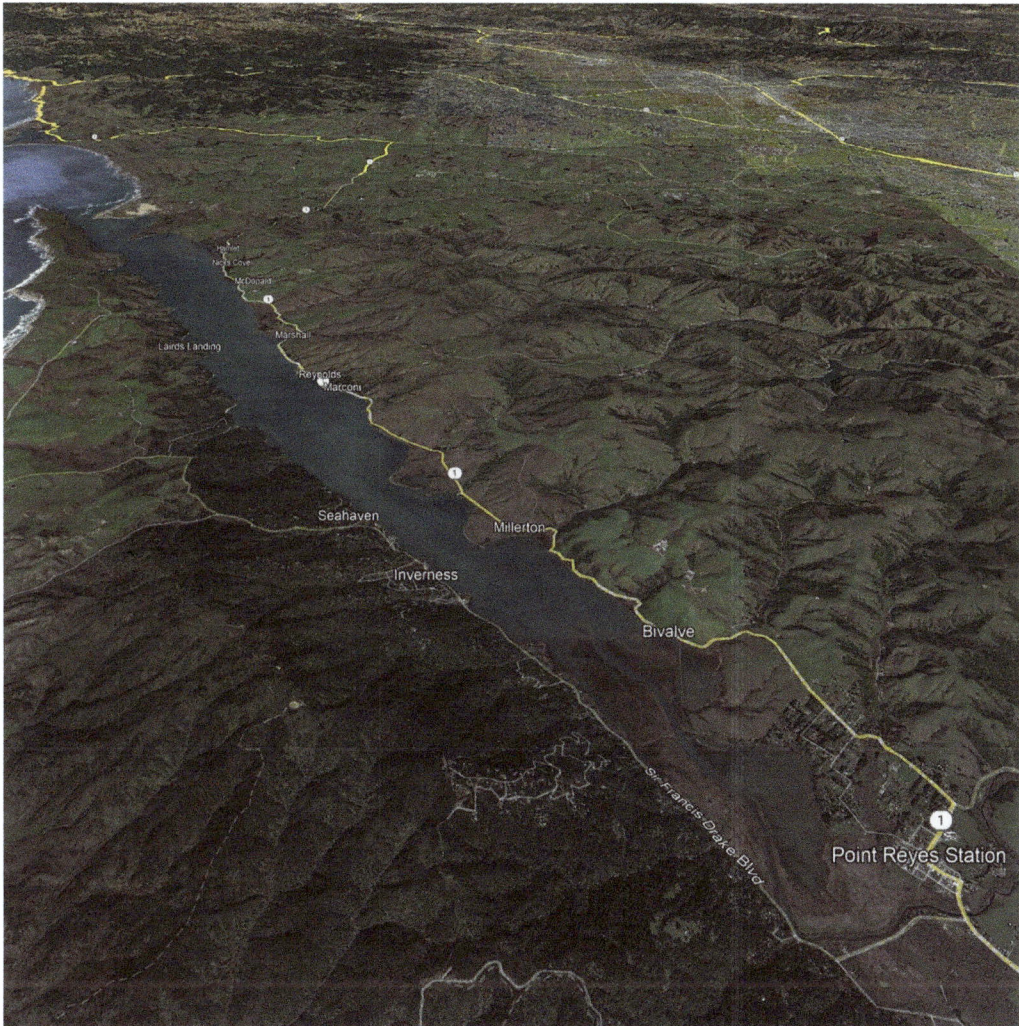

FIGURE 2.8 Composite aerial image of entire Tomales Bay. Designated September 30, 2002; Area: 2,850 ha; Coordinates: 38°09′N 123°23′W; Ramsar site number 1,215.

on the west coast, south of the Columbia River; dunlin and western sandpiper also utilize Tomales Bay as either a wintering ground or a migratory stopover point.

The historic tidal wetlands (aka "Giacomini Marsh") at the southern end of the bay also support viable populations of the federally threatened California red-legged frog (Rana aurora draytonii). Tomales Bay also provides an important nursery for leopard shark (Triakis semifasciata), bat ray (Myliobatis californica), a variety of surfperch (Embiotocidae), rockfish (Scorpaenidae), sculpin (Cottidae), and scale fish (Pleuronectidae), including the Pacific halibut (Hippoglossus stenolepis). Topsmelt and jacksmelt, two abundant species that breed in the eelgrass beds, provide the primary food source for the osprey population that breeds within the Tomales Bay watershed, the largest concentration of osprey in California (Evens 1995).

Noteworthy fauna at the site include:

- Fish: Pacific herring (Clupea harengus pallasi) that spawn in the eelgrass beds with a 20 year average of approximately 5,500 tons, making Tomales Bay one of the major spawning sites in California (Moore and Mello 1995). Additionally, three federally threatened or endangered

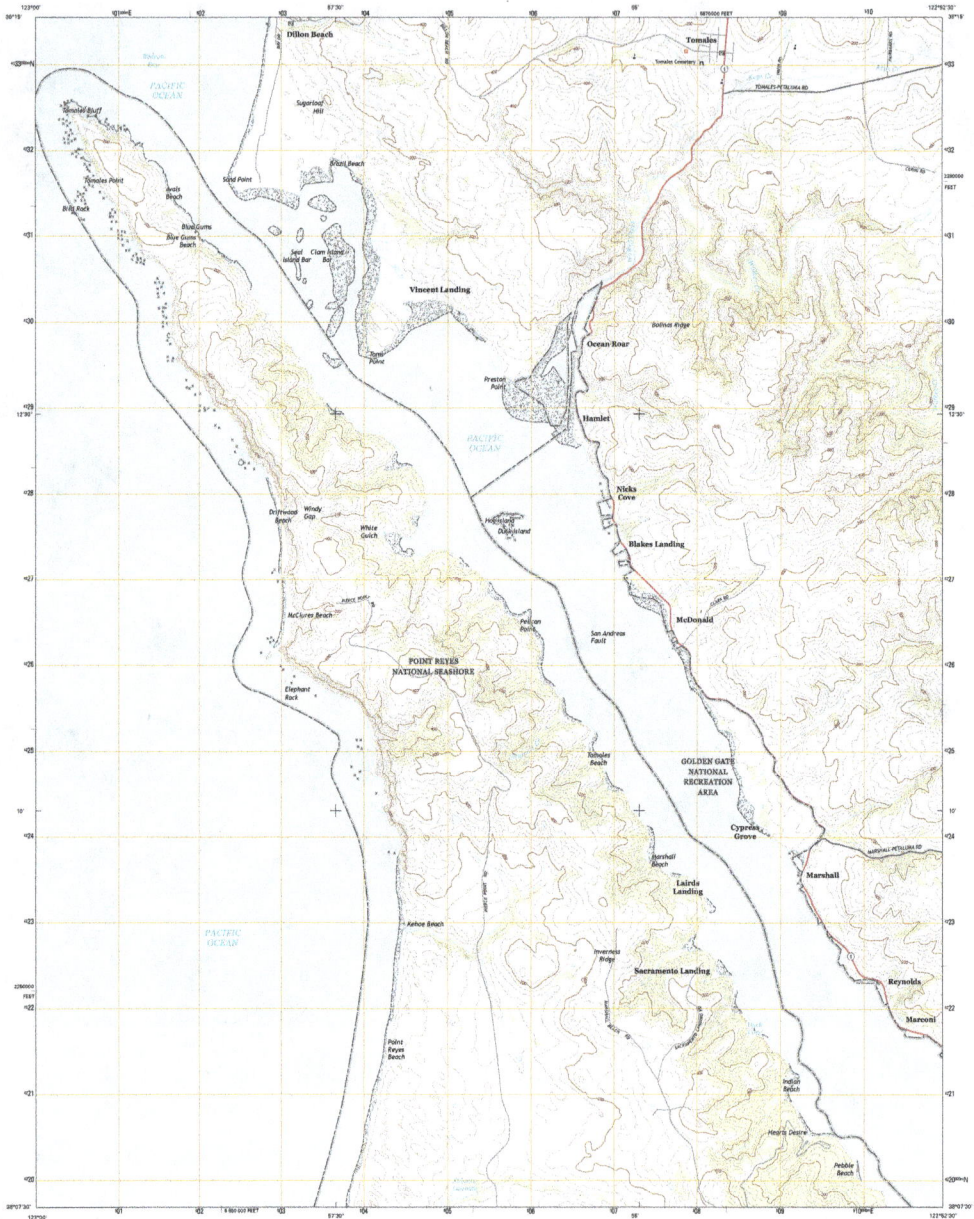

FIGURE 2.9 Map of Tomales Bay, northwest of Marconi, California. Designated September 30, 2002; Area: 2,850 ha; Coordinates: 38°09′N 123°23′W; Ramsar site number 1,215.

taxa breed in the waters of Tomales Bay and its tributaries: tidewater goby (Eucyclogobius newberryi), steelhead (Oncorhynchus mykiss), and coho salmon (Oncorhynchus kisutch);

- Birds: 51 species of waterbirds with an average of 21,943 individuals, with a mean density of 770 birds/km² (Kelly and Tappen 1998); and
- Mammals: Point Reyes peninsula, in this region, supports the largest population of harbor seals (Phoca vitulina) in California and a large proportion of those animals utilize Tomales Bay as a traditional pupping and resting site (Allen 1985, Evens 1993).

The following rare, threatened, and endangered species occur in Tomales Bay: Tomales isopod (Caecidotea tomalensis), California freshwater shrimp (Syncaris pacifica) [federally endangered],

Myrtle's silverspot butterfly (Speyeria zerene myrtleae) [federally endangered], green sturgeon (Acipenser medirostris) [species of special concern], coho salmon (Oncorhynchus kisutch) [state endangered, federally threatened], steelhead (Oncorhynchus mykiss) [federally threatened], tidewater goby (Eucyclogobius newberryi) [federally endangered], California red-legged frog (Rana aurora draytonii) [federally threatened], foothill yellow-legged frog (Rana boylii) [state species of special concern], western pond turtle (Clemmys marmorata) [state species of special concern], brown pelican (Pelecanus occidentalis californicus) [federally endangered], great blue heron (Ardea herodias) [state Sensitive], western least bittern (Ixobrychus exilis) [state species of special concern], Aleutian Canada goose (Branta canadensis leucopareia), hen harrier (Circus cyaneus) [state species of special concern], California black rail (Laterallus jamaicensis coturniculus) [state threatened], California clapper rail (Rallus longirostris obsoletus) [federally threatened], western snowy plover (Charadrius alexandrinus nivosus) [federally threatened], and northern spotted owl (Strix occidentalis caurina) [federally threatened].

Mariculture is an important cultural and economic human activity at the site. Oysters (Crassostrea gigas and C. virginica), California flounder (Paralichthys californicus), herring (Clupea pallasi), and rock crab (Cancer spp.) are all commercially harvested from Tomales Bay. Herring has been harvested in Tomales Bay since the early 1800s, in some years providing up to 75% of the total regional harvest. The commercial halibut take also fluctuates due to environmental variables. The predominant land use in the watershed for over 100 years has been animal agriculture (beef, dairy, and sheep production). Currently there are approximately 25 ranches operating in the watershed, about half dairy and half beef. Ranching families and ranch workers comprise the dominant influence on the social fabric of the communities surrounding the bay. Archeological sites associated with the Coast Miwok native peoples exist at various sites around the shore. All are protected by public entities (National Park Service, California Parks) and indicate that Tomales Bay was historically an important focal point for Miwok culture. More than a hundred village sites have been identified. Most of these are along the west shore of Tomales Bay (Thalman 1993). Indeed, the name "Tomales" probably has its origins in the Miwok word "tamal," meaning 'bay.'

Because the bay waters are within a marine sanctuary and are protected by state and federal laws, all commercial and recreational activities are regulated by the National Oceanographic and Atmospheric Administration (NOAA), the National Park Service (NPS), or the California Department of Fish and Game. Human uses of Tomales Bay include restricted commercial mariculture, commercial and recreational fishing, clamming, and boating (sailing, sea kayaking, motor boating, and windsurfing). Land use practices in the surrounding watershed are dominated by agricultural production (primarily beef and dairy), National Park lands (designated wilderness and low-impact human use areas), rural residential, and municipal watershed activities. Land-based recreational use includes hiking, birdwatching, horse riding, and mountain biking.

BOLINAS LAGOON, CALIFORNIA, USA: OVERVIEW

Located in California, less than 20 km north of San Francisco's Golden Gate Bridge, Bolinas Lagoon (Figures 2.10 and 2.11) is a tidal embayment of open water, mudflat, and marshes, which provides productive and diverse habitats for marine fishes, waterbirds, and marine mammals; it is also part of a much larger protected natural habitat complex in the region. The site is located on the Pacific Flyway, which makes the lagoon an ideal staging ground and stopover site for migratory birds, and the climate provides wintering habitat for a wide array of ducks, geese, and shorebirds. The area supports a number of recreational uses, including a variety of watercraft.

This 445 ha tidal embayment of open water, mudflat, and marsh areas is one of the most popular areas for local residents of, and international visitors to, the San Francisco Bay area, which has a total approximate population of 7.75 million residents. Bolinas Lagoon is approximately 27 km northwest of downtown San Francisco, a tranquil lagoon setting, with relatively few influences from people, pollution, and other disturbances, due to its relatively isolated location. Two rural routes connect to the lagoon from the nearest large towns, Novato and San Rafael. Point Reyes National Sea Shore, a 28,733 ha national park, lies immediately to the north of Bolinas Lagoon and is an expansive area of beaches and hiking trails.

FIGURE 2.10 Composite aerial image of Bolinas Lagoon. Designated September 1, 1998; Area: 445 ha; Coordinates: 37°55′N 122°41′W; Ramsar site number 960.

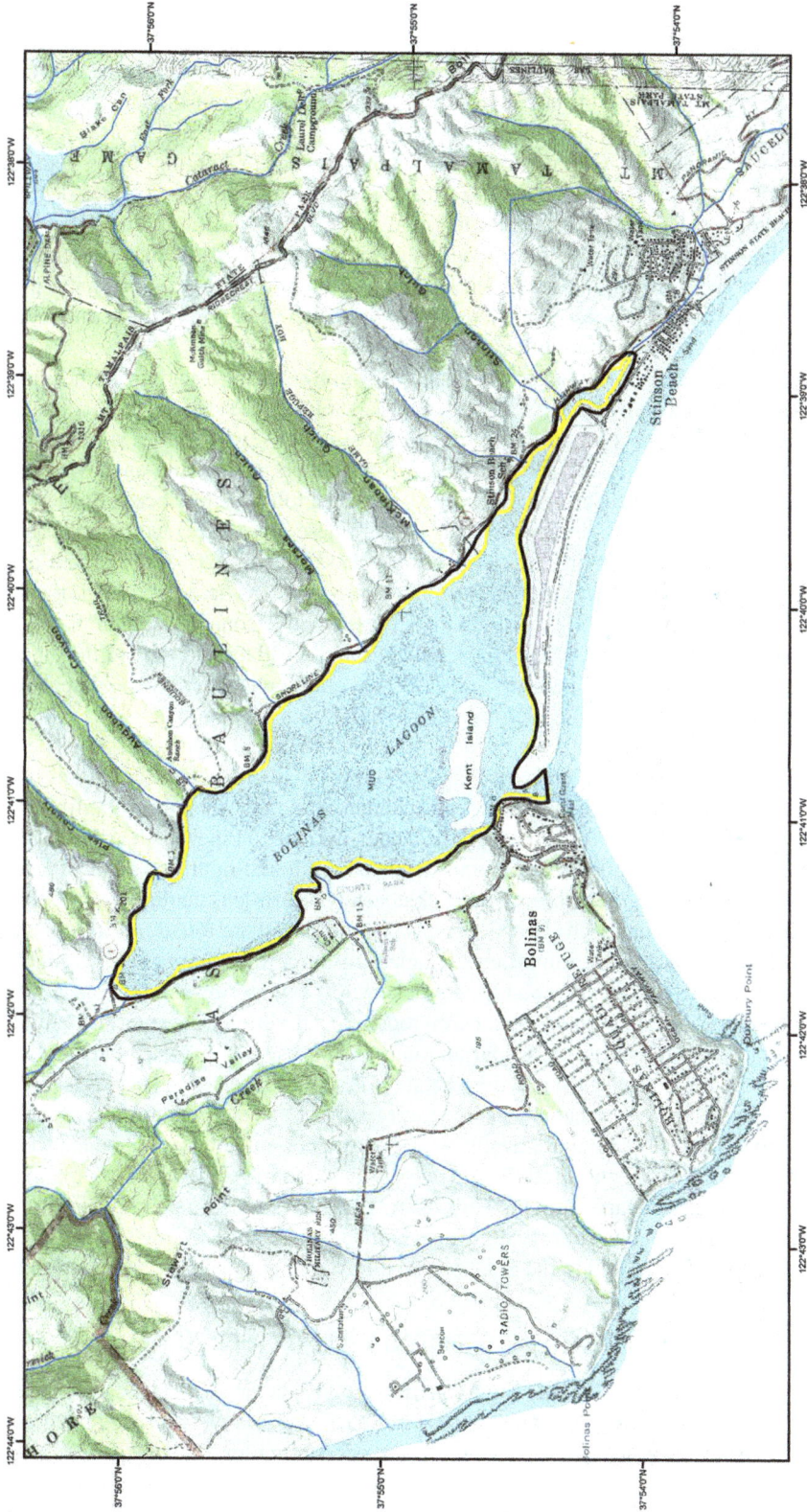

FIGURE 2.11 Map of Bolinas Lagoon. Designated September 1, 1998; Area: 445 ha; Coordinates: 37°55′N 122°41′W; Ramsar site number 960.

SITE DETAILS

Native Americans inhabited the land surrounding Bolinas Lagoon prior to the arrival of Spanish-American settlers, but humans did not significantly impact the watershed until the onset of farming, logging, and cattle grazing in the latter half of the 19th century. In the latter 1800s the area of Bolinas Lagoon and vicinity was dominated by logging and associated sea-faring activities, including timber transportation. Immediately following the end of logging, harvested land was exploited for cordwood production, dairy farming, and cattle grazing that further degraded the land and prevented the watershed from naturally recovering (Philip Williams and Associates, Ltd. 2006). On the far southern edge of Bolinas Lagoon lies the outer sandbar, which presents a barrier to incoming wave energy from the sea. Such an outer beach barrier is a fundamental protective force for this and similar types of wetlands, increasing the retention time of the sediment entering the lagoon within the relatively tranquil waters of the site.

Resident fish species at Bolinas Lagoon include arrow goby, staghorn sculpin, shiner surf perch, and other small channel dwelling species. Some of the schooling and surface-feeding fish like jack-smelt and topsmelt may enter on tidal cycles during most months, while other species (anchovies and herring) are typically episodic and seasonal. Vast numbers of juvenile anchovies migrating northward sometimes enter the lagoon and these fish are often followed by flocks of brown pelicans and terns (Philip Williams and Associates, Ltd. 2006). These episodic events are determined by oceanographic conditions, occurring in warm water periods during late summer and early fall. Pacific herring are seasonal visitors, but Bolinas Lagoon is not considered a spawning ground for this species (Spratt 1981, Suer 1987). Bird numbers can also give some indication of the biomass of fish that enter the Lagoon, with 3,800 brown pelicans and 3,700 terns observed in August 1985; 6,000 terns in August 1985; 6,000 pelicans and 2,500 terns in September 1984; and approximately 2000–3000 terns in September 1984 (Shuford et al. 1989).

Juvenile leopard sharks and bat rays occur on the tidal flats and adults of both species enter Bolinas Lagoon regularly to forage on large clams and probably to breed. Concentrations of leopard sharks in summertime occur on channel edges and sandier tidal flats where they are likely depositing eggs. Anadromous salmonids pass through the lagoon in route to Pine Gulch and other tributary creeks. Juvenile striped bass, coho salmon, and steelhead are found in all the creeks that feed the Lagoon. Three freshwater species [threespine stickleback (Gaslerosleus aculearus), prickly sculpin (Cottus asper), and California roach (Hesperoleucus symmetricus)] are found in Pine Gulch Creek.

Overall, Bolinas Lagoon is an important biological resource that supports: (1) a high species diversity of aquatic birds; (2) an egret and heron rookery; (3) a wintering site for waterfowl, shore-birds, and raptors; (4) a black-crowned night heron roost; (5) a traditional roost for fish-eating flocks of pelicans, cormorants, and terns; (6) a riparian migrant stop-over (Pine Gulch Creek); (7) habitat for 20 species of special concern; (8) breeding habitat for several threatened species (snowy plover and black rail); and (9) foraging habitat for several raptors of special concern (osprey, peregrine falcon, and merlin). In addition to these relatively rare bird species, there are an abundance of other common bird species, such as gulls and the turkey vulture (Cathartes aura). The population of harbor seals (Phoca virulina richardsii) in the Gulf of the Farallones is estimated to comprise 20% of the California population (Allen et al. 1989), which is found at the site sunning and resting. Harbor seals have been closely monitored in the San Francisco Bay area and at Bolinas Lagoon since 1970.

SAN FRANCISCO BAY/ESTUARY, CALIFORNIA, USA: OVERVIEW

The San Francisco Bay/Estuary (SFBE) [Figures 2.12 and 2.13] is the largest estuary on the Pacific Coast of the US, encompassing approximately 160,000 ha. SFBE is widely recognized as one of North America's most ecologically important estuaries, accounting for 77% of California's remaining perennial estuarine wetlands and providing key habitat for a broad suite of flora and fauna and a range of ecological services such as flood protection, water quality maintenance, nutrient filtration and cycling, and carbon sequestration. The site is home to many plant species and over 1,000

FIGURE 2.12 Composite aerial image of San Francisco Bay/Estuary (SFBE). Designated February 2, 2013; Area: 158,711 ha; Coordinates: 37°52′N 122°22′W; Ramsar site number 2,097.

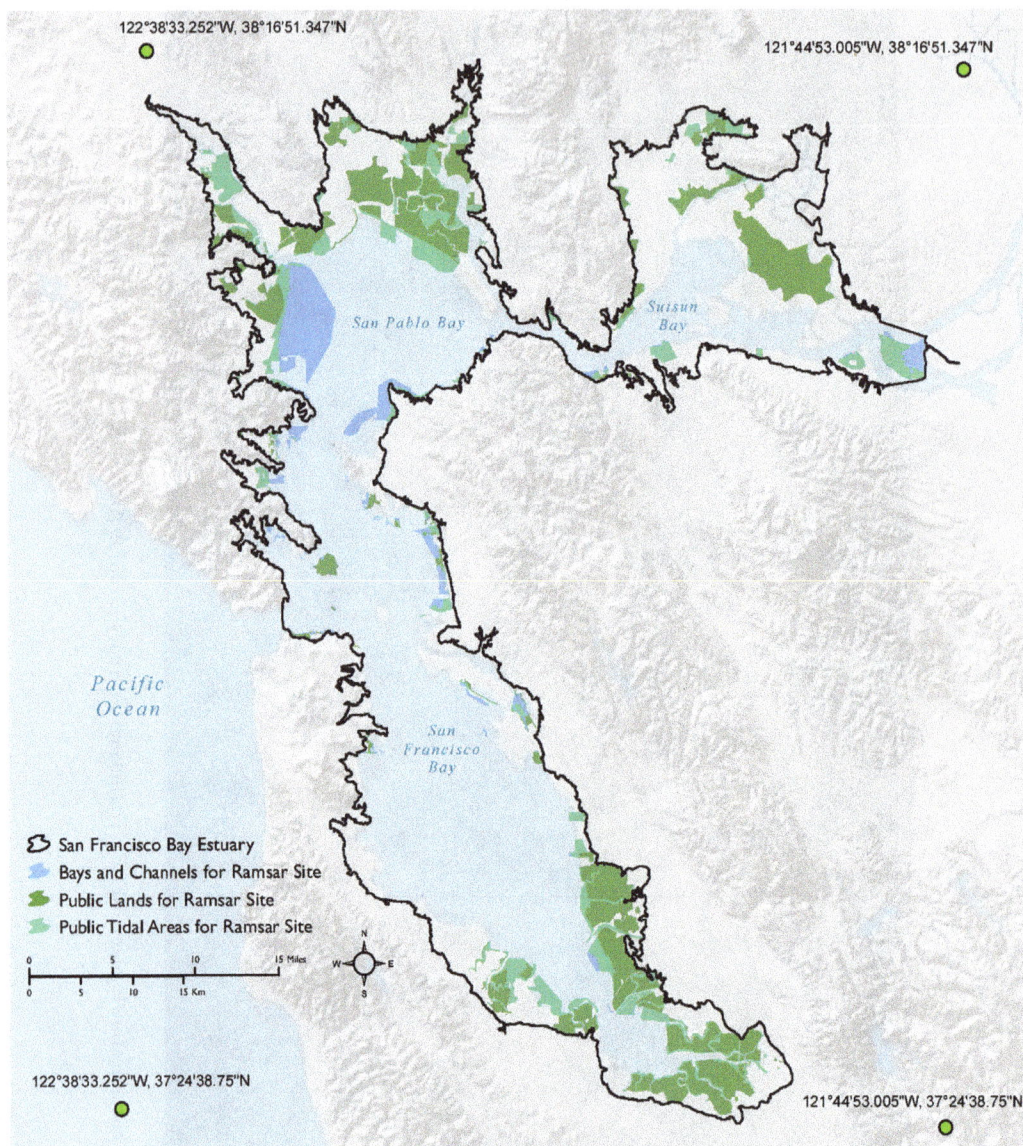

FIGURE 2.13   Map of San Francisco Bay/Estuary (SFBE). Designated February 2, 2013; Area: 158,711 ha; Coordinates: 37°52′N 122°22′W; Ramsar site number 2,097.

species of animals, including endemic and conservation status species. The site is noted for hosting more wintering shorebirds than any other estuary along the US Pacific Coast south of Alaska and is recognized as a site of Hemispheric Importance by the Western Hemisphere Shorebird Reserve Network. The site is also important for over 130 species of resident and migratory marine, estuarine, and anadromous fish species. Development pressures on the site's remaining wetlands and adjacent uplands continue to threaten habitats not owned or managed for conservation. The site is also a renowned international tourism destination.

SFBE is located along the Pacific Coast of Central California, United States. It is bordered by counties: San Francisco, San Mateo, Santa Clara, Alameda, Contra Costa, Solano, Napa, Sonoma, and Marin. Situated mid-site, the most prominent large city within the site is San Francisco, home to a human population of approximately 815,000. SFBE provides for a host of social and economic values through ports and industry, agriculture, fisheries, archaeological and cultural sites, recreation, and research.

SITE DETAILS

At the time of European settlement, tens of thousands of Native American Ohlone people inhabited the SFBE, south to Monterey Bay, in villages and camps. The Muwekma Ohlone Tribe were the original inhabitants of San Francisco and the surrounding Bay Area. The Spanish established a fort (the Presidio) in San Francisco in 1776, and by the mid-1800s, the immigrant human population had swelled as a result of gold mining, which began the major transformations of the SFBE watershed that we see today.

San Francisco is the original urban center of the region, although San Jose is the largest city in the region (population approximately 1 million). Land use in the SFBE region in modern times has seen increasing urban development and industry, agriculture (farming and grazing), and salt extraction. Specifically, nearly 50% of the estuary's watershed has been converted to agriculture; about 4% has been urbanized, and 10% is now industrial sites. Large areas of former marshlands were filled during urbanization, and diked for agriculture or converted to salt evaporation ponds, thus dramatically reducing the overall acreage of tidal wetlands in the watershed today (Cohen 2000). Salt ponds do provide some habitat for shorebirds.

SFBE and associated wetlands support 21 animal and 5 plant species that are listed as threatened or endangered by the United States of America and/or California governments. SFBE is noted for hosting more wintering and migrating shorebirds than any other estuary along the U.S. Pacific Coast south of Alaska (Save the Bay 2007). For this, SFBE is recognized as a Site of Hemispheric Importance by the Western Hemispheric Shorebird Reserve Network. Bay-wide surveys conducted each fall of 2006–2008 averaged over 340,000 shorebirds, including 29 species (Leidy 2007). 589,000–932,000 shorebirds were counted during spring surveys (during the height of migration) conducted between April 1988 and April 1993 (Page and Kjelmyr 1999). Compared to the major wetlands along the Pacific Coast, SFBE supports an average of 55.7% (37.8%–90.1%) of the total number of individuals of 13 key shorebird species. In particular, a significant portion of arctic-breeding dunlin and western sandpipers winter in the SFBE.

The SFBE supports three fish taxa endemic to the bay and nearby waters: delta smelt (Hypomesus transpacificus), San Francisco topsmelt silverside (Atherinops affinis affinis), and tule perch (Hysterocarpus traskii traskii), in addition to four local varieties of chinook salmon (Onchorhynchus tshawytscha). The federally endangered delta smelt occurs only in the San Francisco Bay-Delta Estuary; this species spends much of its lifecycle in the Sacramento and San Joaquin Rivers and deltas that feed into SFBE. Juvenile and adult smelt may also spend time in the northern SFBE, where they have been observed in Suisun and San Pablo Bays, as well as the Napa River (Shuford and Gardali 2008). Although exact population estimates are unknown (Stenzel et al. 2002), relative population levels have been monitored for several decades by federal and state water export facilities (Takekawa 2006). Counts from 2002 to 2007 showed low abundance (USACOE et al. 2001).

The tidal wetlands of SFBE support the entire population of the salt marsh harvest mouse (Reithrodontomys raviventris), a federally-listed endangered species, distributed around the bay's marshes in small, disjunct populations, often in marginal vegetation and almost always in marshes without an upper edge of upland vegetation. The salt marsh wandering shrew (Sorex vagrans haliocetes), a subspecies of the vagrant shrew, is currently found only within the tidal marshes of South San Francisco Bay, and is currently listed as a mammalian species of special concern by the California Department of Fish and Game. The Suisun shrew (Sorex ornatus sinuosis) is currently limited to scattered, isolated remnants of natural tidal salt and brackish marshes surrounding the northern borders of Suisun and San Pablo Bays. The endangered San Francisco garter snake (Thamnophis sirtalis tetrataenia) is endemic to the San Francisco Bay, with most of its population located in marshes and nearby habitats along the bay side and Pacific Ocean side of San Mateo County.

The present San Francisco Bay Estuary is approximately 10,000 years old and is a product of today's high sea level, which presently floods the ancient river drainage from California's Central Valley out to the Pacific Ocean through the Golden Gate. About 90% of the freshwater entering the bay comes from the Sacramento/San Joaquin Delta, with the remaining approximate 10%

originating from local streams and creeks and from wastewater treatment facilities. Today's fresh-water inflow to the bay is about 60% less than the historic flows due to diversions of municipal water (responsible for about 9% of flow reductions) and for California's Central Valley agricultural uses (responsible for about 51% of flow reductions). The average depth of the SFBE is 5.5 m, with deeper water areas at the mouth (107 m) and in the main channels (9–27 m).

The main native habitats in SFBE are primarily open water of varying depths based on tidal conditions and location; tidally influenced mudflats: submerged eelgrass beds; vegetated marshes; sand and salt flats; and sandy and cobble beaches. There are also several rock islands within the SFBE. Pickleweed (Salicornia pacifica) and cordgrass (Spartina foliosa) predominate the marshes in more saline waters, and bulrush (Scirpus spp.) predominates in freshwater dominated marshes. Many parts of the bay have been altered by human activities but still provide habitat for many species; they include diked marshes, agricultural baylands (grazed and farmed), and salt ponds. Upland habitats exist within the bay in the form of islands, some of which are human-made (fill) and some of which are natural and still contain native upland plant communities consisting of grasslands, shrublands, and woodlands.

Noteworthy flora at the site include submerged aquatic plant communities of the shallow subtidal habitats and tidal flats, which are important food sources for estuarine fish, invertebrates, and birds. Submerged plants such as eelgrass (Zostera marina) and certain macroalgae also provide impor-tant cover, spawning, and rearing grounds for invertebrates and estuarine fish, such as migrating salmon and Pacific herring. Eelgrass, surfgrass (Psyllospadix scouleri and P. torreyi), widgeon grass (Ruppia maritima), and sago pondweed (Potamogeton pectinatus) provide important nursery and foraging habitats, dampen wave energy, and aid in sediment capture.

Noteworthy fauna include eight animal species that are endemic to the SFBE and associated wetlands, which provide habitat for a number of near-endemic or range-limited species, subspecies, and varieties of flora and fauna. The bay supports three endemic fish taxa: the federally endan-gered delta smelt (Hypomesus transpacificus), San Francisco topsmelt silverside (Atherinops affinis affinis), and the tule perch (Hysterocarpus traskii traskii), in addition to four local varieties of chi-nook salmon (Onchorhynchus tshawytscha). In addition to the endemic salt marsh harvest mouse (Reithrodontomys raviventris), tidal wetlands of the site support the majority of extant California clapper rails (Rallus longirostris obsoletus).

GRASSLAND ECOLOGICAL AREA, CALIFORNIA, USA: OVERVIEW

Located in the Central Valley in the San Joaquin River Basin, the Grassland Ecological Area (GEA) [Figures 2.14 and 2.15] is the largest remaining contiguous block of freshwater wetlands in California. It consists of semipermanent and permanent marshes, riparian corridors, vernal pool complexes, wet meadows, native uplands and grasslands, featuring alkali sacaton (Sporobolus airoi-des) grassland and endemic delta button celery (Eryngium racemosum). The site is renowned for its wintering waterbirds, which reach several hundred-thousand every winter, including sandhill cranes (Grus canadensis), 19 duck species (northern pintail, Anas acuta; green-winged teal, Anas crecca; northern shoveler, Anas clypeata; canvasback, Aythya valisineria; and others), 6 species of geese, tens of thousands of shorebirds (most abundantly western sandpiper, Calidris mauri; dunlin, Calidris alpina; and long-billed dowitcher, Limnodromus scolopaceus). The site is home to four endangered shrimps as well as the threatened giant garter snake (Thamnophis gigas). Due to flood-control and irrigation projects, the entire hydrology of the valley has been dramatically altered, with more contemporary water quality issues specifically addressed with the Central Valley Project Improvement Act of 1992. Most of the wetlands at the site are managed by the controlled applica-tion of water to the site, using a series of canals and control structures, mimicking historical flood patterns, with pulses of high water flow during winter and spring. The largest potential threat to the site is urban development.

FIGURE 2.14   Composite aerial image of Grassland Ecological Area (GEA). Designated February 2, 2005; Area: 65,000 ha; Coordinates: 37°10′N 120°49′W; Ramsar site number 1,451.

SITE DETAILS

The GEA is located in western Merced County, California. Adjacent to the city of Los Banos (estimated population 46,000), the GEA is 129 km northwest of Fresno (estimated population 545,000), and 80 km south of Modesto (estimated population 219,000). It lies within the San Joaquin River Basin in California's Central Valley and San Francisco Bay Ecoregion. Elevation of the site varies from 18 to 40 m above sea level, with approximately 65%, or 44,111 ha, of the GEA made up of wetlands, including managed wetlands and unmanaged wetland habitats such as sloughs, oxbows,

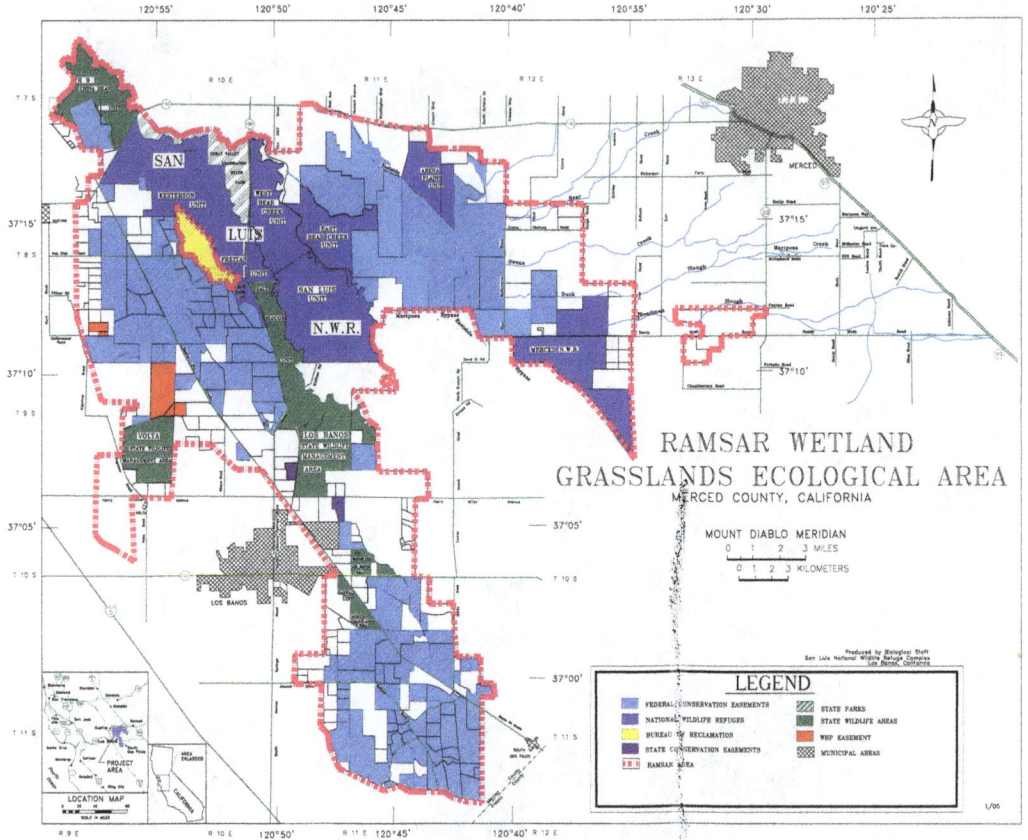

FIGURE 2.15 Map of Grassland Ecological Area (GEA). Designated February 2, 2005; Area: 65,000 ha; Coordinates: 37°10′N 120°49′W; Ramsar site number 1,451.

riparian corridors, vernal pools, and low-lying floodplains subject to periodic inundation. As the largest remaining contiguous block of freshwater wetlands remaining in California, the site consists of a matrix of federal, state, and privately owned seasonal, semipermanent, and permanent marshes, riparian corridors, vernal pool complexes, and grasslands.

The Central Valley of California once consisted of over 1.6 million hectares of wetlands, yet only 5% now remain, mostly due to drainage and conversion to agricultural use. The GEA is thus a unique and critical resource for wildlife. In addition, the GEA is a prime example of a vernal pool complex, with an ecologically distinctive flora and fauna, which is an increasingly rare habitat type (e.g., 90% of California's vernal pools have been destroyed). Riparian habitat in California has been reduced to just 2% of what it was a century ago. The GEA contains riparian habitat along 30 km of the San Joaquin River and about 60 km of other riparian habitat on tributaries to the San Joaquin River, a river used by chinook salmon and other anadromous fish. This segment of the river and associated floodplain are important to water quality as well. As one of the largest remaining vernal pool complexes, the GEA is home to many rare or imperiled species associated with this disappearing habitat, including fairy shrimp (Branchinecta conservatio), longhorn fairy shrimp (Branchinecta longiantenna), vernal pool fairy shrimp (Branchinecta lynchi), vernal pool tadpole shrimp (Lepidurus packardi), California tiger salamander (Ambystoma californiense), and colusa grass (Neostapfia colusana). Other endangered species supported by the GEA include the San Joaquin kit fox (Vulpes macrotis mutica), bald eagle (Haliaeetus leucocephalus), and the western snowy plover (Charadrius alexandrinus nivosus).

The GEA lies within the southern part of the California's Central Valley and is drained by the San Joaquin River and its tributaries. It is considered part of the San Joaquin Valley, which is actually not a valley but the bed of an ancient inland sea filled by alluvial sediments over thousands of years. The San Joaquin Valley meets the Sacramento Valley at Stockton, California where their combined deltas meet at the San Francisco Bay (the east bay area). The Valley is bordered on the east by the Sierra Nevada Range and on the west by the Coast Range. Topographical relief within the GEA is minimal and consists of gentle relief, often associated with swales. Hydrological patterns were historically influenced in the northern grasslands by flood flows caused by melting snow-pack carried by the San Joaquin River and other stream flows, and by rainfall itself. The wide flood plains and shallow water tables of the area promoted recharge of the shallow water table and supported thousands of hectares of marshlands. Historically, the wetlands at the site were primarily riparian in nature and consisted of a complex network of deep slough channels and shallow swales. Based on the severity and timing of flood flows, these wetlands provided a diverse complex of temporary, seasonal, semi-permanent, and permanent wetland habitat types.

The area consists of a number of different ecological communities including riparian cottonwood/willow forest, riparian oak woodlands, freshwater marshes, vernal pools, alkali sinks, and native alkali grasslands. It is a dynamic landscape, with temporal variation both seasonally and annually, as well as spatial variation across its expanse of lands. It serves as critical habitat for waterfowl in the winter, and serves as a migratory corridor for neotropical songbirds, shorebirds, and raptors in the fall and spring, as well as year-round habitat that supports a variety of resident species. The site is one of only a few small remnants of a formerly vast Central Valley wetland complex, and is therefore critical to protect, enhance, and expand.

Noteworthy flora at the site include: vegetative communities within GEA alkali grasslands, such as alkali sacaton (Sporobolus airoides), saltgrass (Distichlis spicata), and iodine bush (Allenrolfea occidentalis); the alkali sink community of iodine bush, seep-weed (Sueda fruitcosa), alkali heath (Frankenia salina), and Atriplex spp.; and the vernal pool communities with goldfields (Lasthenia spp.), tidy tips (Layia spp.), popcorn flower (Plagiobothrys spp.), purple owls-clover (Castilleja exserta), and downingia (Downingia spp.). Rare plants at the GEA include palmate-bracted bird's beak (Cordylanthus palmatus), hispid bird's beak (C. molle hispidus), delta button celery (Eryngium racemosum), and colusa grass (Neostapfia colusana).

Noteworthy fauna at the site include those that depend on the GEA for vital spring and fall migration stopovers, in that several hundred-thousand shorebirds pause during their long international journeys to feed in the invertebrate-rich shallow wetlands. The grasslands of the GEA also host many sensitive or rare animal species including the Aleutian Canada goose (Branta canadensis leucopareia), San Joaquin kit fox, tule elk, bald eagle, peregrine falcon, Swainson's hawk (Buteo swainsoni), and the tri-colored blackbird (Agelaius tricolor). American bald eagles (Haliaeetus leucocephalus) and peregrine falcons (Falco peregrinus) are usually spotted in GEA, every year.

The Yokut People once lived in the Central Valley, and their cultural artifacts and signs have been found throughout the GEA. In present day, the site is a destination for duck hunters and bird watchers, which is accommodated as an eco-tourism destination that supports the local economy.

ELKHORN SLOUGH, CALIFORNIA, USA: OVERVIEW

Elkhorn Slough (Figures 2.16 and 2.17) is a seasonal estuary comprised of intertidal marshes, mudflats, and seasonal brackish pools. This slough harbors eelgrass beds as well as oyster communities, which provide valuable fish nurseries, and the intertidal mudflats nourish migratory shorebirds. These distinctive estuarine communities are among the rarest and among the most threatened habitat type in California, which has lost approximately 91% of its wetlands in the last 100 years. This biologically rich estuary provides habitat for more than 340 species of birds, with more than 20,000 waterbirds, more than 500 species of invertebrates, and 100 species of fish, including the representative bat ray (Myliobatis californica) and leopard shark (Triakis semifasciata). It also provides key habitat for more than 100 individuals of the southern sea otter (Enhydra lutris nereis) that feed, nest, and nurse in the area. The site also provides diverse ecosystem services, such as pollution control, climate regulation, food provisioning, recreational, educational, and research opportunities,

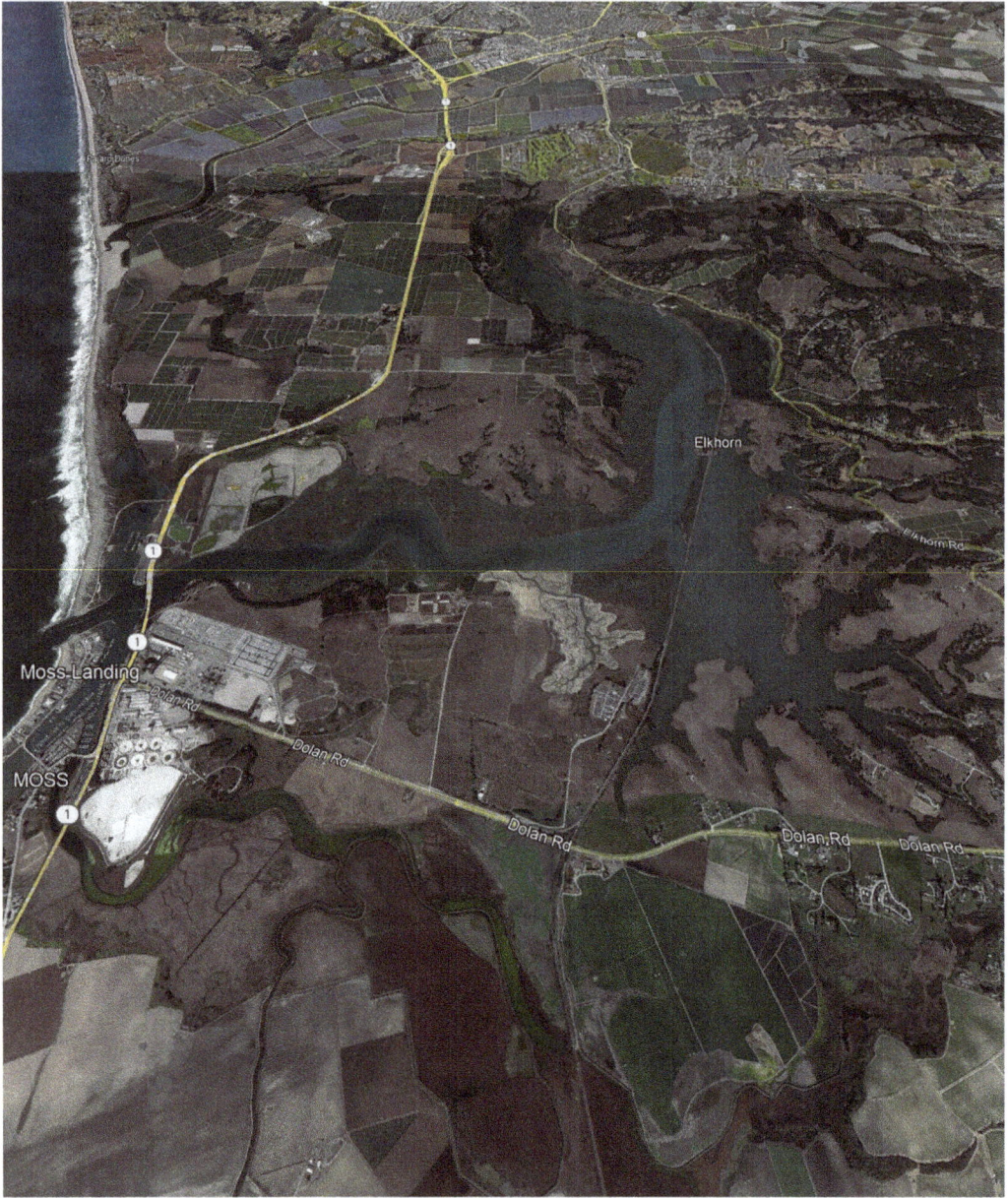

FIGURE 2.16 Composite aerial image of Elkhorn Slough. Designated June 25, 2018; Area: 724 ha; Coordinates: 36°50′N 121°45′W; Ramsar site number 2,345.

attracting thousands of bird watchers every year. The site has also been designated as a Globally Important Bird Area by the National Audubon Society and as a Western Hemisphere Shorebird Reserve Network of Regional Importance.

SITE DETAILS

Near the mouth of the estuary lies the small coastal town of Moss Landing, with a busy harbor that is home to fishing and research boats that work in the ocean offshore, as well as serving as the gateway for kayakers to enter Elkhorn Slough. Carneros Creek is a perennial stream that flows into the

FIGURE 2.17 Map of Elkhorn Slough. Designated June 25, 2018; Area: 724 ha; Coordinates: 36°50′N 121°45′W; Ramsar site number 2,345.

head of the estuary, entering near the northern end of the site. As with most coastal western marshes in the US, Elkhorn Slough is bisected by longshore commercial and industrial railroad traffic.

The Elkhorn Slough estuary supports salt marsh, eelgrass, and oyster communities; all of the biogenic habitats comprised of these foundational species are very rare in California, and have been badly degraded by human activities. Elkhorn Slough hosts a very rich assemblage of indigenous marine and estuarine fish species (Yoklavich et al. 1991, Hughes et al. 2015). About 100 species of fish in 43 different families have been documented in Elkhorn Slough (Caffrey et al. 2002), with the majority of these being endemic species. As the only large estuary in the region, the suite of fish species at the site is globally important as a representation of local diversity. The following are particularly significant and representative species at Elkhorn Slough:

- Elasmobranchs, including leopard sharks (Triakis semifasciata) and bat rays (Myliobatis californica). Sharks and rays are seasonally abundant in the estuary, foraging on clams and worms and leaving distinctive crater-like feeding pits in the mudflats, visible at low tide; they use the estuary as a nursery due to warmer temperatures than offshore, hastening growth and foraging on ample food;
- Gobies (most abundant is the arrow goby, *Clevelandia ios*): these are the most common fish of the estuary, found in every mudflat; and
- Bay pipefish (Syngnathus leptorhynchus): these seahorse relatives are year-round residents that are most abundant in the eelgrass beds of Pacific estuaries, a very limited habitat type; Elkhorn Slough is a rare site in this region that harbors many of these species.

As in other estuarine ecosystems, many fish species spawn in Elkhorn Slough, or use it as a nursery. A number of factors contribute to this habitat use, including an abundant food supply, protection from predation, and the site's character as a thermal refuge with calm waters. An example is the commercially valuable English sole (Parophrys vetulus). A high proportion of the adults of this species caught offshore in the Monterey Bay spent their juvenile period in Elkhorn Slough (Brown 2006), and dissolved oxygen concentrations within Elkhorn Slough correlate with the offshore catch of English sole the following year (Hughes et al. 2015). Elkhorn Slough is designated as Essential Fish Habitat and a Habitat Area of Particular Concern for various fish species life stages managed under the Coastal Pelagic and Pacific Groundfish Fisheries Management Plans of the National Marine Fisheries Service.

Elkhorn Slough is a seasonal estuary and a tidal embayment. During rains, freshwater falls and flows into the slough from the surrounding hills and mixes with salt water carried by tides from Monterey Bay and the Pacific Ocean. These aquatic and terrestrial environments form a complex ecological community that performs many natural and vital functions. The site's roles in these processes are:

- Trapping sediments eroded from the surrounding hills and farms;
- Affording protection from flooding because the slough channels run-off into the bay after heavy storms with the salt marsh acting as a buffer for storm surge;
- Providing habitat and a nursery for fish, with over 80 species of fish known to use the slough waters at some time during their life cycle (including commercial species such as English sole);
- Serving as a way station for tired and hungry birds with over 300 species of birds having been recorded in and around the slough, including resident and migratory birds;
- Supporting habitat for numerous plants and animals, some of which are rare or endangered species; and
- Providing many opportunities for recreation and wildlife viewing.

Kildow and Pendleton (2010) studied the environmental economics of the site but did not undertake a valuation study, however a general estimate of value of the site is in the millions of dollars per year if you include the recreational businesses it sustains (e.g., kayak shops, boat tours, and restaurants), the commercial fisheries it supports offshore (flatfish and crabs), and the other ecosystem services it provides (e.g., shoreline protection from storm and tsunami surges, uptake of nutrients, and carbon sequestration). Elkhorn Slough is visited by tens of thousands of people each year, from all around the USA and beyond, and is a unique site where sea otters can be observed interacting with wetland habitats. The latter use depends upon conservation of the prey items and vegetation types used by sea otters in the estuary.

The site is known as Elkhorn Slough Reserve, which has a Visitor Center open to the public Wednesday-Sunday from 9 am to 5 pm, with exhibits on the estuary and trails. The center hosts school classes in an education lab and provides training for teachers. Associated with these learning opportunities are a number of long-term monitoring efforts, for nutrients, weather, habitat change (geospatial data is collected), and biological indicators (e.g., algal cover in megaplots on mudflats, marsh health along transects, mudflat communities along permanent transects, crabs within traps, oyster recruitment upon tiles, and the monitoring of waterbirds including migratory shorebirds, and breeding birds in heronry). Extensive long-term monitoring programs for the site are described at https://elkhornslough.org/reserve/research/.

TIJUANA RIVER NATIONAL ESTUARINE RESEARCH RESERVE, CALIFORNIA, USA: OVERVIEW

On the border with Mexico facing the city of Tijuana, the Tijuana River National Estuarine Research Reserve (TRNERR) [Figures 2.18 and 2.19] is one of the few unfragmented estuaries and coastal lagoons in Southern California. It is a seasonally marine-dominated estuary receiving freshwater input only during the wet winter period, though its mouth remains open throughout the year. It has several sensitive habitats such as sand dunes and beaches, vernal pools, tidal channels, mudflats and coastal sage scrub. The site is critical habitat for nationally endangered species and subspecies, such as the San Diego fairy shrimp (Branchinecta sandiegonensis), the light-footed clapper rail (Rallus longirostris levipes), the salt marsh bird's beak (Cordylanthus maritimus maritimus), as well as nursery grounds for commercially important fish like the diamond turbot (Hypsopsetta guttulata) and the California flounder (Paralichthys californicus). Dirt roads and border patrol off-road vehicles are a primary cause of impact to the ecology of the site because of lights, noise, and causing erosion, already quite serious due to other erosion and runoff coming from the shared basin with Mexico. The site is isolated from other surrounding habitats, by urban areas, and there are problems with introduced species. A multi-phased restoration program designed to restore tidal exchange and wetland habitats at the site are in place, as well as a management plan. The site is administered jointly by California State Department of Parks and Recreation and the United States Fish and Wildlife Service.

SITE DETAILS

TRNERR is located in the southwest corner of the United States of America and within the extreme southwestern portion of the State of California. The nearest town is Imperial Beach (approximate population 26,000), which borders the TRNERR on the northern boundary of the Reserve. Tijuana (approximate population of 2.3 million) is on the other side of the international border from California, in Mexico. The TRNERR separates Tijuana and San Diego. The site provides 1,011 ha of protected coastal wetland habitat for endangered plant, fish, and animal species. This sensitive coastal wetland is home to over 370 bird species that include nine species that are federally listed as either threatened or endangered, one (state listed) threatened species, and three regionally rare species present across most of estuary's habitats.

The site is the second largest home for the federally listed endangered light-footed clapper rail, supporting over 22% of the total estimated population in the biogeographic region. The site is the only binational watershed in California, sharing its watershed with Mexico, which serves as a major stopover for migrating birds along the Pacific Flyway. The site contains six wetland types (intertidal marshes, estuarine waters, sand shores, salt flats, seasonal rivers, and seasonal and permanent freshwater pools), and does not suffer habitat fragmentation faced by other coastal lagoons in Southern California, due to its unique socio-political location. The site is an excellent example of the wetland habitat that, in the past, was a prominent feature in Southern California's coastal zone, where today more than 95% of those wetlands have been converted for coastal development or severely degraded. This unique area is among the most biologically productive coastal wetland ecosystems on earth, and has thus been recognized by several national environmental organizations for its ecological significance.

The site receives drainage from a 448,327 ha watershed, of which 73% lies within Mexico, and supports vulnerable, endangered, and critically endangered species and threatened habitat. Over

FIGURE 2.18   Composite aerial image of Tijuana River National Estuarine Research Reserve (TRNERR). Designated February 2, 2005; Area: 1,021 ha; Coordinates: 32°33′N 117°07′W; Ramsar site number: 1,452.

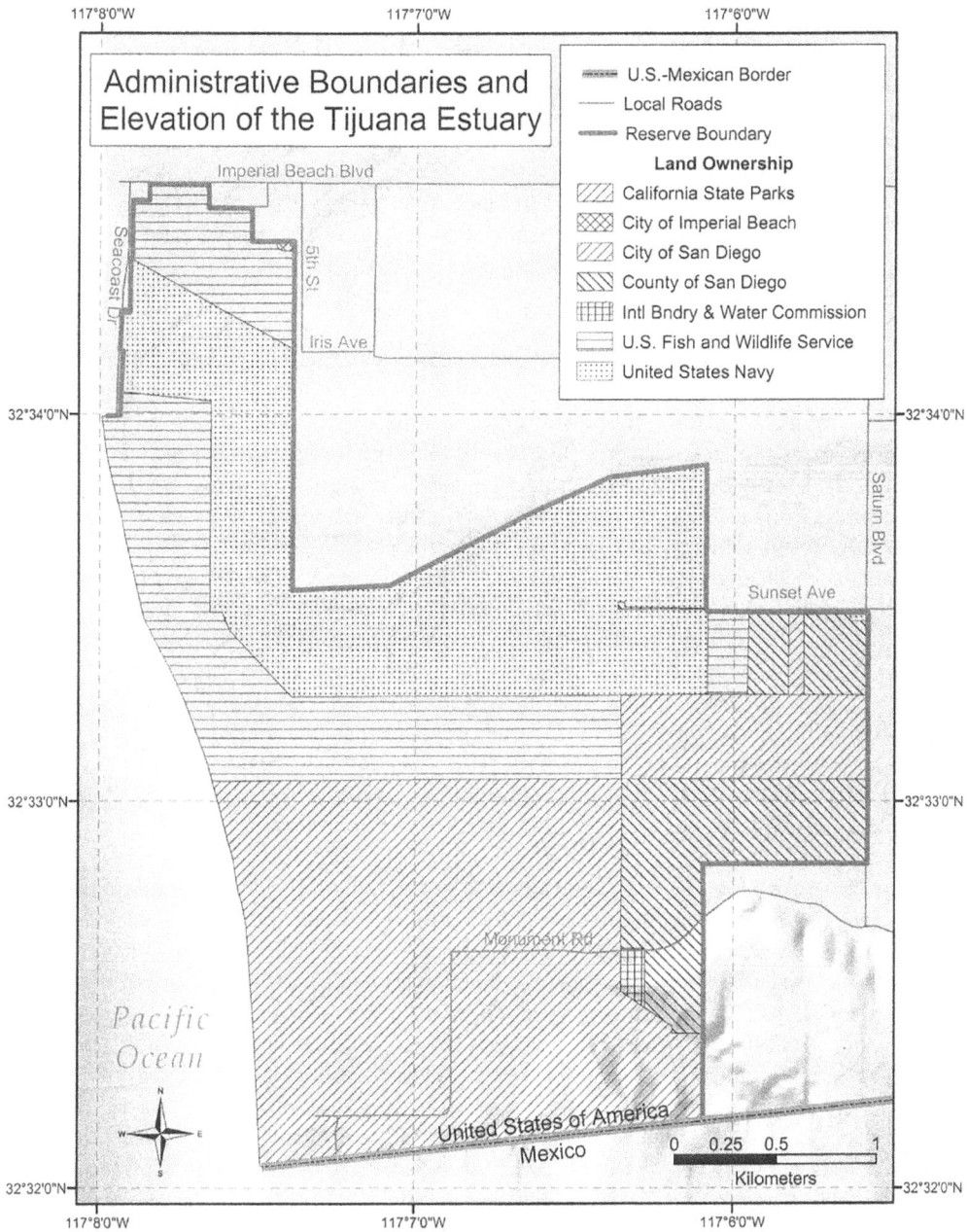

FIGURE 2.19   Map of Tijuana River National Estuarine Research Reserve (TRNERR). Designated February 2, 2005; Area: 1,021 ha; Coordinates: 32°33'N 117°07'W; Ramsar site number: 1,452.

95% of the site is designated as critical habitat that supports six bird species, listed as federally threatened or endangered (light-footed clapper rail, *Rallus longirostris levipes*; least Bell's vireo, *Vireo bellii pusillus*; California least tern, *Sternula antillarum browni*; brown pelican, *Pelecanus occidentalis*; southwestern willow flycatcher, *Empidonax traillii extimus*; and one federally listed endangered plant subspecies, i.e. salt marsh bird's beak, *Cordylanthus maritimus maritimus*).

Located within TRNERR's 1,021 ha ecosystem is the Tijuana Slough National Wildlife Refuge (NWR). Comprised of 428 ha, the NWR was established in 1980, in support of work necessary to implement aspects of the United States Endangered Species Act (1973), which was incorporated into the TRNERR in 1982 with the expressed purpose of, "conserving fish or wildlife, which are listed as endangered species or threatened species…or plants, which are listed as endangered species or threatened species." A few small vernal pools can be found at the site; these shallow pools, which hold several centimeters of water during the wet months, support the San Diego fairy shrimp (Branchinecta sandiegonensis), a federally endangered species also listed in the Red List (IUCN 2007).

The Tijuana Estuary is located at the end of the Tijuana River Valley, where it meets the Pacific Ocean. The Tijuana River Valley is a historic river valley naturally formed by both ocean and riverine processes over long periods of time. Deteriorated water quality represents a significant problem for the TRNERR, especially in the river channel and along the ocean beach, and has necessitated short-term closure of parts of the TRNERR for public health and safety.

Main vegetation types at the site include:

- Cordgrass (Spartina foliosa) forming robust stands along tidal channels in the northern reaches of the Reserve;
- Succulents, including the annual dwarf glasswort (Salicornia bigelovii) and saltwort (Batis maritima) as dominants, as well as pickleweed (Salicornia virginica) and sea blite (Suaeda californica). At higher elevations, these succulents grade into a dense matted cover of shoregrass (Monanthochloe littoralis). At the highest elevations, another species of glasswort (Salicornia subterminalis) becomes co-dominant with shoregrass;
- Filamentous bluegreen and green algae and dozens of species of diatoms form mats, up to 1 cm thickness in most soils and which occur at all intertidal elevations;
- Coastal sage scrub communities that are comprised mostly of *Eriogonum, Rhus laurina, Rhus integrifolia, Baccharis consanguinea, Artemisia californica, Lotus scoparius, Encelia farinosa*, and *Simmondsia chinensis*, in varying combinations;
- Mule fat (Baccharis viminea), found within the upland habitat of the site. In the northeastern and southern regions of the site, mule fat is the dominant plant species. In other areas of upland habitat at the site, mule fat can be found in mixed plant communities, along with *Juncus* and *Haplopappus venetus, Salix*, and *Tamarix*.

Bird populations have been an important factor in the special protective status attributed to the Tijuana Estuary at the TRNERR. Wintering waterfowl at the site include pintail (Anas acuta), cinnamon teal (Anas cyanoptera), American wigeon (Anas americana), surf scoter (Melanitta perspicillata), and ruddy duck (Oxyura jamaicensis). The site's wetlands are important habitats for a large number of shorebirds that account for the majority of the migratory bird population. Four species, i.e., willet (Catoptrophorus semipalmatus), short-billed dowitcher (Limnodromus griseus), western sandpiper (Calidris mauri), and marbled godwit (Limosa fedoa) account for most of the shorebird populations throughout the year. Abundance and species composition of the migratory birds at the site fluctuate seasonally, with its intertidal sand and mudflats supporting the largest numbers of individuals and species.

Fish assemblages at the site are sampled in the estuary, and often include topsmelt silverside (Atherinops affinis), longjaw mudsucker (Gillichthys mirabilis), arrow goby (Clevelandia ios), and the California killifish (Fundulus parvipinnis). Adult flathead grey mullet (Mugil cephalus) are also common. Abundance varies widely from year to year, but total density peaks in the summer and declines in the winter. The tidal channels at the site function as a nursery for important recreational fish, such as the diamond turbot (Hypsopsetta guttulata) and the California flounder (Paralichthys californicus), as well as the croaker family, topsmelt silverside (Atherinops affinis), and northern anchovy (Engraulis mordax). Other game fish that have been found in the estuary include: kelp bass (Paralabrax clathratus), sand bass (Paralabrax maculatofasciatus), opaleye (Girella nigricans), and white croaker (Genyonemus lineatus). While these fish species are not harvested commercially, they are fished recreationally offshore by both private and commercial recreational fishing boats. The site also provides for public use that involves environmental education, interpretive programs, special events, recreation (e.g., wildlife observation, foot trails, equestrian trails, beach use, surfing, and photography), and scientific research.

KAWAINUI AND HĀMĀKUA MARSH COMPLEX, HAWAI'I, USA: OVERVIEW

Sacred to Native Hawaiians, Kawainui Marsh (Figures 2.20 and 2.21), the largest remaining emergent wetland in Hawai'i and Hawai'i's largest ancient freshwater fishpond, is located in what was once the center of a caldera of the Ko'olau shield volcano. The Hawaiian Islands are the most isolated human-inhabited high islands in the world, and constitute a distinct biogeographic region. The marsh provides primary habitat for four of Hawai'i's endemic and endangered waterbirds, including Laysan duck and Hawaiian goose or nēnē, and contains archaeological and cultural resources, including ancient walled taro water gardens (lo'i) where fish were also cultivated. Kawainui Marsh stores surface water, providing flood protection for adjacent Kailua town, the largest town on the windward side of O'ahu. Hāmākua Marsh is a smaller wetland (Figures 2.20 and 2.21), historically connected to and immediately downstream of Kawainui Marsh, which also provides significant habitat for several of Hawai'i's endemic and endangered waterbirds.

Kawainui Marsh and the surrounding area was a significant prehistoric settlement as evidenced by Hawaiian legend, extensive agricultural systems, ceremonial sites, burial sites, and habitation areas. This area once supported one of the largest native Hawaiian settlements, contains some of the oldest known Hawaiian agricultural sites, and Kawainui Marsh and surrounding environs have provided significant information about Hawaiian culture, particularly having to do with the relationship of the early Hawaiians within the environment, collectively referred to as 'Āina in the Hawaiian Language, of the windward valley (Handy et al. 1972, Kelly and Clark 1980, Kelly and Nakamura 1981, Drigot and Seto 1982).

SITE DETAILS

Kawainui Marsh has a distinctive grass community of California grass (Brachiaria mutica), an exotic species that was introduced to Hawai'i for cattle grazing, which covers more than half of Kawainui Marsh and is spreading (USACOE 1998). Other plant species found in the grass community include honohono (Commelina diffusa), arrowhead (Sagittaria latifolia), and scattered stands of cattail (Typha latifolia and Typha angustifolia). In areas of persistent emergent herbaceous vegetation dominated by California grass, vegetation can reach 2–3 m tall, except for where grazed or mowed. Other emergent vegetation species found within the grass communities of the marsh include Hilo grass (Paspalum conjugatum) and *Pluchea* spp. Other typical vegetation at the site include bulrush (Schoenoplectus californicus), neki or swamp fern (Schoenoplectus lacustris), sawgrass (Cladium leptostachyum), and paperbark (Melaleuca quinquenervia) [USACOE 1998], and specifically, Indian pluchea (Pluchea indica) [USFWS 1997]. In the higher elevation areas of the marsh, plant species such as *Mimosa pudica* var. *unijuga*, wedelia (Wedelia trilobata), and kamole (Ludwigia octovalvis) have been present. In disturbed areas, bamboo (Phyllostachys nigra) can also be found (USFWS 1997). The tree and shrub community exists on the outer edges of the grass and bulrush plant communities, and along the slopes above the marsh, consisting primarily of river tamarind (Leucaena leucocephala), guava (Psidium guajava), Chinese banyan (Ficus microcarpa) and monkeypod (Samanea saman). Hau trees (Hibiscus tiliaceus) are the dominant species growing along the banks of Kahanaiki and Maunawili Streams (USACOE 1998).

Hāmākua Marsh is a 9 ha, brackish, inland marsh with predominantly persistent emergent herbaceous vegetation. The marsh is located between a slow-moving, brackish waterway (Kawainui Stream) and dry lowland slopes. A broad band of Indian pluchea (Pluchea indica) marks the transition from upland to wetland along the landward (southwestern) portion of the wetland (Ducks Unlimited 1993), but has been variously controlled/removed by the State Division of Forestry and Wildlife as part of their management of the area (Smith 2002) over the years. Saltwort (Batis maritima) forms an herbaceous zone inside of the *Pluchea*. Ponded water or mudflats border pickleweed, with emergent vegetation dominated by seashore paspalum (Paspalum vaginatum), water hyssop (Bacopa monnieri), and widgeon grass (Ruppia maritima). Common indigenous plant species found

FIGURE 2.20 Composite aerial image of Kawainui and Hamākua Marsh complex. Designated February 2, 2005; Area: 414 ha; Coordinates: 21°24′N 157°45′W; Ramsar site number 1,460.

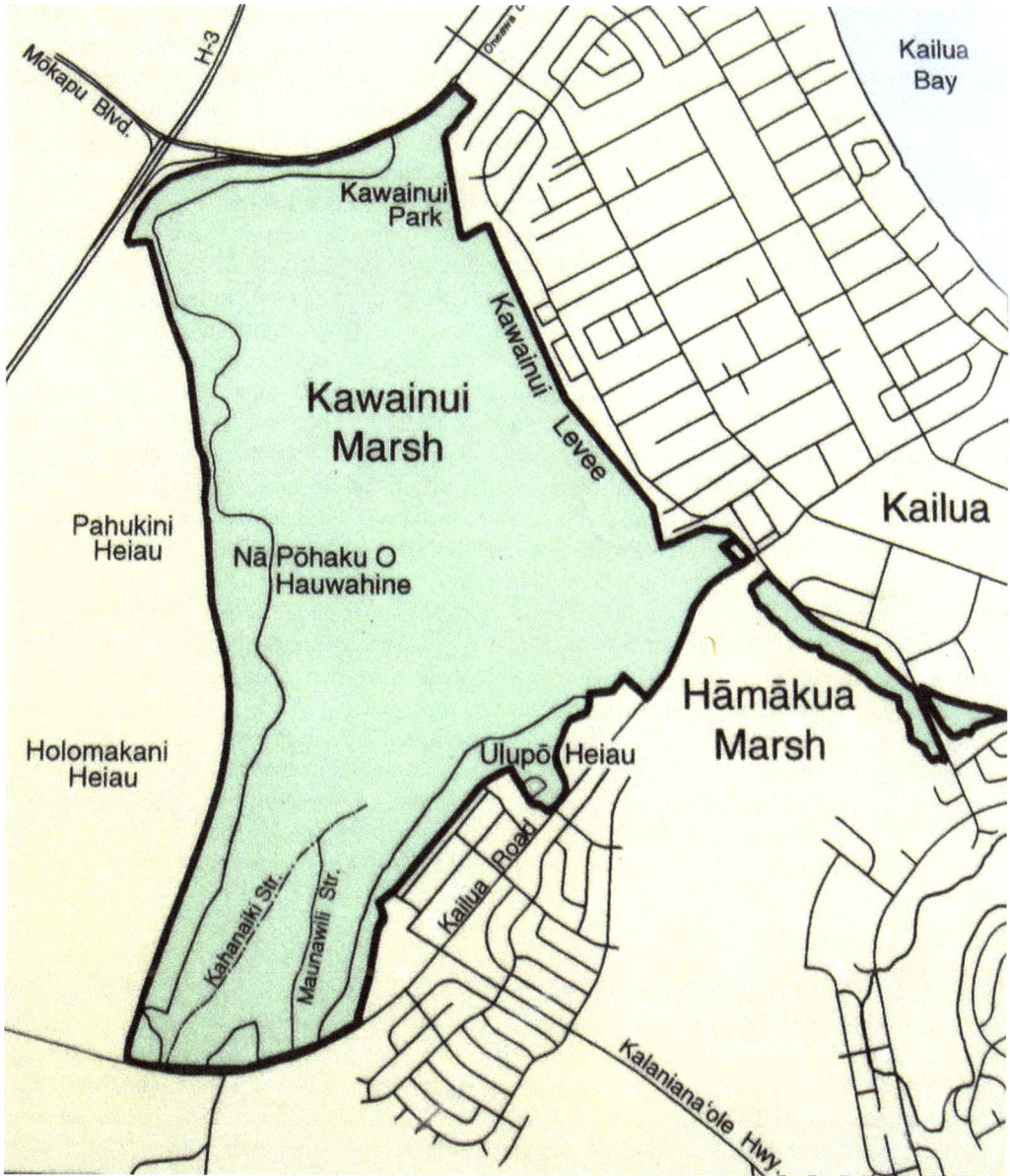

FIGURE 2.21 Map of Kawainui and Hāmākua Marsh complex. Designated February 2, 2005; Area: 414 ha; Coordinates: 21°24′N 157°45′W; Ramsar site number 1,460.

throughout the marsh include bulrush (Bolboschoenus maritimus), water hyssop, ditchgrass, and makai sedge (Scirpus maritimus var. paludosus) [Ducks Unlimited 1993].

The Hawaiian duck, a distinctive species, have undergone extensive hybridization with mallard duck (Anas platyrhynchos) on Oʻahu, with the near disappearance of Hawaiian duck alleles (Browne et al. 1993). Other birds of special note at the site are the alae keʻokeʻo or Hawaiian coot (Fulica alai), the aeʻo or the Hawaiian stilt (Himantopus mexicanus knudseni), and the aukuʻu or black-crowned night heron (Nycticorax nycticorax hoactli) [Conant 1981, USFWS 1997]. The only indigenous non-wetland dependent bird species that regularly frequents the Kawainui and Hāmākua Marshes is the ʻiwa or great frigatebird (Fregata minor palmerstoni) [Conant 1981, USACOE 1998]. The cattle egret (Bubulcus ibis) is abundant in the marshes, and is the marshes' predominant alien waterbird (Conant 1981).

Kawainui and Hāmākua Marshes also provide habitat for migratory waterfowl and wintering shorebirds (Ducks Unlimited 1993). Migratory waterfowl, such as the northern pintail (Anas acuta), northern shoveler (Anas clypeata), mallard duck (Anas platyrhynchos), Canada goose (Branta canadensis), lesser scaup (Aythya affinis), green-winged teal (Anas crecca), American wigeon (Anas americana), and redhead (Aythya americana) are found within the small ponds in the larger open water areas of Kawainui Marsh during winter months (Conant 1981, USFWS 1997). Feral mallards are regular, year-round inhabitants. Migratory shorebirds reported at Kawainui Marsh include the Pacific golden plover (Pluvialis fulva), ruddy turnstone (Arenaria interpres), sanderling (Calidris alba), and wandering tattler (Heteroscelus incanus) [Conant 1981, USFWS 1997].

Introduced species dominate the aquatic fauna in Kawainui Marsh. Common introduced fish and invertebrates found in the marsh are Mozambique tilapia (Tilapia mossambica), mosquito fish (Gambusia affinis), guppies (Poecilia spp.), carp (Cyprinus carpio), Chinese catfish (Clarias fuscus), swordtail (Xiphophorus helleri), smallmouth bass (Micorpterus dolomieui), Tahitian prawn (Macrobrachium lar), crayfish (Procambarus clarkia), damselfly (Ischnura ramburii), apple snails (Pomacea spp.), and pond snails (Melanoides spp.) [USFWS 1997]. Oriental rice eel (Monopterus albus) and the bullfrog (Rana spp.) also inhabit the marsh. In the upper reaches of the marsh, Kahanaiki and Maunawili Streams are also dominated by introduced species of fish and crustaceans. Native species, including shrimp (Atyoida bisculata) and Hawaiian river shrimp (Macrobrachium grandimanus), are present in small numbers. Numerous native species inhabit Oneawa Canal, below the marsh, including amphidromous gobies and eleotrids *Awaous guamensis* (an indigenous goby), *Stenogobius hawaiiensis* (an endemic goby), *Eleotris sandwicensis* (an endemic eleotrid), as well as aholehole (Kuhlia sandwicensis), mullet (Mugil cephalus), and barracuda (Sphyraena barracuda).

Both Kawainui and Hāmākua marshes are wetland types that have undergone significant decline in area and quality (USFWS 2002). The Kawai Nui Heritage Foundation and 'Ahahui Malama I Ka Lokahi have been organizing volunteers since 1997 to remove alien vegetation at the site, to create open water areas, and to plant native vegetation in a portion of Kawainui Marsh. The Kawai Nui Heritage Foundation and 'Ahahui Malama I Ka Lokahi periodically sponsor tours on the natural and cultural history of Kawainui Marsh and nearby sites, bringing up to as many as 200 visitors at a time.

PALMYRA ATOLL NATIONAL WILDLIFE REFUGE, EQUATORIAL ISLANDS, USA: OVERVIEW

Palmyra Atoll National Wildlife Refuge (Figures 2.22 and 2.23) is comprised of coral reefs, permanent shallow marine waters, intertidal forested wetlands, submerged lands, and ocean waters extending out to 22 km from the islands' edges, and is located approximately 1,545 km south of Honolulu, Hawai'i in the equatorial Pacific. A National Wildlife Refuge (NWR) since 2001, the site supports a variety of species with different conservation statuses under the National Endangered Species Act and the IUCN Red List, such as the Hawaiian monk seal (Monachus schauinslandi), hawksbill sea turtle (Eretmochelys imbricata), and green sea turtle (Chelonia mydas). It is also an important feeding and nesting ground for seabirds like the red-footed booby (Sula sula), with the third largest colony in the world, and it sustains approximately 5% of the total population of the bristle-thighed curlew (Numenius tahitiensis). As a National Wildlife Refuge, the site is closed to public use without a permit issued by the manager, but scientific research is coordinated between the US Fish and Wildlife Service and The Nature Conservancy, along with a Palmyra Atoll Research Consortium. Threats include the presence of invasive species like the scale (Pulvinaria urbicola), which was once responsible for a decline in the *Pisonia grandis* forest coverage. Current restoration projects are responsible for the regrowth of these forests, and associated return of a number of bird species.

SITE DETAILS

Palmyra Atoll NWR is located at the northern end of the Line Islands in the equatorial Pacific. To the southeast of Palmyra Atoll NWR are eight Line Islands that belong to the Republic of Kiribati;

FIGURE 2.22 Composite aerial image of Palmyra Atoll national wildlife refuge. Designated April 1, 2011; Area: 204,127 ha; Coordinates: 05°52′N 162°06′W; Ramsar site number 1,971.

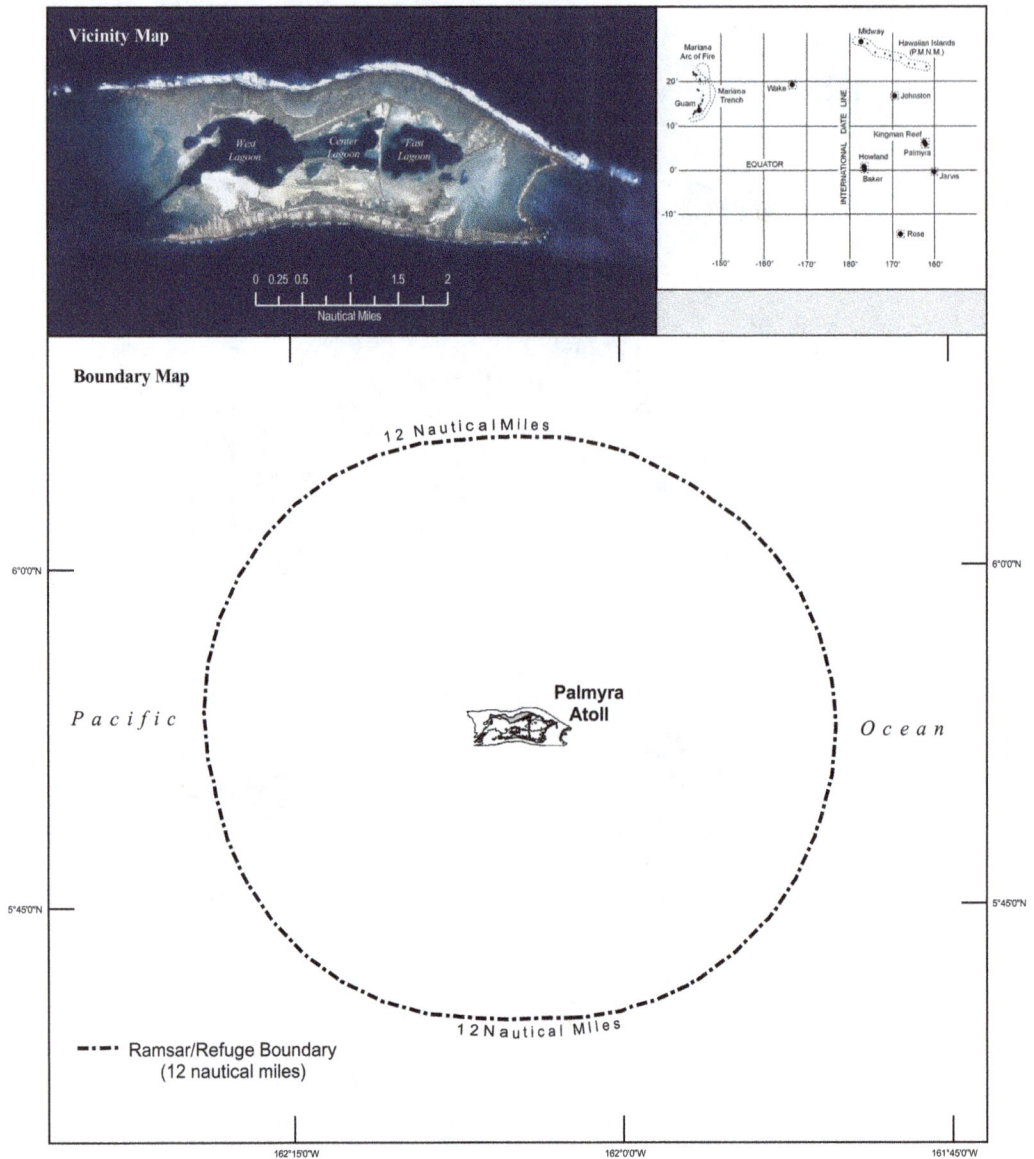

FIGURE 2.23 Map of Palmyra Atoll national wildlife refuge. Designated April 1, 2011; Area: 204,127 ha; Coordinates: 05°52′N 162°06′W; Ramsar site number 1,971.

to the north is Kingman Reef. A terrestrial 130 ha portion of the site, on Cooper Island, is owned and operated by The Nature Conservancy. The atoll, consisting of a ring of interconnected islands and islets surrounding three small sub-lagoons, is at the center of a large shallow coral platform approximately 19.3 km long and 4.8 km wide.

The entirety of the atoll was formed by the growth of coral reefs along the rim of an ancient volcano, and subsequent erosion of volcanic lands. At first, the coral adhered to the volcanic island, forming a fringing reef. As the volcanic lands began to erode, the reef separated from the shore, creating a barrier reef. With an average annual rainfall of more than 4 m, Palmyra is a wet atoll within the Inter-Tropical Convergence Zone (Mueller-Dombois and Fosberg 1998). This high rainfall, coupled with high nutrient influx in the form of guano from thousands of resident seabirds and migratory shorebirds, supports thickly vegetated rain forest that is unique even in comparison with nearby, more arid islands such as in Kiribati (Wester 1985).

There are a number of threatened or endangered species (by the Endangered Species Act of 1973) that have been documented at Palmyra Atoll: the critically endangered (IUCN 2010) Hawaiian monk seal (Monachus schauinslandi), occasionally seen off Palmyra NWR; sperm whales (Physeter macrocephalus), classified as vulnerable on the IUCN Red List; and bottlenose dolphins (Tursiops sp.) and spinner dolphins (Stenella longirostris), who are occasional visitors at Palmyra NWR. The site has been noted as a significant foraging habitat for the green sea turtle (Chelonia mydas), and the hawksbill sea turtle (Eretmochelys imbricata), although nesting activity at Palmyra is relatively rare. The bristle-thighed curlew (Numenius tahitiensis) is registered as vulnerable by Birdlife International and within the IUCN Red List, and 'Yellow' by the Audubon Watchlist and American Bird conservancy.

The site is an important marine feeding ground for several seabirds since it is the only nesting area available within 116.5 million ha of the surrounding ocean (Flint 1992). Located on the boundary between the North Equatorial Countercurrent and other ocean currents in the vicinity, more than a million nesting seabirds visit the atoll. The site supports the third largest red-footed booby (Sula sula) colony in the world. The atoll also supports healthy populations of brown noddy (Anous stolidus) and brown booby (Sula leucogaster). Palmyra shelters approximately 20,000 black noddies (Anous minutus), the largest nesting colony in the central Pacific. The sooty tern (Onychoprion fuscatus), commonly found throughout the tropical Pacific Ocean, is an abundant resident of Palmyra Atoll. A 1993 survey estimated a total of 750,000 sooty tern nests at Palmyra. Other seabird nesters include the great frigatebird (Fregata minor), sooty tern (Onychoprion fuscatus), white tern (Gygis alba), red-tailed tropicbird (Phaethon rubricauda), and masked booby (Sula dactylatra).

Palmyra has one of the largest remaining undisturbed *Pisonia* forests in the Pacific. Increased human population and development has caused this vegetation community to degrade or disappear throughout its historic range. Even *Pisonia* forests in protected areas, such as the Rose Atoll NWR, lack the diversity and moisture present at Palmyra. This community supports several native ferns including bird's nest fern (Asplenium nidus) and lau'ae fern (Phymatosorus scolopendria), as well as the introduced sword fern (Nephrolepis hirsutula).

A population of melon-headed whales (Peponocephala electra) are resident off Palmyra NWR (Brainard et al. 2005). Rare occurrences of the sea cucumber (Holothuria atra), the echinoid *Echinothrix calamaris*, and the asteroid *Echinaster luzonicus* have been recorded on the fore reef. The holothurians *Holothuria atra*, *Holothuria edulis*, and *Stichopus chloronotus* are present in shallow habitats of the lagoon. The giant clam (Tridacna maxima) is present in low numbers on the southern reef and at the southeastern back reef and the shallow "coral gardens" pool habitat (Dalebout et al. 1992).

Early visitors characterized the atoll as one of the "Pearls" of the Pacific and Polynesia, and a "necklace of emerald islets" (Boddam-Whetham 1876, Bryan 1940, Wright 1948). Beginning in 1939, the U.S. Navy dredged a 9 m deep and 60 m wide ship channel through the shallow perimeter reef between the southwest ocean reef and West Lagoon, dredged the reef separating the West and Center Lagoons to a depth of 3 m for a seaplane runway, and connected all but a few islets by constructing elevated road causeways on the shallow perimeter reefs around most of West and Center Lagoons. Corals, crustose coralline algae, and other typical lagoon reef species are almost entirely absent on lagoon reefs, a lasting legacy of earlier military dredging and filling operations. The most unique and healthy habitats at the site are the two shallowest and easternmost of the reef pools off the southeast and northeast corners of the shallow eastern reef flat. These are dominated by crustose coralline algae and many species of *Acropora*, *Montipora*, *Astreopora*, and *Pocillopora*, and are commonly referred to as the "coral gardens." Other healthy reef substrates occur along the entire north ocean-facing reef slopes of the atoll where many species of corals, algae, reef fish, and invertebrates are supported. Halimeda sands are prevalent in the fore reef habitats of the site, nearly absent from comparable habitats such as Kingman Reef.

NWR designation closes the site to all public uses unless the Refuge Manager evaluates each via a permit application and issues a special use permit prior to entry. Commercial fishing and other commercial activities are prohibited within refuge, and compatible uses normally include nondestructive conservation research, monitoring, birding, wildlife photography, environmental education, and other low impact uses. Due to the concern over alien species, all visitors landing at any refuge island must comply with quarantine procedures, since substantial efforts are constantly being made to rid the islands of alien and undesirable invasive species. Palmyra has historically been a popular stopover for yachts sailing in the central Pacific because of safe anchorage and plentiful freshwater. Yachts are still permitted by the NWR to visit Palmyra, but are limited.

REFERENCES

Allen, S. 1985. *Harbor seals at Point Reyes*. Point Reyes Bird Observatory Newsletter 68.

Allen, S. G., Huber, H.R., Ribic, C.A., and D. G. Ainley. 1989. Population dynamics of harbor seals in the Gulf of the Farallones, California. *California Department of Fish & Game Report*. 75(4): 224–232.

Bird Studies Canada. 2010. British Columbia Coastal Waterbirds Survey. Online database. (https://www.birdscanada.org/birdmon/default/datasets.jsp?code=BCCWS)

Brown, J. A. 2006. Using the chemical composition of otoliths to evaluate the nursery role of estuaries for English sole Pleuronectes vetulus populations. *Marine Ecology Progress Series*. 306: 269–281.

Browne, R., Griffin, C., Chang, P., Hubley, M., and A. Martin. 1993. Genetic divergence among populations of the Hawaiian duck, Laysan duck, and mallard. *The Auk*. 100: 49–56.

Butler, R.W. and R.W. Campbell. 1986. *The Birds of the Fraser River Delta: Populations, Ecology and International Significance*. CWS Occasional Paper No. 86, Ottawa.

Caffrey, J., Brown, M., Tyler, W.B., and M. Silberstein. 2002. *Changes in a California Estuary: A profile of Elkhorn Slough*. Elkhorn Slough Foundation. 280pp.

Cohen, A. 2000. *An introduction to the San Francisco*. San Francisco Estuary Institute, Oakland, California. 42pp.

Conant, S. 1981. *A survey of the waterbirds of Kawainui marsh*. State of Hawai'i Department of Planning and Economic Development. 63pp.

Dalebout, M., Baker, C.S., Robertson, R.M., Chivers, S.J., Perrin, W.F., Mead, J.G., Grace, R.V., and E. Flint. 1992. *Survey of the terrestrial biota of Palmyra Atoll: 18 February to 9 March 1992*. Unpublished trip report, U.S. Fish and Wildlife Service. Honolulu, HI. 19pp.

Drigot, D. and M. Seto. 1982. *Ho'ona'auao No Kawai Nui (Educating About Kawai Nui). A multi-media educational guide*. The Environmental Center, University of Hawai'i, Honolulu, 196pp.

Ducks Unlimited. 1993. *Hawaiian Islands wetlands conservation plan*. Ducks Unlimited, Inc., Sacramento, CA.

Evens, J. 1993. *Natural history of the point Reyes Peninsula*. Revised Edition. Point Reyes National Seashore Association. 224pp.

Evens, J. 1995. *Population size and reproductive success of osprey at Kent Lake, Marin County, California*. Final Report for Avocet Research Associates to Marin Municipal Water District. November 15, 1995.

Evens, J. and G. Page. 1986. Predation on black rails during high tides in salt marshes. *Condor*. 88: 107–109.

Handy, E., Handy, E., and M. Pukui. 1972. *Native planters in old Hawai'i: Their life, lore, and environment*. Bernice P. Bishop Museum Bulletin 233. Bishop Museum Press: Honolulu, HI.

Hughes, B. B., Levey, M.D., Fountain, M.C., Carlisle, A.B., Chavez, F.P., and M.G. Gleason. 2015. Climate mediates hypoxic stress on fish diversity and nursery function at the land-sea interface. *Proceedings of the National Academy of Sciences*. 112(26): 8025–8030.

IUCN. 2007. *IUCN Red List of Threatened Species*. (https://www.iucnredlist.org/)

Kelly, J. 2001. Distribution and abundance of winter shorebirds on Tomales Bay, California: Implications for conservation. *Western Birds*. 32: 145–166.

Kelly, M. and J. Clark. 1980. *Kawainui Marsh, Oahu: Historical and archaeological studies*. Bishop Museum, Department of Anthropology, Honolulu, HI. 82pp.

Kelly, M. and B. Nakamura. 1981. *Historical study of Kawainui Marsh area, Island of Oahu*. Bernice P. Bishop Museum. Prepared for Hawai'i Department of Planning and Economic Development, Honolulu, HI.

Kelly, J. and S. Tappen. 1998. Distribution, abundance, and implications for conservation of winter waterbird on Tomales Bay, California. *Western Birds*. 29: 103120.

Kildow, J. and L. Pendleton. 2010. *Elkhorn Slough restoration policy and economics report*. Elkhorn Slough Foundation. 174pp.

Leidy, R. 2007. Ecology, assemblage structure, distribution, and status of fishes in streams tributary to the San Francisco Estuary, California. San Francisco Estuary Institute, Oakland, CA. 194pp.

Moore, T.O. and J.J. Mello. 1995. Pacific Herring (Clupea harengus pallasi), studies and fisheries management in Tomales Bay, 1992-93, with notes on Humboldt Bay and Crescent City area landings. California Department of Fish and Game Marine Resources Division. Admin. Report 95-5.

Mueller-Dombois, D. and F.R. Fosberg. 1998. *Vegetation of the tropical Pacific islands.* Springer, New York. 737pp.

Page, G., Stenzel, L., and J. Kjelmyr. 1999. Overview of Shorebird abundance and distribution in wetlands of the Pacific coast of the contiguous United States. *The Condor.* 101(3): 461–471.

Philip Williams and Associates, Ltd. 2006. *Projecting the future evolution of Bolinas Lagoon.* Report to Marin County Open Space District. 217pp.

Save the Bay. 2007. *Greening the Bay. Financing wetland restoration in San Francisco Bay.* San Francisco Estuary Institute, Oakland, CA. 20pp.

Shuford, W. D. and T. Gardali, Eds. 2008. California bird species of special concern. *A Ranked Assessment of Species, Subspecies, and Distinct Populations of Birds of Immediate Conservation Concern in California.* Department of Fish and Game, Sacramento. 65pp.

Shuford, W. D., Page, G.W., Evens, J.G., and L.E. Stenzel. 1989. Seasonal abundance of waterbirds at point Reyes: A coastal California perspective. *Western Birds.* 20: 137–265.

Smith, D. 2002. *Hamakua Marsh ecosystem restoration and community development project.* Hawai'i State Division of Forestry and Wildlife, Honolulu, HI. 7pp.

Spratt, J. D. 1981. Status of the Pacific Herring, Clupea harengus paliasii, resource in California, 1972 to 1980. Department of Fish and Game: Fish Bulletin 171.

Stenzel, L., Hickey, C., Kjelmyr, J., and G. Page. 2002. Abundance and distribution of shorebirds in the San Francisco Bay area. *Western Birds.* 33(2): 69–98.

Suer, A. L. 1987. *The Hening of San Francisco and Tomales Bays.* The Ocean Institute, San Francisco, CA. 64pp.

Takekawa, J., Woo, I., Spautz, H., Nur, N., Grenier, L., Malamud-Roam, K., Nordby, C., Cohen, A, Malamud-Roam., F., and S. Wainrwright-de La Cruz. 2006. Environmental threats to Tidal-Marsh vertebrates of the San Francisco Bay Estuary. *Studies in Avian Biology.* 32: 176–197.

Thalman, S.B. 1993. *The coast Miwok Indians of the point Reyes area.* Point Reyes National Seashore Association. 40pp.

USACOE (U.S. Army Corps of Engineers). 1998. Kawainui Marsh environmental restoration project. Final Ecosystem Restoration Report and Environmental Assessment. Honolulu, HI.

USACOE (U.S. Army Corps of Engineers), U.S. Environmental Protection Agency, San Francisco Bay Conservation and Development Commission, and San Francisco Bay Regional Water Quality Control Board. 2001. *Long-Term Management Strategy for the Placement of Dredged Material in the San Francisco Bay Region.* Management Plan. 174pp.

USFWS (U.S. Fish and Wildlife Service). 1991. Determination of endangered status for three plants, Blennosperma bakeri (Sonoma Sunshine or Baker's Stickyseed), Lasthenia burkei (Burke's goldfields), and Limnanthes vinculans (Sebastopol Meadowfoam). *Federal Register.* 56(23): 61173–61182.

USFWS (U.S. Fish and Wildlife Service). 1997. Final fish and wildlife coordination act report for the Kawainui Marsh environmental restoration project Oahu, Hawai'i. Pacific Islands Office, Honolulu, HI.

USFWS (U.S. Fish and Wildlife Service). 2002. *Draft Revised Recovery Plan for Hawaiian Waterbirds, Second Revision.* Portland, OR. 140pp.

Wester, L. 1985. Checklist of the vascular plants of the Northern Line Islands. *Atoll Research Bulletin.* 287: 1–38.

Yoklavich, M. M., Cailliet, G.M., Barry, J.P., Ambrose, D.A., and B.S. Antrim. 1991. Temporal and spatial patterns in abundance and diversity of fish assemblages in Elkhorn Slough, California. *Estuaries.* 14: 465–480.

3 Mexico

States of Baja California and Baja California Sur

The Ramsar wetlands along the western coasts of Mexico vary widely in their biophysical environmental conditions, as well as their species composition. The Baja California Peninsula is also quite unique to this continental region in that it includes some of the hottest and driest environments of the continent, surrounded by some of the most biologically diverse and rich marine ecosystems. Consequently, the wetlands of this peninsular area (including its islands) mediate many of these processes between the arid uplands and the open water ocean environments in both the Pacific Ocean and the Gulf of California. The 16 coastal wetlands within this diverse peninsular area (Figure 3.1) are:

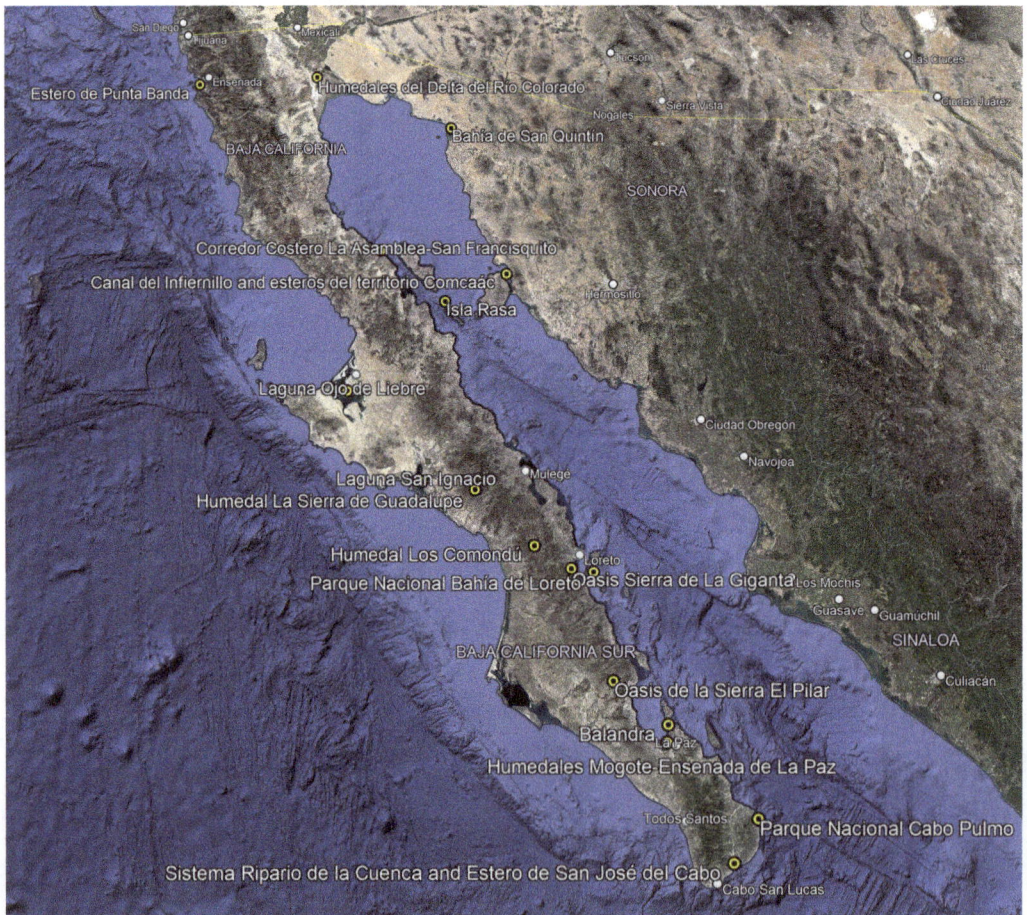

FIGURE 3.1   Referenced Ramsar wetlands of coastal western Mexico, shown in their general location among the states of Baja California and Baja California Sur.

DOI: 10.1201/9781003046394-3

- Humedales del Delta del Río Colorado
- Estero de Punta Banda
- Bahía de San Quintín
- Corredor Costero la Asamblea-San Francisquito
- Isla Rasa
- Laguna Ojo de Liebre
- Laguna San Ignacio
- Humedal la Sierra de Guadalupe
- Humedal los Comondú
- Oasis Sierra de la Giganta
- Parque Nacional Bahía de Loreto
- Oasis de la Sierra el Pilar
- Balandra
- Humedales Mogote-Ensenada la Paz
- Parque Nacional Cabo Pulmo
- Sistema Ripario de la Cuenca and Estero de San José del Cabo

HUMEDALES DEL DELTA DEL RÍO COLORADO, BAJA CALIFORNIA AND SONORA: OVERVIEW

Humedales del Delta del Río Colorado (Figures 3.2 and 3.3) is a system of natural and artificial wetlands consisting of intertidal wetlands, brackish deltas, riverine environments, and permanent freshwater lakes and ponds, residing within one of the largest hydrographic basins in North America, the Colorado River Basin. Site vegetation consists of diverse aquatic plants and coastal halophytic species, with numerous endemic, rare, or threatened species. Over 400 species of flora are present at the site, which are replete with resident and migratory waterbirds. Human activities at the site include fishing, hunting, agriculture, scientific research, environmental education, eco-tourism, and ranching. The agriculturally dominated landscape of the site is the result of intensive landscape modifications that have had tremendous impacts on groundwater resource use; wetland areas are mostly natural, although they currently depend on the surplus water released from the Morelos Dam, and in a few artificial cases are dependent on agricultural return water or the transport of irrigation water.

SITE DETAILS

Most of the wetlands at the site are located within the Agricultural Valley of Mexicali. As in most of the State of Baja California, lakes, rivers, streams, and springs are scarce. Average annual rainfall in this region of the country is 0 to 125 mm. The vast majority of water resources in the valley come from underground aquifer resources and from existing surface water storage brought about by dams, i.e., wholly 88% of water resources of the State within the Mexicali Valley; of this percentage, the Colorado River contributes 57%.

The site is located within northeastern Mexico, and is made up exclusively of the land on the 'left bank' of the Colorado River, within the State of Sonora, and on the 'right bank' of the Colorado River within the State of Baja California. Although this is the final reach of the Colorado River, no significant runoff is generated in this area because it originates in the United States where the vast majority of its waters are consumed, nearly entirely, by agricultural activities and for urban use. Within the valley, there are two other rivers: (1) the Hardy River, with a length of 26 km, which is formed by agricultural return water from the Mexicali Valley, which flows into the southern Colorado River, and (2) the Nuevo River, formed by agricultural return water and domestic and industrial wastewater from Mexicali. Thus, the site consists of two basins, one located entirely in Baja California (Bacanora-Monumentos Sub-basin) and the other almost entirely in the State of Sonora (Bacanora-Mejorada Sub-basin).

FIGURE 3.2 Composite aerial image of Humedales del Delta del Río Colorado. Designated March 20, 1996; Area: 250,000 ha; Coordinates: 31°50′N 114°59′W; Ramsar Site number: 814.

The Colorado River Delta (Delta del Río Colorado) is located at the northern end of the Gulf of California, in a desert area. In the delta, the Colorado River divides into two main branches and several secondary ones that form two islands, Montague and Pelicano Islands, and a large flood-plain. The high rates of evaporation (approximately 0.9 m/year) and virtually no fresh water input from the Colorado River, as well as low precipitation (annual average of 68 mm), results in what is referred to as the Upper Gulf of California "anti-estuary," characterized by large intervals of up to 8 m between the minimum and maximum tide levels within an intertidal zone that occupies more than 7,000 km². This area of the Upper Gulf of California and the Colorado River Delta were origi-nally declared Biosphere Reserves by Mexico in 1993.

The wetlands of the site provide ecosystem services such as climate change mitigation on a regional scale (e.g., its carbon storage and a physical buffer capabilities against the erosive forces of the sea), providing seasonal resting habitat for migratory birds on their route through the Pacific Flyway, reducing salinity in the soils of surrounding agricultural areas, and purification of water and the maintenance of water quality.

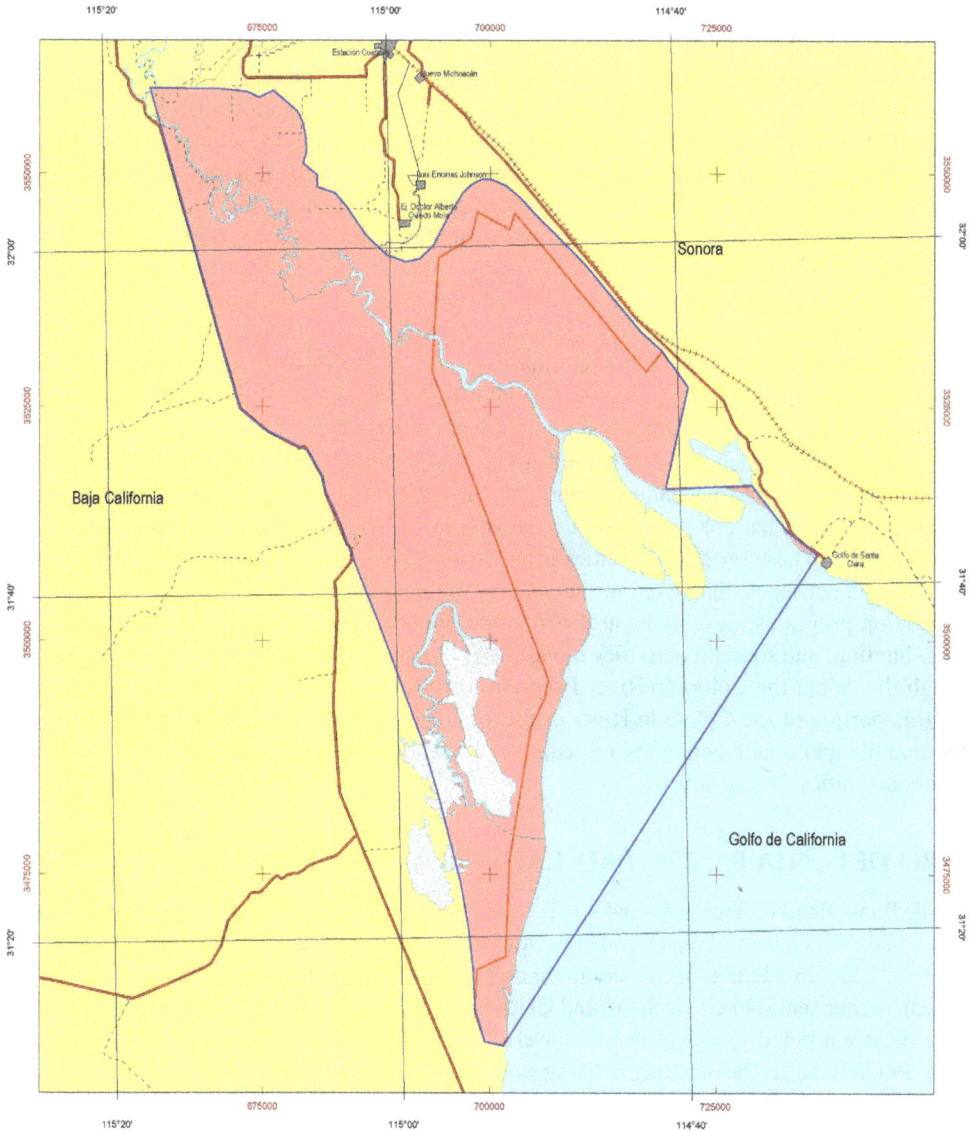

FIGURE 3.3 Map of Humedales del Delta del Río Colorado. Designated March 20, 1996; Area: 250,000 ha; Coordinates: 31°50′N 114°59′W; Ramsar Site number: 814.

Main flora of the site include emergent vegetation, mainly consisting of southern cattail (Typha domingensis), with patches of sedges (Scirpus maritimus and S. americanus). Halophytic vegetation includes saltgrass, and the species *Allenrolfea occidentalis*, *Suaeda torreyana*, and *Atriplex canescens*. The wetlands at the site are of vital importance for many species of vertebrates, especially for amphibians, fish, and waterfowl. There are over 350 bird species recorded at the site, such as ducks (Anas cyanoptera, Anas clypeata, Anas platyrhynchos, and Anas americana), and herons (Ardea alba, Egretta thula, Ixobrychus exilis, and Ardea herodias).

The site's wetlands are of great cultural importance, once settled by the Cucapá Tribe, who were dependent upon the wetlands of the area, and subsequently affected by the establishment of the Mexicali Valley when the draining of thousands of hectares of wetlands occurred, which eventually brought the completion of the Morelos Dam in 1950.

Human disturbance factors at the site include a number of thermoelectric plants that utilize geo-thermal resources of the area, as well as intensive agricultural activities, illegal hunting of migratory birds, aquaculture, extensive livestock farming, fishing, and eco-tourism. The general extraction of water in the Colorado River Basin within some of states in the southern United States has resulted in a regional lack of water, due to management that favors human consumption and excludes eco-logical service requirements. Similarly, throughout the entirety or the Colorado River Basin, and in the Mexicali Valley, urban and agricultural growth constitute an enormous threat to these wetlands, and their capability to protect the regional areas from the impacts of global climate change, such as drought within the region.

Environmental education programs are carried out in the area, focused on the Colorado River and the Hardy River, with local communities and colonies settled in the vicinity of these rivers. There are also ecological restoration programs at the site, focused on riparian vegetation restoration, which involves the planting of poplar trees (Populus fremontii), willow trees (Salix gooddingii), and honey mesquite (Prosopis glandulosa) along the river corridors.

Environmental outreach and education activities have been carried out by personnel from both the Sonoran Institute and Pronatura Sonora, within schools of the region, both for this area and for all the wetlands that are located along the riparian corridor of the Colorado River Delta and the Hardy River. These types of activities promote eco-tourism activities within wetlands. Among the restoration activities conducted at the site are the creation of recreational areas, environmen-tal education programs, development of economic opportunities based on natural resources (i.e., fishing, hunting, and tourism activities along the river), as well as bird/mammal/reptile watching. Through the Adopt the Colorado River Program for both children and adults, volunteer activities along this portion of the Colorado River occur, offering community benefits such as maintaining water quantity and quality, enjoying recreational areas, and activities such as fishing, hunting, and other opportunities for eco-tourism.

ESTERO DE PUNTA BANDA, BAJA CALIFORNIA: OVERVIEW

Estero de Punta Banda (Figures 3.4 and 3.5) is comprised of intertidal marshes, mud and sand flats, and seagrass beds, which have completely disappeared along the coastal areas from Ensenada to Southern California, USA. In addition to the common dolphin (Delphinus delphis), gray whale (Eschrichtius robustus), harbor seal (Phoca vitulina), and California sea lion (Zalophus californianus), the fauna of the site include a rich diversity of benthic invertebrates that support the entire food web of the site's estuary. Punta Banda is the breeding, feeding, and nursing ground for at least 150 species of fish, many of which have commercial importance. The site has also been used for at least the past 2,000 years by the Kumiai People, hunter-gatherers whose language gave origin to many of the languages now spo-ken in the area. Since the 1980s, the construction of oil exploration and housing infrastructure in the area has affected a considerable portion of the site, and still threatens its ecology.

SITE DETAILS

The muddy lowlands, coastal dunes, tidal channels, and seagrass areas of the site are varied in their biology, ranging from terrestrial flora (coastal dunes) to marine flora of phytoplankton and benthic algae, to a diversity of seed producing plants. The fauna of the site include highly diverse benthic invertebrates, which support the food web of the estuary, as well as fish that breed and feed at the site. The site is a priority site for conservation by Mexico's National Commission for the Knowledge and Use of Biodiversity (CONABIO), at the state and national level, due to its high level of ende-mism, species richness, and key ecosystem functions. The site has also been designated a BirdLife Category 5, "Important Bird and Biodiversity Area" (IBA).

The site is L-shaped and measures approximately 8 km long by an average of 0.35 km wide, and abuts the edge of the Maneadero Valley. The region that makes up the landward portion of

FIGURE 3.4 Composite aerial image of Estero de Punta Banda. Designated February 2, 2006; Area: 2,393 ha; Coordinates: 31°44′N 116°38′W; Ramsar Site number: 1,604.

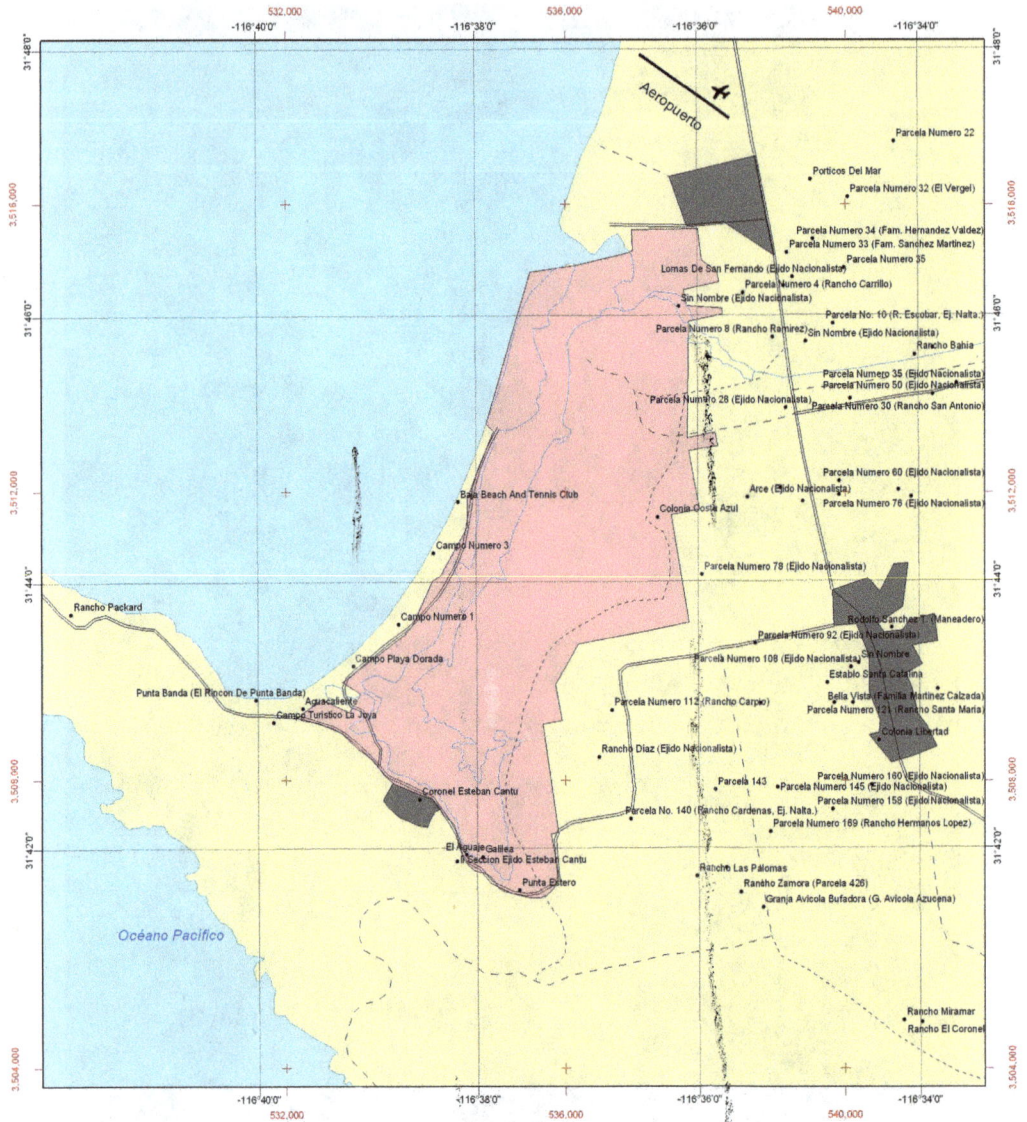

FIGURE 3.5 Map of Estero de Punta Banda. Designated February 2, 2006; Area: 2,393 ha; Coordinates: 31°44′N 116°38′W; Ramsar Site number: 1,604.

the estuary is alluvium, with the remainder of the site a sand bar and sandy coastline, with an interior comprised of marshes and alluvial soil of the Maneadero Valley. The site is located within Mexico's Hydrological Region No. 1 (Baja California, Northwest Ensenada), in the southern part of the Tijuana River-Arroyo Maneadero Basin. The site has intermittent surface flow, via Arroyo Maneadero; the Maneadero Valley aquifer originates just to the east of the site.

The site consists of beaches with a low slope (0.5–1%), and low dunes (1–2 m). As a coastal lagoon, the site receives fresh water during the winter rains and its depth decreases from the mouth of the estuary to upland edge, registering maximum depths of 12.5 m at the mouth, and approximately 1 m depth at the upstream end. The mouth of the lagoon varies in extension according to the increase and decrease in the volume of sediments carried by the coastal current, via longshore flow.

The terrestrial flora on the site's sand bar are comprised primarily of coastal dune ground vegetation and, to a greater extent, coastal scrub genera, such as *Acalypha*, *Artemisia*, *Agave*, *Euphorbia*,

Lycium, and *Machaerocereus*. The sand bar supports approximately 23 species of plants, of which the most common species is sand verbena (Abronia maritima). Vegetation behind the dunes is dominated by *Juncus acutus* var. *sphaerocarpus, Carpobrotus edulis, C. aequilaterus*, and *Haplopappus venetus* var. *vernonioides*. Other flora in the estuary are comprised of phytoplankton, benthic algae, and seed producing plants. Phytoplankton are represented by 28 genera: 18 diatoms, nine dinoflagellates, and one silicoflagellate. A total of 253 species of benthic diatoms have been reported at the site, of which the most common are: *Nitzschia punctata, Denticula subtilis*, and *Amphora* sp. Benthic algae are represented by 39 genera and 47 species, of which 11 genera and 13 species are Rhodophytes; four genera and six species are Chlorophytes; and four genera and five species are Phaeophytes. Twenty genera and 23 species of marine phanerogams have been reported at the site, with eelgrass (Zostera marina) being the dominant species. Salt marsh bird's beak (Cordylanthus maritimus maritimus) has also been recorded at the site. Other common species in the estuary are cordon grass (Spartina foliosa), saline grass (Monantochloe littoralis), brine grass (Salicornia pacifica), saltwort (Batis maritima), alkali heath (Frankenia grandifolia), and wooly seablite (Suaeda californica var. californica). Among the invasive species at the site are *Carpobrotus edulis, C. aequilaterus*, and *Mesembryanthemum crystallinum*, all of which are fast-growing and aggressive exotics that have been displacing the native vegetation of the area.

The site represents an important habitat for birds, with up to 144 reported species, of which 63 inhabit open water areas, 62 on beaches, and of which 11 are birds of prey and eight are primarily terrestrial. During the winter, the estuary, located in the Pacific Migration Corridor, is used by around 4,000 western sandpiper (Calidris mauri), among other wintering birds, such as the American widgeon (Anas americana), the lesser scaup (Aythya affinis), and the northern pintail (Anas acuta); the western snowy plover (Charadrius alexandrinus nivosus) also nests in the estuary.

More than 50% of the fish species found at the site are of commercial importance, with the most economically important fish larvae (for their abundance and temporal persistence) inclusive of the families Gobidae and Atherinidae. In addition to their economic importance, the predominance of larval and juvenile forms of fish in the estuary represents a food source for larger individuals, for fishing birds, and even for marine mammals.

The Punta Banda Estuary has been used by humans for at least the last 2,000 years. The region has been recognized as Kumiai territory by various historians, and at least 3,000 years ago the Kumiai People arrived in Southern California and northern Baja California. These groups, called Yumans, came from the desert area located in the northeast of the region. Linguistic and archaeological studies have revealed that these indigenous people were hunter-gatherers who spoke a common language, which is still used by some members of the native communities in the area of the site, today. Within the estuary vicinity, indigenous people could find a great variety of animals and plants that they used as food and raw material, as well as sufficient fresh water in the surrounding areas. There, they fished, hunted, and collected plant material (rush, tule, and willow) with which they made baskets and rafts. In the Punta Banda Estuary, there is at least one shell dump, hypothesized to be of Yuman origin. The settlement is characterized by the presence of the remains of mollusc shells and debris from lithic (flaked rock) material carvings. At these sites, the shells of bivalve molluscs, such as clams and oysters, were observed, as well as the remains of snails. These archeological finds encompass a 50 m radius, within which five areas of irregularly scattered similar remains have been observed.

In modern times, the coastal area of the Punta Banda Estuary is used by tourists and local communities as a place for recreating. Sport fishing is common, carried out from small boats, small yachts, or from the coast. There are several tourist camps at the site, which offer sport fishing trips for visiting tourists, as well as tourist camps owned by small family businesses, where cabins, camping sites, and boats are rented.

BAHÍA DE SAN QUINTÍN, BAJA CALIFORNIA: OVERVIEW

Located in a transitional biogeographic region between temperate and subtropical zones, often referred to as Mediterranean climate type, the Bahía de San Quintín (Figures 3.6 and 3.7) site supports the largest breeding populations of one species, and five subspecies, of threatened or endangered birds:

FIGURE 3.6   Composite aerial image of Bahía de San Quintín. Designated February 2, 2008; Area: 5,438 ha; Coordinates: 30°25′N 115°58′W; Ramsar Site number 1,775.

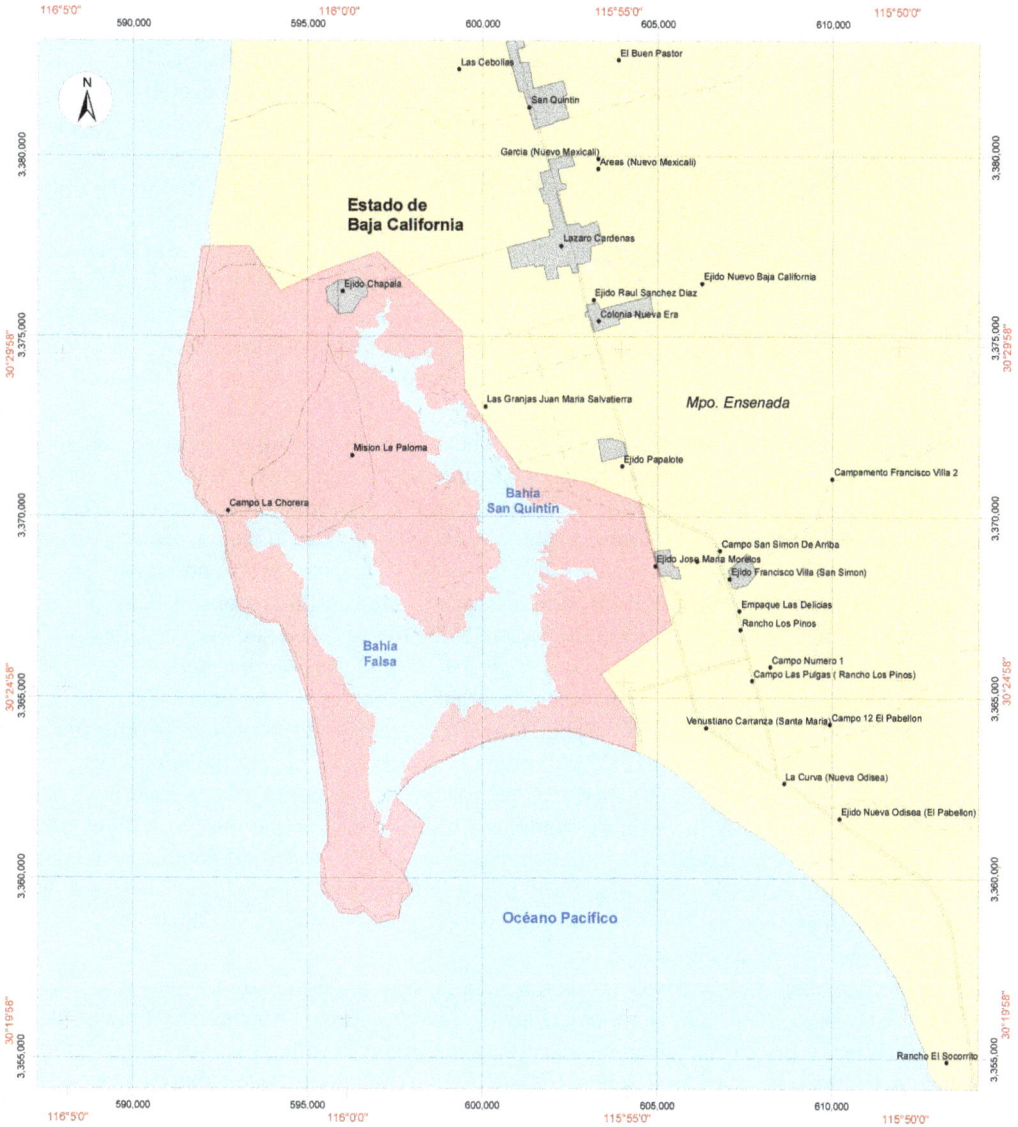

FIGURE 3.7 Map of Bahía de San Quintín. Designated February 2, 2008; Area: 5,438 ha; Coordinates: 30°25′N 115°58′W; Ramsar Site number 1,775.

light-footed clapper rail (Rallus longirostris levipes), black rail (Laterallus jamaicensis), California least tern (Sternula antillarum browni), the Belding's savannah sparrow (Passerculus sandwichensis beldingi), the California gnatcatcher (Polioptila californica), and the western snowy plover (Charadrius alexandrinus nivosus) [Massey and Palacios 1994]. The site is the only place on the Pacific coast of Baja California where black crake, an endangered species, have been sighted recently. The site also supports populations of the short-eared owl (Asio flammeus) and peregrine falcon (Falco peregrinus) [Aguirre et al. 1999], species under special protection per NOM-059-SEMARNAT-2001 and/or -2010. The site also supports the ornate shrew (Sorex ornatus) [Mellink 2002], with agricultural activity in the region likely restricting the range of distribution of this shrew species (Aguirre et al. 1999). More than 25,000 migratory waterbirds hibernate in the site's wetlands, which support 30–50% of the total *Branta bernicla nigricans* population during those periods of migration. *Astragalus harrisonii*, *Chorizanthe chaetophora*, *Chorizanthe interposita*, and *Chorizanthe jonesiana* are all endemic flora species of the Mediterranean climate area, which are similarly supported by the site.

SITE DETAILS

Bahía de San Quintín is one of only two wetlands on the western coast of Baja California, representing what in the past was a more common ecosystem in the Californian Biogeographic Region. Its biodiversity derives from its high productivity, owing to a wide variety of habitats in a healthy state of conservation, and the Mediterranean climate of this biogeographic region. Accordingly, flora and fauna with tropical and temperate affinities converge in this area, giving the region a unique character (Aguirre et al. 1999). Due to the aforementioned characteristics, the site is considered rare or uncommon for this biogeographic region. In the regional context, Bahía de San Quintín is the only natural area in the Mediterranean climate region of Alta and Baja California that remains relatively intact in its ecological functions as a coastal lagoon, while also being a scenic area of great beauty for people. Many of the priority conservation species and subspecies at the site depend to a large extent on the health and continued conservation status of this wetland for their survival (Aguirre et al. 1999).

The site supports at least ten species of plants endemic to this Mediterranean region, such as *Astragalus harrisonii* (Fabaceae), *Chorizanthe chaetophora* (Polygonaceae), *Chorizanthe interposita, Chorizanthe jonesiana, Chorizanthe turbinata, Dudleya anthonyi* (Crassulaceae), *Eriogonum fastigiatum, Hazardia berberidis* (Compositae), *Oenothera wigginsii* (Onagraceae), and *Senecio californicus* var. *ammophilus* (Compositae). The meadows of marine plants and marsh vegetation at the site are the best preserved of all of those existing in the coastal lagoons of Baja California. Several species of dune-forming plants at the site are at the limit of their southern distribution, such as marine rocket (Cakile maritima), *Ambrosia chamissonis*, and *Carpobrotus chilensis.*

Bahía de San Quintín is important in the biogeographical context of the Baja California region, as it is the habitat, both breeding and wintering, for various species and subspecies of birds that seek refuge at the site, including more than 25,000 migratory shorebirds during the winter (Page et al. 1997). The site is also an important supportive ecosystem for Pacific migratory waterfowl, shorebirds, and songbirds, such as the western sandpiper (Calidris mauri) and the black brant (Branta bernicla nigricans), of which it is known that approximately 50% of its total population winters in the area (Massey and Palacios 1994). The site is also a wintering area for birds of prey, such as the burrowing owl (Athene cunicularia), the short-eared owl (Asio flammeus), and the peregrine falcon (Falco peregrinus) [Aguirre et al. 1999].

The site's channels, when exposed at low tide, can be seen to be bordered by eelgrass (Zostera marina), with the other algal flora of the bay relatively inconspicuous. Other notable flora of the site are: saltmarsh bird's beak (Cordylanthus maritimus maritimus), Anthony's live forever (Dudleya anthonyi), and cordgrass (Spartina foliosa). *Cordylanthus maritimus maritimus* currently only grows in seven locations in the Californian region, with Bahía de San Quintín being one of these critically important few, where a very large population of this species has been recorded (Martínez-Fragoso 1992, Aguirre et al. 1999).

Notable vertebrates at the site include the endemic legless lizard (Anniella geronimensis), the endemic San Quintín meadow rat (Microtus californicus aequivocatus), the endemic San Quintín kangaroo rat (Dipodomys gravipes), the locally common California sea lion (Zalophus californianus), the harbor seal (Phoca vitulina), and the common bottlenose dolphin (Tursiops truncatus).

There is evidence indicating that in the precolonial period the site was frequented and used by regional indigenous peoples. Based on the shell piles and some lithic artifacts and projectiles that have been found in at the site, it is inferred that Bahía de San Quintín was a hunting area for migratory waterfowl and sea turtles. Additionally, various molluscs and fish are believed to have been commonly taken from its inner margins. During those earlier times, the bay served as a port of refuge and a fishing area for the tule rafts that indigenous people used, inside and outside the bay, for their fishing activities (Martínez-Fragoso 1992). Approximately 3,000 years ago, the dunes were very important to indigenous society, as evidenced by seven archeological sites consisting of shell-piling areas (concheros), areas with the remains of lithic tools, and seasonal campsites that have been described and are found in the Punta Mazo area (Serrano-González 1998).

The only thing that is known about the colonial period is that the salt pans located north of San Quintín were regularly exploited by missionaries (Martínez-Fragoso 1992). Toward the beginning of the 19th century, the bay was frequented by North American ships in smuggling transactions and, later, by whaling ships. During that same century the bay was also commercially exploited for its abundant population of sea otters (Enhydra lutris), which were completely exterminated by American and Russian hunters before 1850. It was not until around 1895 that the process of colonization and establishment of fixed communities began at the site. In those years, some North American corporations, and later English companies, began a large-scale project of colonization and development of local agriculture. During this time, a dock was built to export the harvests of a flour mill in the area, which formed the central part of town. The town was connected to the north by a railway line. Additionally, rainfed agricultural was established on the eastern and southern plains of the bay, since the late 19th century.

In the last 40 years, the region of the bay that is currently known as Molino Viejo has been the center of operation for the exploitation of abalone and lobster on Isla San Martín. The region's fishing areas are supplied in this area, where the catches are landed. It is also worth noting that Molino Viejo has been the base field laboratory for almost all the scientific explorations that have been carried out in San Quintín (Martínez Fragoso 1992).

Although it is difficult to achieve sustainable development for many projects, where natural resources are cultivated, sustainable use has been practiced daily in Bahía de San Quintín for 30 years. For example, the culturing of bivalve molluscs has proven to be a profitable and non-polluting use of the bay, with a high degree of sustainability. Another extractive use of resources in the area of the site is the extraction of volcanic rock known as "morusa" or "volcanic slag," from the nearby Kenton Volcano and Picacho Vizcaíno, from which red and black ash is obtained, respectively, for the North American market, mainly in the State of California for the construction industry, and for agrochemistry, soil improvement, and as ornamentation for landscaping roads and properties. Other extractive uses in the vicinity of the site are extraction of building stone and gravel; salt mining; and the harvesting of abalone (Haliotis sp.), the ladybug clam (Agropecten circularis), the sand clam (Chione succinta), the sea cucumber (Isostichopus fuscus), sea urchins (Strongylocentrotus sp.), and the rugose pen shell (Pinna rugosa) [Gobierno del Estado de Baja California 2006].

CORREDOR COSTERO LA ASAMBLEA-SAN FRANCISQUITO, BAJA CALIFORNIA: OVERVIEW

Corredor Costero La Asamblea-San Francisquito (CCLASF) [Figures 3.8 and 3.9] provides diverse marine and coastal habitat conditions for abundant fauna and flora. Among the species present at the site are the marine turtles Chelonia mydas, Caretta caretta, Lepidochelys olivacea, Eretmochelys imbricata, and Dermochelys coriacea, as well as the whale shark (Rhincodon typus). The site contains numerous seagrass beds and coral formations, but also bears the impact of the introduction of exotic species, the collection of eggs, and various other human activities.

The site is located on the eastern coast of the State of Baja California, which is comprised of 21 closed or semi-closed coastal water bodies (estuaries, marshes, small coastal lagoons, and hypersaline pools), 22 sandy beaches and dune areas, and 17 islands that make up the Bahía de Los Angeles (BLA) archipelago, and numerous coastal formations and island reefs. The site is home to flora and fauna that are characterized by a high degree of specialization, with 71 plant species and 87 animal species recorded, of which 21 are under some category of protection under NOM-059-SEMARNAT-2001 and/or -2010 and/or listed by the International Convention on threatened Species of Wild Fauna and Flora (CITES 2005). Of the 3,452 marine species that have been reported for the Gulf of California, at least 35% have been recorded at the site, for a total of 87 plant species and 1200 marine faunal species, including 66 endemic species and 68 species under some category of protection under NOM-059-SEMARNAT-2001 and/or -2010 and in the CITES (2005). Of the 117 plant species identified in the BLA Archipelago, 11 species are endemic to Baja California and

FIGURE 3.8 Composite aerial image of Corredor Costero La Asamblea-San Francisquito (CCLASF). Designated November 27, 2005; Area: 44,304 ha; Coordinates: 29°27′N 113°50′W. Ramsar Site number: 1,595.

two (Xylorhiza frutescens and Mammillaria insularis) are also endemic to the islands of the BLA. Along the southwestern coast of Isla Coronado, a small bay of mangroves (Rhizophora mangle) exist, the northernmost extent of this species for Mexico. Of the 26 species of fauna reported in the BLA archipelago, six species are endemic to the islands and 13 are protected under NOM-059-SEMARNAT-2001 and/or -2010 and in the CITES (2005).

SITE DETAILS

The CCLASF brings together a mosaic of estuaries, marshes, small coastal lagoons, hypersaline pools, sandy beaches, dune areas, islands, islets, coastal and insular reef formations, soft seabeds, muddy shoals, mangroves, and kelp beds, supporting the most productive marine ecosystem in all of Mexico, adjoining an extremely arid desert. This combination of environmental conditions, ecosystems, and ecological relationships gives CCLASF its unique character, which is recognized

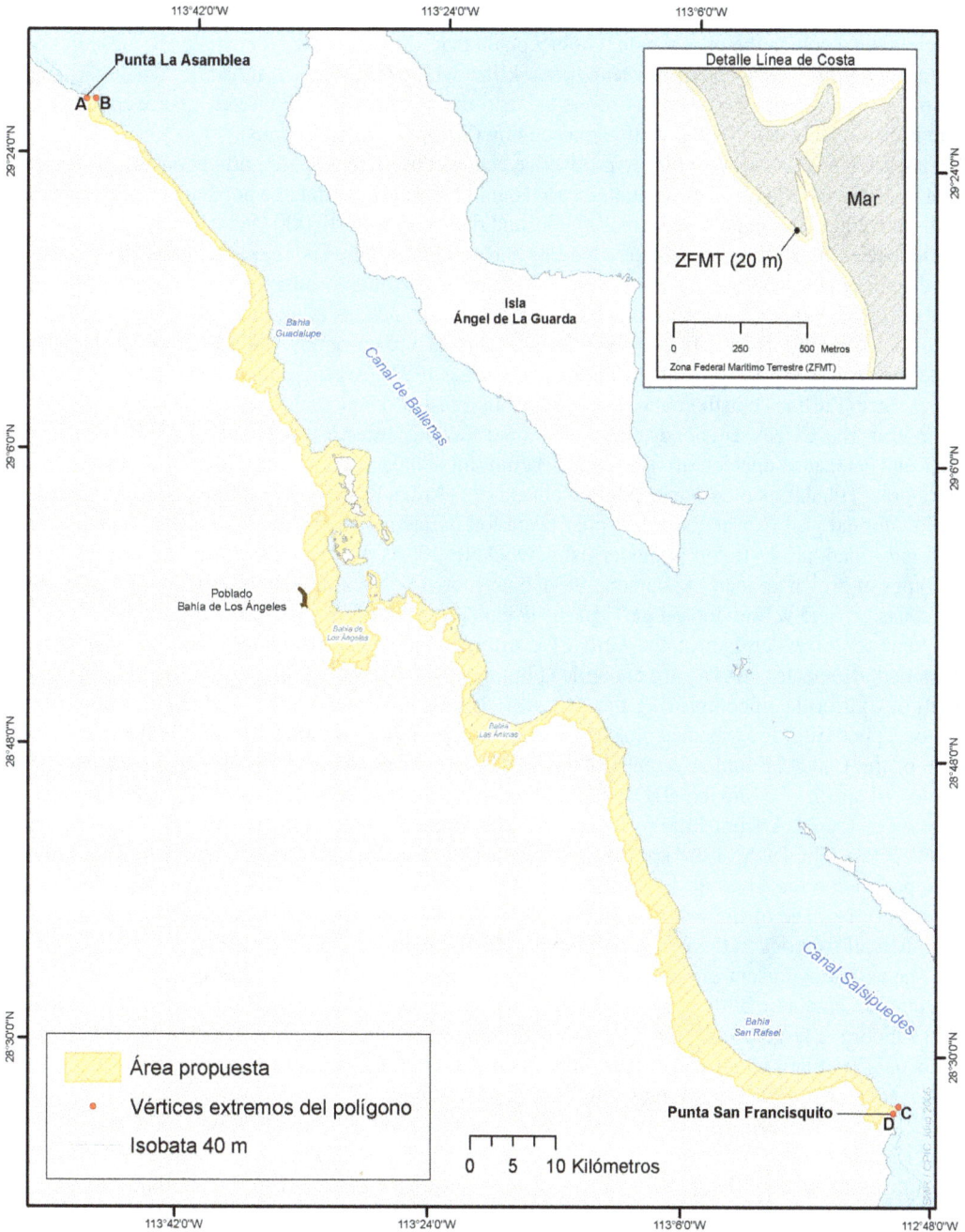

FIGURE 3.9 Map of Corredor Costero La Asamblea-San Francisquito (CCLASF). Designated November 27, 2005; Area: 44,304 ha; Coordinates: 29°27′N 113°50′W. Ramsar Site number: 1,595.

as a priority area for conservation by The National Commission for the Knowledge and Use of Biodiversity (CONABIO) and the Coalition for the Sustainability of the Gulf of California.

The CCLASF is critical habitat for imperiled sea organisms such as the green sea turtle (Chelonia mydas), loggerhead turtle (Caretta caretta), olive ridley sea turtle (Lepidochelys olivacea), hawksbill sea turtle (Eretmochelys imbricata), and leatherback sea turtle (Dermochelys coriacea), as well as the whale shark (Rhincodon typus). The area has also been documented as an important feeding and refuge area for fin whale (Balaenoptera physalus), Eden's whale (B. edeni), the common dolphin (Delphinus delphis), the pilot whale (Globicephala macrorhynha), the blue whale (B. musculus), the humpback whale (Megaptera novaeangliae), killer whale (Orcinus orca), minke whale (B. acuto-rostrata), gray whale (Eschrichtius robustus), and the sperm whale (Physeter macrocephalus). The site is also widely used by the California sea lion (Zalophus californianus).

The CCLASF is part of a highly productive ocean ecosystem that supports populations of important seabirds. On Isla Rasa, bird counts carried out in 1999 indicate that the populations of Heermann's gull were 260,000, elegant tern were 200,000, and royal tern were 10,000 (Velarde and Ezcurra 2002); these species all use the CCLASF as a feeding and perching areas. The region is also home to the largest colonies in the world of brown pelican (Pelecanus occidentalis), with populations reported between 6,000 and 18,000 pairs in the archipelago of San Lorenzo Islands (Anderson 1983). 110,000 pairs of blue-footed booby (Sula nebouxii) and 74,000 pairs of brown booby (Sula leucogaster) have been reported on Isla San Pedro Martir (Tershy and Breese 1997). All these taxa, as well as the Mexican shearwater (Puffinus opisthomelas) are endemic or quasi-endemic to this region (Howell et al. 2001).

Within the CCLASF islands there are noted nesting areas for substantial numbers of brown pelican (Pelecanus occidentalis), Brandt's cormorant (Phalacrocorax penicillatus), double-breasted cormorant (Phalacrocorax auritus), great blue heron (Ardea herodias), blue egret (Egretta caerulea), yellow-footed gull (Larus livens), osprey (Pandion haliaetus), peregrine falcon (Falco peregrinus), and red-tailed hawk (Buteo jamaicensis) [CONANP 2000]. The site also hosts the world's largest colonies of brown pelican (Anderson 1983), blue-footed booby (Sula nebouxii), brown booby (Sula leucogaster), and yellow-footed gull (Larus livens) [Tershy and Breese 1997].

Analyses of endemism in the Gulf of California indicates that of the 782 bony fish species recorded, 81 species (10.4%) are endemic (Findley et al. 1996). Of these, 53 are found in the Upper Gulf of California, encompassing the CCLASF. It should be noted that out of ten endemic species of the Upper Gulf, four species of scorpionfish (Scorpaenidae) occur exclusively in the marine portion of the CCLASF and adjacent waters. The whale shark (Rhincodon typus), a threatened and protected species in Mexico (DOF 2002, 2010) and in the world (CITES 2005), uses the internal portion of the BLA from June to November as a zone of shelter and feeding (Enríquez-Andrade et al. 2003). The BLA is the spawning habitat of the California flounder (Paralichthys californicus) population that lives in the Gulf of California, which has significant commercial importance. Migratory passage of the yellowtail amberjack (Seriola lalandi), a species used by both sport and commercial fisheries in the BLA (Torreblanca 2003) occurs in the CCLASF.

The avifauna present in the CCLASF are made up of a minimum of 258 species, of which 126 are marine or aquatic birds (Anderson 1983). The site is part of a highly productive ocean ecosystem that supports a large proportion of the world's population of the black storm petrel (Oceanodroma melania, 70%) and lesser storm petrel (O. microsoma, 90%), which nest on Isla Partida (Velarde and Anderson 1994); Heermann's gull (Larus heermanni, 90%;); the royal tern (Sterna maxima, 70%); the elegant tern (S. elegans, 95%), which nest specifically on Isla Rasa; and Craveri's murrelet (Synthliboramphus craveri, 90%), which nests on several islands in the region (Velarde 1989, Velarde and Anderson 1994). Among the BLA islands there are seven species of reptiles and three lizards, including the western zebra-tailed lizard (Callisaurus draconoides), common spotted sided lizard (Uta stansburiana), Xantus' leaf-toed gecko (Phyllodactylus xanti), the Angel Island chuckwalla that is endemic to the Great Islands Region (Sauromalus hispidus), as well as the night snake (Hypsiglena torquata) and the speckled rattlesnake (Crotalus mitchellii) [Grismer, 1999]. Of these reptiles, Uta stansburiana and Sauromalus hispidus are the most well-known, as they are present among 16 of the islands of the site (Grismer 1999). Coronado Island stands out as having all of the species present in this archipelago, on that one island, with six of the islands in the archipelago devoid of observed reptiles. All of the reptiles mentioned above are protected per NOM-059-SEMARNAT-2001 and/or -2010.

The first inhabitants of the CCLASF were the Cochimíes people who lived within the site between 2,000 and 4,000 years ago in the surroundings of the aguaje (spring) that is within the BLA;

they were engaged in hunting, gathering, and fishing. They were known to use simple utensils such as small nets, traps, and turtle shells for collecting seeds, and harpoons for hunting sea turtles, deer, and bighorn sheep. They also hunted small sea lions and took advantage of dolphins and stranded whales. The presence of remains of organisms such as rays, sharks, puffer fish, crabs, sea turtles, mussels, groupers, birds, gastropods, and bivalves that have been found in the shell piles among the CCLASF provide evidence of the resources that this indigenous group extracted from BLA (Aschman 1959). In 1759, the Jesuit missionaries founded the San Borja Mission, 38 km from BLA. The Mission is currently a tourist attraction and is part of the historical heritage of the State of Baja California.

Whether through extraction, exploitation, direct use, or through non-consumptive uses, most of the economic activities carried out in the CCLASF depend on the natural resources of the area. 40% of families in the vicinity of the site receive their livings from fishing; 10% from jobs in transport services (i.e., drivers); 7.7% of families in the vicinity of the site receive income directly from tourism (e.g., employees in hotels, restaurants and camps, sport fishing guides, or naturalist guides); 6.2% of families in the vicinity offer mechanical services for income; 17.7% of people in the vicinity of the site engage in other activities (e.g., teachers, police, and bricklayers) for income; and 6.9% of those in the vicinity "do not work" (Brandstein 1998).

The islands of the BLA region face environmental impacts from the reduction of native floral and faunal populations due to the introduction of exotic species; collection of seabird eggs; wildlife disturbance (particularly breeding seabird and California sea lion colonies); looting of native and endemic flora and fauna (e.g., collection of seeds and plant shoots, and the capture of reptiles); alteration and/or degradation of habitats (deforestation and soil erosion due to the removal of vegetation and/or stones, for the creation of trails and for the establishment of new camping areas on some islands); contamination of the coastal zone by garbage and chemical residues and hydrocarbons from boats; and potential increase in the flow of visitors without proper planning occurring in advance of such activities.

ISLA RASA, BAJA CALIFORNIA: OVERVIEW

Isla Rasa (Figures 3.10 and 3.11) is an island of volcanic origin with three coastal lagoons, located in a zone of high marine productivity. Over 80 species of terrestrial and marine bird species have been observed on and near the island. The site hosts a large proportion of both populations of Heermann's gull (Larus heermanni) and the elegant tern (Sterna elegans), with hundreds of thousands of individuals of each species. The island also hosts two endemic reptile species: leaf-toed gecko (Phyllodactylus tinklei) and the common side-blotched lizard (Uta stansburiana), but no native mammals. The introduced black rat (Rattus rattus) and the house mouse (Mus musculus), introduced in the late 19th and early 20th centuries, were successfully eradicated from the island in 1995.

SITE DETAILS

Located in the Gulf of California, the town closest to the site is Bahía de los Ángeles (BLA), with approximately 1,000 permanent residents, approximately 56 km to the northwest; the island is approximately 93 km directly offshore of the predominantly uninhabited peninsular area. The island is located in the northern half of the Gulf of California, southeast of Isla Ángel de la Guarda, and northwest of Isla Salsipuedes. The site is one of the richest fishing grounds of the region and an important source of fish production for Mexico, particularly for its Monterey sardine (Sardinops caeruleus). The coasts of the island, which are rocky, and other noted lagoons, serve as resting and feeding sites for a large number of migratory birds, mainly shorebirds. The shorelines of the lagoons on the island, in their intertidal zone, consist of patches of halophytes with abundant Salicornia pacifica, Sesuvium verrucosum, and Batis maritima. In general, the island has very scarce vegetational cover, as with the Sonoran Desert overall, with a predominance of dwarf saltbush (Atriplex barclayana), two species of cholla (Opuntia cholla

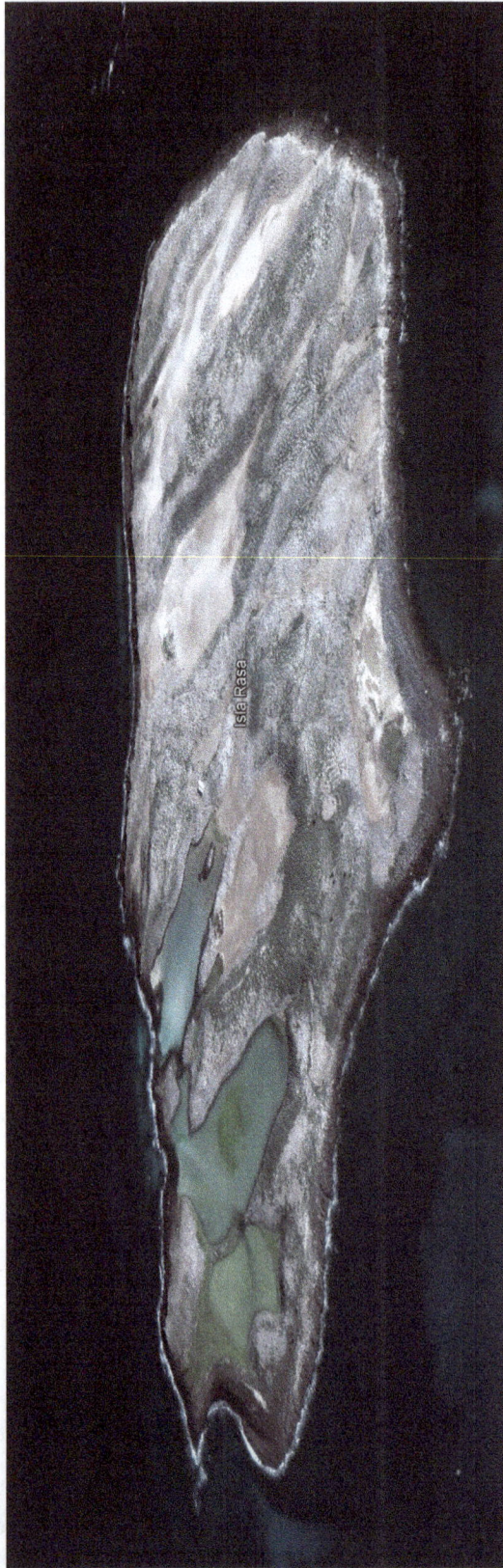

FIGURE 3.10   Composite aerial image of Isla Rasa. Designated February 2, 2006; Area: 66 ha; Coordinates: 28°49′N 112°59′W; Ramsar Site number: 1,603.

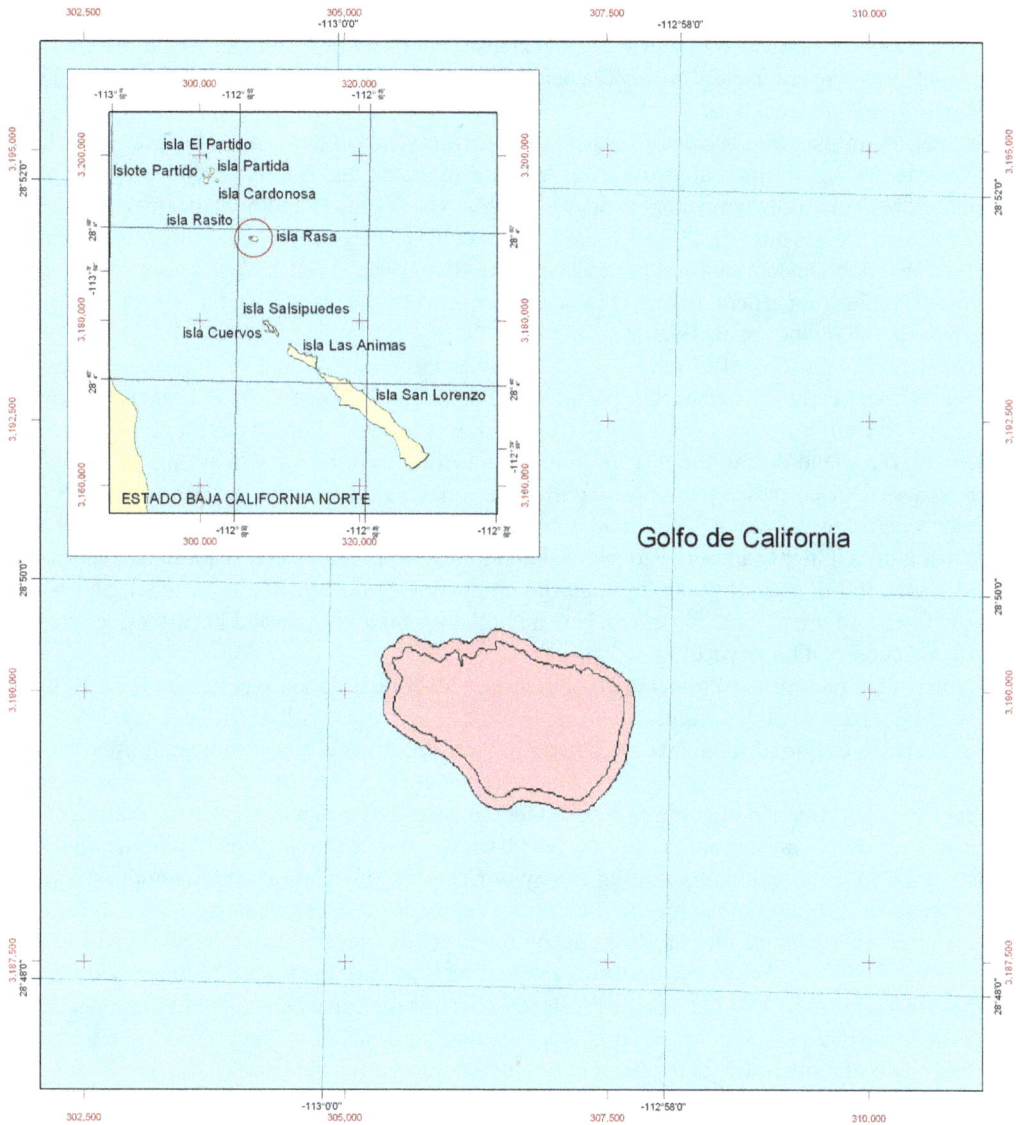

FIGURE 3.11 Map of Isla Rasa. Designated February 2, 2006; Area: 66 ha; Coordinates: 28°49′N 112°59′W; Ramsar Site number: 1,603.

and O. bigelovii), and the other cacti species *Pachycereus pringlei*, *Stenocereus gummosus*, and *Lophocereus schotii*, as well as *Cressa truxillensis* and *Lycium brevipes*. The site's intertidal plant species include *Batis maritima*, *Sesuvium verrucosum*, and *Salicornia pacifica*. The dominant animals on the island are seabirds that nest during spring and early summer. The island is the nesting site of approximately 95% of the world's population of Heermann's gull (Larus heermanni) and elegant tern (Sterna elegans), with 260,000 and 200,000 individuals, respectively, resident. In addition, the island is a nesting colony of royal tern (Sterna maxima) that fluctuates annually at the site between 10,000 and 17,000 individuals. Also, nesting on the island in small numbers, is the yellow-footed gull (Larus livens). In total, Isla Rasa supports approximately 475,000 individuals of the species of seabirds mentioned above.

Rasa Island originates from the Holocene, its approximate age estimated at 10,000 years, and it is composed of basalt rock pavements. Its soils are basically regosols and lithosols, with valleys of different dimensions, and hills that have a maximum height of 33 m. The valleys are full of guano that seabirds have deposited throughout its geological history and the site lacks fresh water, but has three coastal lagoons, one of which has an artificial extension of its natural entrance to the sea, therefore filling and emptying completely during the periods of high and low tide. Tides in the vicinity fluctuate between −0.5 m and 3.0 m.

In total, more than 80 species of birds have been recorded on the island (Velarde 1989, Cody and Velarde 2002), of which approximately half are migrants and the other half are residents of the region, but only a few are island residents, such as the osprey (Pandion haliaetus), which feed mainly on fish species near the coastal areas of the island. There is tremendous diversity within the intertidal invertebrate communities of the island, but these populations have not been well studied, with the exception of a general listing of intertidal decapod crustaceans from the islands of the Gulf of California (Villalobos et al. 1989).

The island has few historical references, however guano was exploited on it for several years in the late 19th and early 20th centuries by the Mexican Guano Company of San Francisco, which it transported to California and other parts of the world. It is believed that the largest proportion of the workers on the island during the time of guano production were indigenous people of the region, and prisoners. During these operations, the invasive rodents *Rattus rattus* and *Mus musculus* were likely introduced to the island. The island provides several ecosystem services, such as high scenic value that makes it highly attractive to national and foreign tourists, as well as for the dissemination of knowledge to the general public through the production of documentaries, articles, and books. The site is also of great value for researchers and others who are interested in furthering environmental and conservation sciences.

The island is presently a Protected Natural Area without extraction, yet human activity in the surrounding waters includes commercial and sport fishing, which is high in intensity. There is a risk of overexploitation of some resources in the region, specifically sea cucumber, lobster, various species of molluscs, and some of the rocky reef fish that are exploited in the area. Approximately a century ago, the extraction of guano was extremely intense on the island and the associated modification of the natural landscape was extreme, which came with the introduction of rats and mice; this set in motion the extirpation of a nesting colony of Craveri's murrelet (Synthliboramphus craveri), a small seabird that nests in cavities in the ground and between rocks. Although there are no permanent human inhabitants on Isla Rasa, uncontrolled human presence due to occasional visitors (approximately 500 visitors per year, during the nesting season of the birds) severely affects reproductive success of the birds that nest there. Because of this disturbance and the high proportion of the world's population of species of birds that nest there, the island is a key site for conservation. The fragility of the site is thus at the heart of the site's conservation philosophy: a single season (i.e., a single day for any individual nesting bird/egg/chick) without protection and with human disturbance can mean the reproductive failure of hundreds of thousands of nesting pairs of birds. Another adverse factor on the site is over-extraction of fish species that birds feed upon. The island's bird species feed on small pelagic fish, mainly Monterrey sardine (Sardinops caeruleus), and northern anchovy (Engraulis mordax). These two fish species, particularly the sardine, are also quite economically important and have been subject to overexploitation by fishers. It should be noted that the Monterrey sardine has suffered major population declines due to violations of catch size, i.e., the capture of fish that have not yet had a chance to reproduce. Isla Rasa is also located within the Protection Zone of Flora and Fauna Islands of the Gulf of California, which is a federal protection that has a general management program requirement for the entire protected area, including a conservation focus on the research, monitoring, protection, visitor presence (i.e., guidance), and the establishment of zero-access areas or restricted areas for nestlings at the site.

The island's organized, visiting tourists are held to strict controls, however there are isolated visitors who come to the island in small boats (sailboats, private yachts, and pangas) who are unaware of the impacts that their mere presence in the interior of the island brings to the fragile bird populations. Therefore, general information has been distributed to as many visitors as possible, prepared by interested conservation groups, but there are no regularly organized programs to prevent all incursions. At times the island is also visited or occupied by Mexican Navy personnel, and those activities are regulated in terms of the circulation of biosecurity information or other information about avoiding disturbance to the nesting colonies of birds, or other types of disturbances such as the consumption of native marine species.

LAGUNA OJO DE LIEBRE, BAJA CALIFORNIA SUR: OVERVIEW

A hypersaline coastal lagoon, Laguna Ojo de Liebre (Figures 3.12 and 3.13) is one of the main refuges for gray whales (Eschrichtius robustus), where they meet for pairing, breeding, and raising their young. The site also provides shelter for a wide variety of birds along the river and on the islands within the lagoon. A total of 94 species have been noted at the site, which contains a lagoon, coastal and marine ecosystems, as well as tidal channels and intertidal areas. The site surroundings also contain unstable dunes, saltpeter areas, halophilous shrubs, and herbaceous marsh vegetation. Waste disposal, a consequence of the fishing activities, and the over-exploitation of the area for fishing are the main ecological threats to the site. The lagoon has been declared a sheltered area for whales, since 1972.

SITE DETAILS

The Ojo de Liebre Lagoon is a hypersaline coastal lagoon and one of the main refuges for the gray whale, which migrates from the Bering Strait to the western coast of the Baja California Peninsula. The site is a prime area for mating and birthing of the gray whale (Eschrichtius robustus), which is a species of special protection in the area per NOM-059-SEMARNAT-2001 and/or -2010. From 1952 to 1979 the average number of gray whales was 1,002 individuals; from 1980–1989 their numbers increased to 1,545, and in 1990 their numbers increased to 1,551 individuals. Sea lion (Zalophus californianus) and the dolphin (Delphinus delphis) are two species of marine mammals that are present in the site's lagoon and are also under special protection per NOM-059-SEMARNAT-2001 and/or -2010. The endangered green sea turtle (Chelonia mydas) is also present in the Ojo de Liebre Lagoon, where it breeds and feeds. The black brant (Branta bernicla nigricans), threatened in Mexico, lives in large concentrations in the lagoon during the winter season. More than 35,000 brant (Branta bernicla) [Castellanos and Llinas 1991] and more than 270,000 shorebirds (Page et al. 1997) winter at the site.

In general, the climate of the site is very dry, with scarce precipitation throughout the year. The annual global infiltration and the recharge of the aquifers is very low, with only about 70 mm of annual precipitation. There are no streams or rivers that drain directly into the lagoon.

The main flora types at the site are coastal dune communities with halophytic scrub, including Abronia gracilis, Atriplex canescens, Dalea maritima, Plantago insularis, and Oenothera primiveris; large mudflats where, due to the influence of high tides, seawater reaches several kilometers inland, comprised of mainly the genera Atriplex, Salicornia, Allenrolfea, Suaeda, and Limonium; and seasonably variable and distributed microalgae, comprised of approximately 85 species in the Ojo de Liebre Lagoon, of which 40 are from the Rodophyta, 29 from the Chlorophyta, and 15 from the Phaeophyta. Species that contribute the highest annual biomass at the site are Spyridia filamentosa and Polysiphonia pacifica. Abundance observations of phytoplankton at the site determined that diatoms were the most abundant in the entire lagoon, with the most representative genera being Nitzschia, Rhizosolenia, and Cocconeis. Cordgrass (Spartina foliosa) is the main species that remains exposed in the floodplain, with eelgrass (Zostera marina) growing as far as the tide allows, up to 7 m deep in the channels. Other than eelgrass, characteristic species of submerged areas are Phyllospadix scouleri and Ruppia maritima.

There are 94 species of birds recorded for the site, with the following most abundant: Phalaropus lobatus, Calidris mauri, Podiceps nigricollis, Limosa fedoa, and Branta bernicla. The Ojo de Liebre Lagoon complex also supports nesting habitat for several species of birds of prey and seabirds, as well as harbors hundreds of species of migratory birds. A total of 20 species of ducks winter in the area. A comprehensive study of fish at the site noted 36 families, 50 genera, and 59 species observed; Fundulus parvipinnis and Atherinops affinis were the most abundant. A total of 111 species of benthic invertebrates have also been reported in different studies at the site. In the lagoon you can find large banks of ladybug clams (Argopecten circularis), lion's paw scallop

FIGURE 3.12 Composite aerial image of Laguna Ojo de Liebre. Designated February 02, 2004; Area: 36,600 ha; Coordinates: 27°45'N 114°05'W; Ramsar Site number: 1,339.

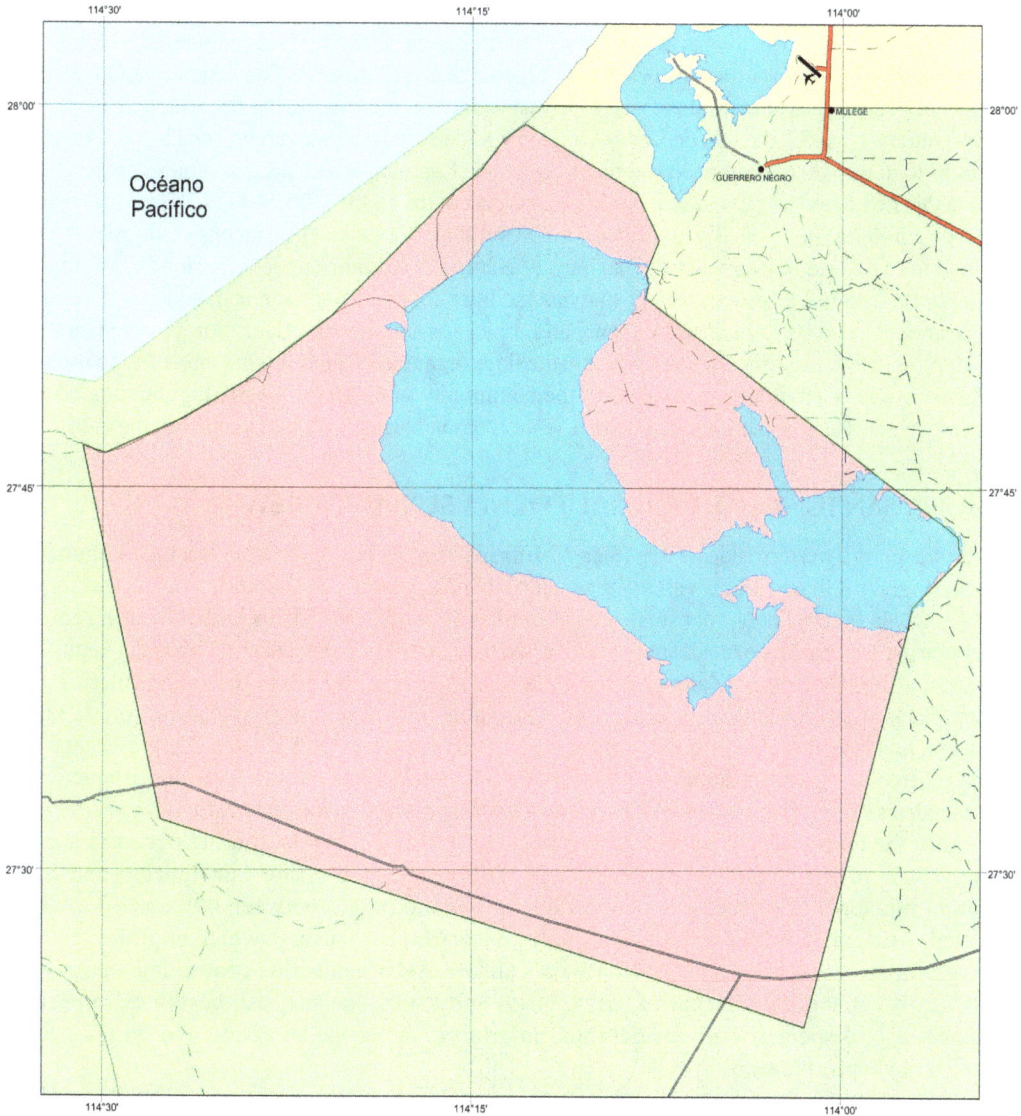

FIGURE 3.13 Map of Laguna Ojo de Liebre. Designated February 02, 2004; Area: 36,600 ha; Coordinates: 27°45′N 114°05′W; Ramsar Site number: 1,339.

(Lyropecten subnodosus), chocolate clam (Megapitaria spp.), rugose pen shell (Pinna rugosa), mangrove cockle (Anadara tuberculosa), flying clam (Pecten voqdesi), the less abundant Panama pearl oyster (Pinctada mazatlanica), mother-of-pearl shell (Pteria sterna), panocha snails (Astrea undosa and A. turbanica), black-and-white murex (Muricanthus nigritus), and Eastern Pacific giant conch (Strombus galeatus). There are also other invertebrates at the site, such as lobster (Panulirus sp.) and crab (Callinectes sp.). In total, 31 species of mammals can be found in the lagoon area.

The hydrology of the site is impinged upon by agricultural use in the area, and within the municipality there are 137 wells and 3 springs. Among cattle ranches, there are agrarian communities dedicated to breeding cattle, sheep, and goats. Forestry is an incipient activity whose use is of dead wood, being used in an artisanal way, and as firewood for domestic use. Mining at the site consists mainly of non-metallic minerals (salt production) in Guerrero Negro, plaster production in Santa Rosalía, and metallic mining (copper and cobalt) in Santa Rosalía. There is a geothermal project located in the Volcán de las Vírgenes, with the purpose of generating electric power for the region. In recent years, tourism has become a rapidly growing activity with the most visited sites within the lagoon complexes, the refuges of the gray whale, cave painting areas, and the Mission founded by the Jesuits in San Ignacio, as well as other historical monuments of Santa Rosalía. The beaches, and adventure tourism in general, are attractions of most interest to foreign tourists during the winter season.

On January 14, 1972, the Ojo de Liebre Lagoon was declared a whale refuge, which was then modi-fied in 1980 to include the Guerrero Negro and Manuela Lagoons, which are a complex comprising the Laguna Ojo de Liebre. The Ojo de Liebre Lagoon is part of the MAB-UNESCO El Vizcaíno Biosphere Reserve (2,546,790 ha), which was decreed on November 30, 1988. The lagoons of the site are thus included in the World Heritage listings, as Sites of Natural Importance under criterion IV. Laguna Ojo de Liebre Lagoon has a Visitors Center that has facilities for education/outreach, and two dormitories for researchers, as well as parking, signage, and a pier for small boats. Education programs for con-servation include tourist services for whale watching, organized training workshops for eco-tourism guides, teaching of environmental issues in the community of Guerrero Negro, highlighting ecologi-cally important issues of the Laguna Ojo de Liebre Lagoon for residents and visitors to the area.

LAGUNA SAN IGNACIO, BAJA CALIFORNIA SUR: OVERVIEW

On the very arid western side of the Baja California Peninsula, the coastal-brackish Laguna San Ignacio (Figures 3.14 and 3.15), with its large intertidal flats, provides particularly important refuge for the gray whale (Eschrichtius robustus), which number 300–400 individuals each winter in this area. Other marine mammals are resident at the site as well, including elephant seals (Mirounga angustiros-tris) and California sea lions (Zalophus californianus). The site is also one of the main wintering areas for the black brant (Branta bernicla nigricans), contains the northernmost mangrove forests on the continent, and is an important nursery for fish. The site is part of the El Vizcaíno World Heritage Site, which receives a significant number of tourists for whale-watching and other eco-tourism activities. The site also suffers from the negative impacts of overfishing and inadequate waste disposal.

Within the mid-northern portion of the Baja California Peninsula, the site occupies approx-imately 80% of the Municipality of Mulegé. The nearest towns are San Ignacio and Santa Rosalía (municipal seat). The average altitude of the site ranges between 0 m and 10 m above sea level. As a major refuge for the gray whale (Eschrichtius robustus), which migrates from the Bering Strait to the western coast of the Baja California Peninsula, this coastal lagoon provides areas for this species to concentrate, give birth, and mate. The site also has an extraordinary presence of bird species, both resident and migratory, along the shoreline and on the islands found in the central lagoon.

SITE DETAILS

This lagoon is located in the San Ignacio-San Raymundo Basin and San Ignacio Sub-basin; the channels in the San Ignacio Lagoon are intermittent in nature and eventually discharge to the Pacific Ocean, but are commonly lost to infiltration before reaching the coastline, and thus contains mainly torrential flows within ephemeral streams. The main streams that drain to the lagoon area are: Arroyo El Granado, Arroyo La Higuera, Arroyo Palo Verde, Arroyo Salvioso, San Ignacio Stream, Arroyo Patrocinio, Arroyo El Batequi, Arroyo San Ángel, Arroyo Covarrubias, San Juan Stream, San Hilario Stream, and Santa Rosa Stream. Mean annual runoff is from 0 mm to 20 mm, so that the annual global infiltration and recharge of the aquifers is of low magnitude or nil. The low annual precipitation, 50 mm to 70 mm, and the high potential evapotranspiration in the region, result in an average annual deficit of water.

The San Ignacio Lagoon contains marine coastal lagoon ecosystems. The lagoon has a natural channel depth of 6 m and is approximately 680 m wide, which crosses the shoals to enter the lagoon. The upland end of the lagoon has large flooded areas, transitioning to the interior area of the lagoon where there are extensive marshes, as well as stands of mangroves toward the entrance to the mouth

FIGURE 3.14 Composite aerial image of Laguna San Ignacio. Designated February 02, 2004; Area: 17,500 ha; Coordinates: 26°45'N 113°07'W; Ramsar Site number 1,341.

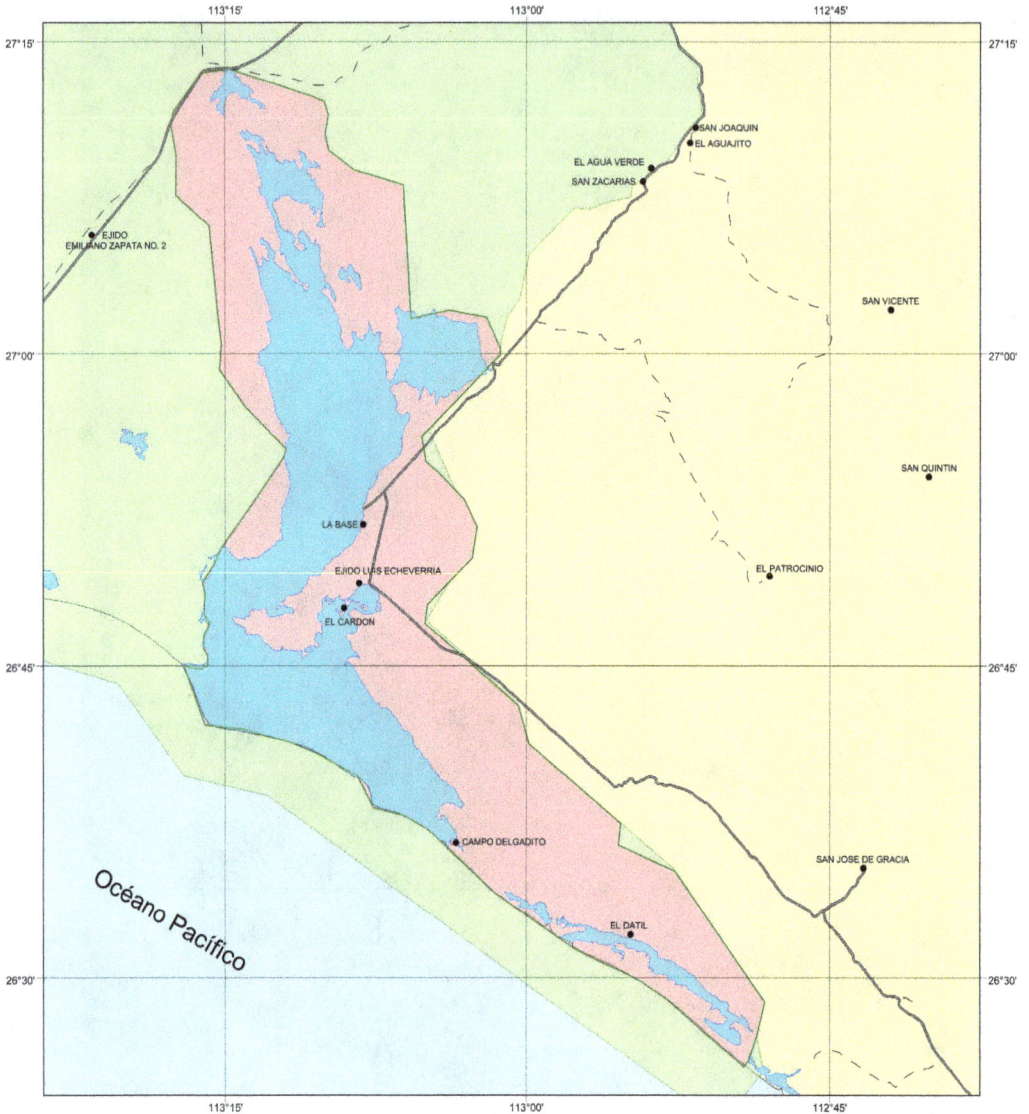

FIGURE 3.15 Map of Laguna San Ignacio. Designated February 02, 2004; Area: 17,500 ha; Coordinates: 26°45′N 113°07′W; Ramsar Site number 1,341.

of the lagoon. The site's surrounding terrestrial vegetation is composed of fleshy stemmed scrub, microphyllous scrub, and other halophilic vegetation. Vegetation in San Ignacio Lagoon is as follows:

- Coastal dunes, dominated by *Abronia gracilis*, *Atriplex canescens*, *Dalea maritima*, *Plantago insularis*, and *Oenothera primiveris*;
- Mudflats dominated by species of the genera *Atriplex*, *Salicornia*, *Allenrolfea*, *Suaeda*, and *Limonium*;
- Mangroves, with dominant species *Rhizophora mangle* and *Laguncularia racemosa*; and
- Marine submerged areas, dominated by *Zostera marina*, *Phyllospadix scouleri*, and *Ruppia maritima*.

Fauna of the site include approximately 173 bird species. Seabirds that dominate the site are of orders Gaviformes (Gavia pacifica and Gavia immer), the Pelecaniformes (Pelecanus occidentalis), and the

Anseriformes, among which the brant (Branta bernicla) stands out. Bird species that reproduce in the area are *Pelecanus occidentalis*, *Phalacrocorax auritus*, *Egretta thula*, *Egretta tricolor*, *Egretta rufescens*, *Nycticorax nycticorax*, *Larus occidentalis*, *Charadrius alexandrinus*, *Terna maxima*, *Hydroprogne caspia*, *Falco peregrinus*, and *Pandion haliaetus*. Other bird species include: *Buteo jamaicensis*, *Falco sparverius*, and *Callipepla californica*. 107 fish species are reported within the lagoon, of which 12 species are considered rare, with 34 fish species from temperate regions, and 27 fish species from tropical origins; a total of 244 species of benthic invertebrates have been identified at the site. Two species of amphibians [the spotted toad (Bufo punctatus) and Couch's spadefoot toad (Scaphiopus couchii)], 37 species of reptiles, and 40 species of mammals have been observed at the site.

Active commercial fishing in the vicinity of the site generates an approximate income of 60 million dollars (US), with an approximate annual catch of 15-million kilograms, supporting between 9,000 and 12,000 jobs. Fishers are organized into a number of Fishing Cooperatives, grouped into three Federations; harvesting of fish is solely authorized for local residents, and the aquaculture of local species is promoted and encouraged. Other activities in the vicinity are natural grazing, as well as farm-based agriculture. The predominantly irrigated agricultural areas in the municipality are El Valle de Vizcaíno, El Valle from Mulegé, and San Bruno, utilizing 137 wells and 3 springs. Forestry is a developing activity, often utilizing dead wood for artisanal work. Mining in the area includes salt production and plaster production, as well as copper and cobalt mining.

Eco-tourism is rapidly growing in the vicinity of the site, with highest numbers of visiting tourists in the lagoon complexes viewing gray whales; viewing cave paintings; and experiencing the local Mission founded by Jesuits in San Ignacio, and associated historical monuments of Santa Rosalía. The beaches, wildlife, and other types of adventure tourism in the area is very attractive for international tourists, especially during the winter months.

A number of key conservation measures are supportive of the ecology of the site, such as:

- Agreement for the protection of migratory birds signed by the governments of Mexico and the United States on January 9, 1937 that established closed seasons for the hunting of some species at the site and that also recommended the creation of refuge zones;
- Mexico's entry into the International Whaling Commission in 1949, protecting and promoting the rational use of whaling resources, worldwide;
- Establishment of the site as a Reserve Zone and Refuge for Migratory Birds and Wild Fauna in September 1972; and
- Establishment of the El Vizcaíno Biosphere Reserve, decreed on November 30, 1988, which subsequently led to the publication of the management program for the El Vizcaíno Biosphere Reserve in June 2000.

The environmental education program of the El Vizcaíno Biosphere Reserve includes activities specific to the San Ignacio Lagoon, such as:

- Advice on the contents of environmental issues that are taught to visitors to this lagoon by the providers of tourist services for whale watching;
- Organization of training workshops for eco-tourism guides that are established and operate in Laguna San Ignacio, in coordination with the federal and state tourism agencies;
- Teaching of environmental topics in the schools of communities that are located near the San Ignacio Lagoon; and
- Promotion of workshops and events related to conservation, for the communities that are near the San Ignacio Lagoon.

HUMEDAL LA SIERRA DE GUADALUPE, BAJA CALIFORNIA SUR: OVERVIEW

Humedal La Sierra de Guadalupe (Figures 3.16 and 3.17) comprises four intermittent riverine ecosystems that form several oases and wetlands at the marine interface. The site is located on the

FIGURE 3.16 Composite aerial image of Humedal La Sierra de Guadalupe. Designated February 02, 2008; Area: 348,087 ha; Coordinates: 26°40′N 112°30′W; Ramsar Site number: 1,815.

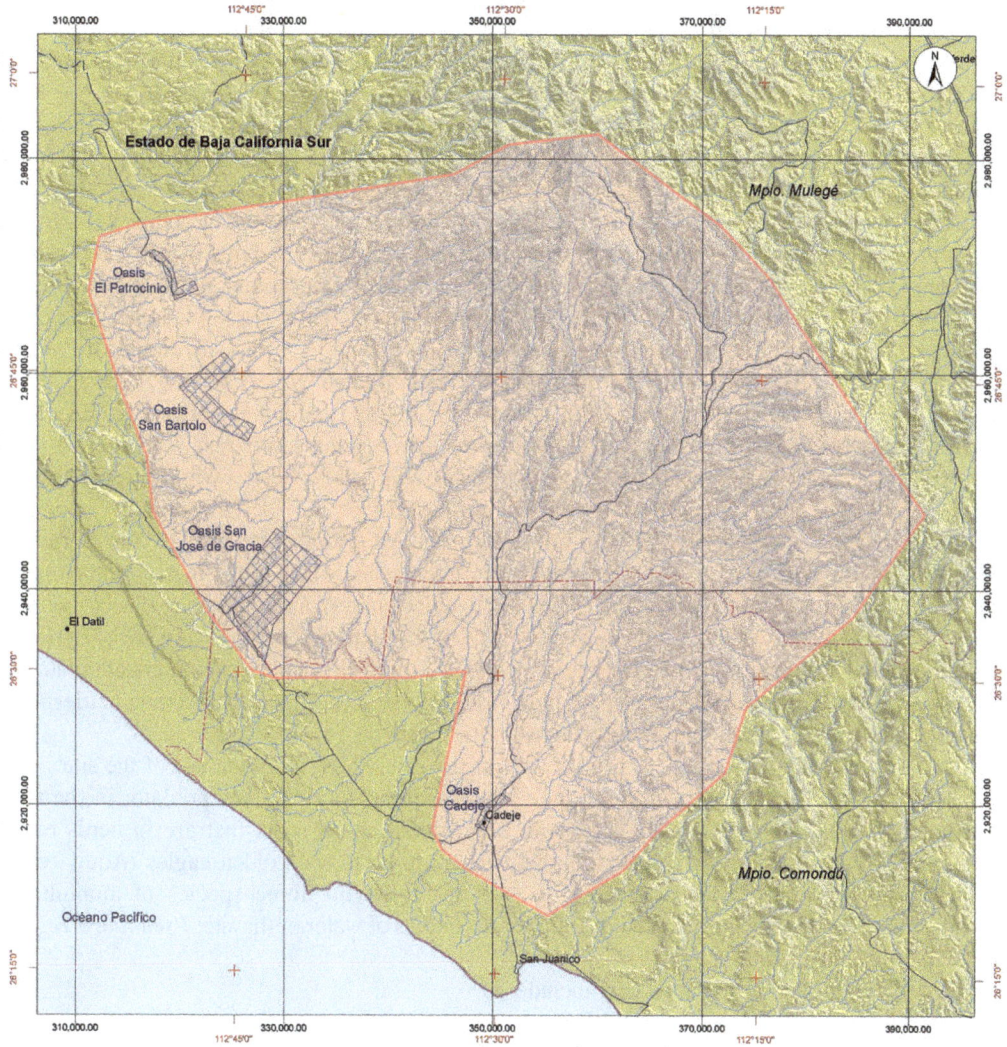

FIGURE 3.17 Map of Humedal La Sierra de Guadalupe. Designated February 02, 2008; Area: 348,087 ha; Coordinates: 26°40′N 112°30′W; Ramsar Site number: 1,815.

Baja California Peninsula, situated between the Pacific Ocean and the Gulf of California coasts, serving as an important and sole freshwater source in this predominantly desert landscape, for a variety of local and migratory fauna. Two endemic floral species have been recorded for the site: *Washingtonia robusta* and *Prosopis articulata*, as well as eight endemic reptile species, including *Bipes biporus*. Drought conditions as well as anthropogenic impacts from water extraction in the area are the biggest ecological threats to the site. The site is within the municipality of Mulegé, approximate population of 64,000, and ranges in altitude from 0 m to 270 m, with an average altitude of 135 m within the regional mountainous areas, with peaks in the range of 1,400 m.

SITE DETAILS

As the sole source of fresh water for the region, the site is critical habitat for a number of species that are attracted to the site's lower-lying areas, which are fed by freshwater from overland flow or springs, and which maintain water in coastal areas such as Cadejé Estuary (a marine coastal wetland with sandy beaches and a delta estuary to the east of the site, in the Bahía Concepción). The runoff of numerous streams forming in the elevated portion of the mountains of the site are gradually integrated into approximately a dozen streams that reach the coast of the Gulf of Mulegé and the Pacific (i.e., the estuaries of the San Ignacio and Cadejé Lagoons).

The Sierra de Guadalupe mountain range, the main orographic feature in the north-central part from Baja California Sur, captures water from precipitation and delivers it to wetlands that reside sporadically along major streams and secondary water courses, generally located below 300 m in elevation. The larger of these wetlands are permanent, and they contain woody vegetation dominated by palm and willow trees. Vegetation in wetlands of the site is subject to dynamic processes related to the seasonal and inter annual abundance of water, as well as natural and anthropogenic disturbances, which drive changes in the vegetational communities of these 'oasis' wetlands. The oases of site contain plant communities with groves of date palms and royal palms, surrounded by vegetation that is characteristic of the streams of the region, predominated by pearlberry (Vallesia glabra), sweet acacia (Acacia farnesiana), and other shrubby plants and perennial herbs (e.g., Baccharis salicifolia, B. sarothroides, Heliotropium curassavicum, and Mimulus guttatus). A total of 28 plant species have been recorded for the area, including two endemic species (Washingtonia robusta and Prosopis articulata), two exotic species (Phoenix dactylifera and Ricinus), and one imperiled species per NOM-059-SEMARNAT-2010, ironwood (Olneya tesota). The oasis of San Bartolo is one of the wetlands that maintains a strong agricultural presence where citrus (Citrus spp.), mango (Mangifera indica), avocado (Persea americana), maize (Zea mays), and sugarcane (Saccarum officinarum) are cultivated. Surrounding areas consist of fleshy stemmed scrub, white stick (Lysiloma candidum), candelabra cactus (Mirtyllocactus cochal), southern elephant tree (Bursera epinnata and B. hindsiana), cholla (Cylindropuntia spp.), palo Adán (Fouquieria diguetii), Arizona nettlespurge (Jatropha cinerea), and organ pipe cactus (Stenocereus thurberi).

Rebman et al. (1998) observed 48 bird species at the site, which all make use of the site's wetlands to one or another degree. Cougar (Puma concolor improcera) is a key predator in the area, whose abundance seems to be associated with high predation on goats that are in herds raised by residents of the area. In the winter months, it is common to see golden eagles (Aquila chrysaetos canadensis) at the site. Hollingsworth lists the following three species of amphibians, dependent upon both of the permanent and temporary bodies of water of the site: *Pseudacris regilla*, *Bufo punctatus*, and *Scaphiopus couchii*.

Hydrophytic flora observed at the site include:

Asteraceae: *Baccharis salicifolia* and *B. sarothroides*;
Boraginaceae: *Heliotropium curassavicum*;
Campanulaceae: *Lobelia laxiflora*;
Crassulaceae: *Sedum alamosanum*;
Salicaceae: *Populus monticola, P. trichocarpa*, and *Salix bonplandiana* var. *bonplandiana*; and
Scrophulariaceae: *Mimulus dentilobus, M. floribundus, M. glabratus* var. *glabratus, M. guttatus*, and *Bacopa monnieri*
Monocot flora observed at the site include:
Arecaceae: *Phoenix dactylifera* and *Washingtonia robusta*;
Cyperaceae: *Juncus acutus, J. mexicanus, Fuirena simplex, Scirpus koilolepis*, and *Schoenoplectus americanus*;
Potamogetonaceae: *Ruppia maritima, Potomegeton pectinatus*, and *Zannichellia palustris*;
Pontederiaceae: *Heteranthera limosa*;
Typhaceae: *Typha domingensis* and *T. latifolia*;
Ferns observed at the site include *Adiantum concinnum* and *Thelypteris puberula* var. *sonorensis*; and
Green alga observed at the site include Naiadaceae: *Najas marina*.

Birds observed at the site include: *Ardea herodias, Pyrocephalus rubinus flammeus, Butorides virescens, Myiarchus cinerascens, Cathartes aura, Corvus corax clarionensis, Accipiter striatus velox, Auriparus flaviceps, Accipiter cooperii, Campylorhynchus brunneicapillus affinis, Parabuteo unicinctus superior, Salpinctes obsoletus obsoletus, Buteo albonotatus, Catherpes mexicanus, Buteo jamaicensis calurus, Troglodytes aedon parkmanii, Aquila chrysaetos canadensis, Cistothorus palustris plesius, Falco sparverius, Regulus calendula, Falco peregrinus anatum, Polioptila caerulea obscura, Callipepla californica achrustera, Vireo plumbeus plumbeus, Rallus limicola limicola, Vermivora celata, Dendroica petechia, Fulica americana americana, Dendroica pensylvanica, Zenaida asiatica, Dendroica coronata auduboni, Zenaida macroura, Dendroica nigrescens, Columbina passerina pallescens, Dendroica townsendi, Geococcyx californianus, Mniotilta varia, Otus kennicottii, Seiurus noveboracensis, Bubo virginianus, Oporornis tolmiei tolmiei, Chordeiles acutipennis, Geothllypis trichas occidentalis, Hylocharis xantusii, Geothllypis beldingi goldmani, Calypte costae, Wilsonia pusilla chryseola, Melanerpes uropygialis, Myioborus pictus pictus, Sphyrapicus nuchalis, Icteria virens auricollis, Picoides scalaris, Piranga ludoviciana, Colaptes auratus, Cardinalis cardinalis seftoni, Contopus sordidulus, Cardinalis sinuatus peninsulae, Empidonax difficilis difficilis, Pheucticus melanocephalus, Sayornis nigricans semiatra, Passerina amoena, Pyrocephalus rubinus flammeus, Pipilo chlorurus, Myarchus cinerascens, Spizella passerina, Sayornis nigricans semiatra, Spizella pallida, Pyrocephalus rubinus flammeus, Spizella breweri breweri, Myarchus cinerascens, Chondestes grammacus strigatus, Sayornis nigricans semiatra, Amphispiza bilineata, Polioptila californica, Melospiza lincolnii lincolnii, Catharus guttatus slevini, Zonotrichia leucophrys gambelli, Mimus polyglottos polyglottos, Zonotrichia leucophrys oriantha, Toxostoma cinereum cinereum, Agelaius phoeniceus, Bombycilla cedrorum, Icterus cucullatus, Phainopepla nitens lepida, Carpodacus mexicanus ruberrimus, Vireo vicinior, Carduelis psaltria hesperophilus, Vireo cassini cassini,* and *Passer domesticus domesticus.*

Lizards observed at the site include: *Dipsosaurus dorsalis, Petrosaurus repens, Uta stansburiana, Urosaurus nigricaudus, Sceloporus orcutti, Sceloporus zosteromus, Phrynosoma coronatum, Callisaurus draconoides, Crotaphytus vestigium, Phyllodactylus xanti, Coleonyx variegatus, Eumeces lagunensis, Elgaria paucicarinata, Aspidoscelis tigris,* and *Aspidoscelis hyperythrus.*

Snakes observed at the site include: *Leptotyphlops humilis, Eridiphas slevini, Bogertophis rosaliae, Thamnophis hammondii, Pituophis vertebralis, Salvadora hexalepis, Sonora semiannulata, Tantilla planiceps, Hypsiglena torquata, Chilomeniscus cinctus, Crotalus ruber,* and *Crotalus mitchellii.*

Mammals observed at the site include:

Rodentia, Sciuridae: *Ammospermophilus leucurus* and *Spermophilus atricapillus*;
Rodentia, Heteromidae: *Chaetodipus spinatus* and *Chaetodipus arenarius*;
Rodentia, Cricetidae: *Peromyscus eremicus, Peromyscus eva,* and *Neotoma lepida*
Chiroptera, Phyllostomidae: *Macrotus californicus*;
Chiroptera, Vespertilionidae: *Antrozous pallidus, Lasiurus borealis, Lasiurus ega, Eptesicus fuscus, Myotis californicus, Pipistrellus hesperus,* and *Corynorhinus townsendii*;
Chiroptera, Mollosidae: *Tadarida brasiliensis*;
Lagomorpha, Leporidae: *Sylvilagus audubonii confinis*;
Carnivora, Procyonidae: *Bassariscus astutus palmarius* and *Procyon lotor grinelli*;
Carnivora: Mustelidae: *Spilogale gracilis leucoparia*;
Carnivora, Felidae: *Puma concolor improcera*; and
Carnivora, Canidae: *Canis latrans* and *Urocyon cinereoargenteus californicus.*

Low-density, extremely isolated human settlements exist at the site, all of that maintain ancestral traditions for this mountainous region. There are cave paintings and petroglyphs nearby that remind of more ancient times, and current residential use, agricultural use, and everyday use of the water resources of the site are likely quite similar to how they have been for many years. Prolonged dry periods and these anthropogenic impacts, especially the extraction of water resources for human subsistence agriculture, as well as irrigation of cultivated food in orchards and gardens, are the main impacts to the site presently, and in the future. There are no major conservation efforts in the area other than efforts to fence off bodies of water, and prohibitions from using mechanized water pumps for any agriculture other than for subsistence. Due to the relative isolation of the area, there are relatively few visitors for tourism, nor are there any major research or education/outreach activities at the site.

HUMEDAL LOS COMONDÚ, BAJA CALIFORNIA SUR: OVERVIEW

Situated in Sierra La Giganta, Humedal Los Comondú (Figures 3.18 and 3.19) is at an average altitude of 478 m. Distinguished by three large oases, the site supports threatened species such as *Geothlypis beldingi*, *Vireo bellii*, *Polioptila californica*, and endemic species like the Xantus' hummingbird (Hylocharis xantusi). The site also supports 36 species of neotropical migratory birds, providing them with feeding habitat and shelter. Primary land use at the site is grazing and agriculture, along with urbanization, which all require water extraction. The greatest ecological threats to the integrity of the site include hurricanes, introduction of exotic flora like common reed (Phragmites australis), *Phoenix dactylifera*, buffelgrass (Pennisetum ciliare), and rubber vine (Cryptostegia grandiflora), and introduced fish species like the guppy (Poecilia reticulata) and the redbelly tilapia (Tilapia zillii), which supplant native fish species.

SITE DETAILS

There are 29 recorded reptile species at the site, 18 of which are imperiled; the pond slider (Trachemys scripta) stands out in this regard, as the sole turtle species present on the peninsula, and is listed in protected status. The vegetation of the site supports nesting *Geothlypis beldingi*, which is endemic to the southwestern region of the Baja California Peninsula. The endemic hummingbird *Hylocharis xantusi* is also resident at the site. Ten endemic reptile species are recorded at the site, such as *Gambelia copeii*, *Eumeces lagunensis*, and *Elgaria velazquezia*, all highly dependent upon the mesic environments of the site. Within the site's La Purísima Oasis, 64 unique species of arachnids, one endemic fish species, and 4 endemic hydrophytic species have been recorded. Within the site's San José Oasis, 12 species of scorpions that are endemic to Baja California have been recorded. At a larger landscape scale, the oases that make up the site serve as a network of resting and feeding sites for 36 neotropical migratory bird species. The site's oases are also used as breeding areas by the endangered passerine *Geothlypis beldingi*, which uses the microhabitat of reed beds, within these oases.

The mountain range where Los Comondú Wetland is located covers the central part of Baja California and is notable for its rugged conditions and elevations ranging from 800 m to 1600 m above sea level. This central region comes close to the coast of the Gulf, where it gives rise to narrow valleys. Its gentle slopes that face the Pacific give rise to an enormous plain known as Llanos de Magdalena, or Llanos Purísma-Iray. Riparian wetlands appear sporadically throughout and along the channels that exist in this plain, with some of the area's streams in the range of 100 km long. Much of this region of the Sierra de La Giganta is made up of repeated layers of clastic volcanic stone and conglomerates of the Miocene (Comondú Formation), formed by volcanic sandstones and conglomerates, lahars, or mudflows of pyroclastic material, as well as lava flows. To the west the Comondú sandstones decrease in thickness along the gentle slope that descends to the Pacific, disappearing under the alluvium, and depositing some of their derived sediments, carried by currents, before reaching the coast. On this west side of the site, Mission San Javier receives 300 mm of rain (García 1973), with an average annual temperature of 19° C to 22° C; winter temperature is relatively moderate with few days of frost. The Magdalena region belongs to the Pacific slope and occupies an area of approximately 29,000 km², to the north within the Arroyo Mezquital Basin, and to the south it ends before Cabo San Lucas. The Arroyo de La Purísima Basin is approximately 5,000 km² and has a mushroom shape, with the mushroom head quite distant from the Pacific

FIGURE 3.18 Composite aerial image of Humedal Los Comondú. Designated February 02, 2008; Area: 460,959 ha; Coordinates: 26°05′N 111°48′W; Ramsar Site number: 1,761.

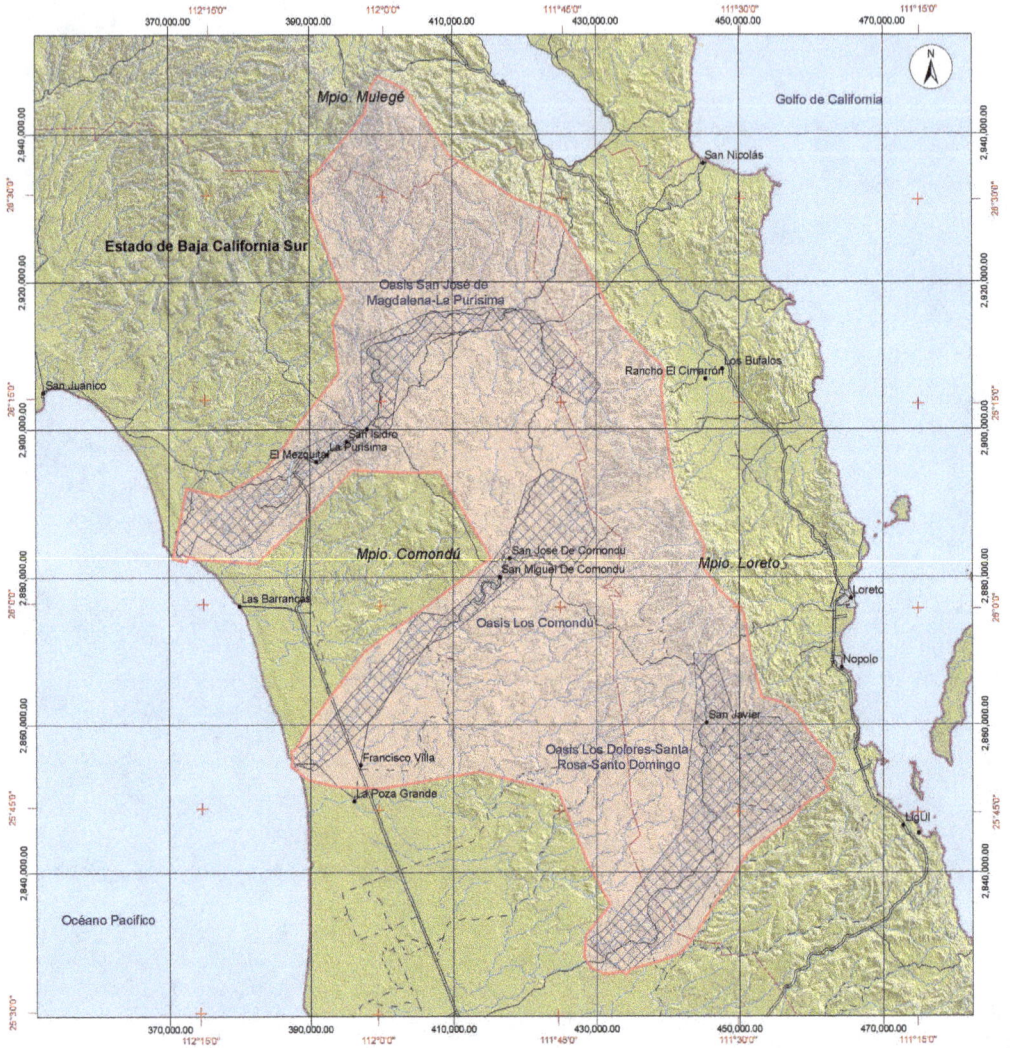

FIGURE 3.19 Map of Humedal Los Comondú. Designated February 02, 2008; Area: 460,959 ha; Coordinates: 26°05′N 111°48′W; Ramsar Site number: 1,761.

Ocean, where it collects water from a large number of small tributaries that descend from the neighboring mountains to Bahía Concepción, converging into two main tributaries, one of them the Arroyo Comondú Viejo.

The Comondú Wetland is within the Magdalena Basin, within the following sub-basins: Arroyo Comondú, Arroyo San Gregorio, and Río Codegomo, with a variety of basaltic plateaus and ravines, branching valleys, saline plains with dunes, and low hilly areas. In general, the site is important for seasonal water supply to the wetlands of the area, as well as for the recharge of the aquifer of micro-region.

Main species of flora at the site include two associations; the first corresponds to groves of *Washingtonia robusta* (Mexican fan palm) that reach a height of up to 18 m, with accompanying shrubs like *Vallesia glabra* (pearlberry), *Acacia farnesiana* (sweet acacia), *Ambrosia ambrosioides* (chicura), and *Cercidium praecox* (palo brea). The second association is made up of two shrubby grasses commonly called reeds (Phragmites communis and Arundo donax) that establish themselves on the shore of the bodies of water at the site and are accompanied by other aquatic herbs, mainly from the Cyperaceae family and other grasses, including, species of the genera *Scirpus*,

Eleocharis, Typha, Juncus, Bulbostylis, Cynodon, Sporobolus, and *Muhlenbergia*. In addition to the native associations there are cultivation areas, within which there are dates (Phoenix dactylifera), mango (Mangifera indica), and various citrus (Citrus spp.). Vegetation in the vicinity of the site also includes fleshy stemmed scrub, with other common species including the palo Adán (Fouquieria diguetii), matacora (Jatropha cuneata), elephant tree (Jatropha cinerea), mesquite (Prosopis articulata), blue palo brea (Cercidium floridum), and red elephant tree (Bursera hindsiana). Oasis flora of significance are *Olneya tesota* (Fabaceae), *Amoreuxia palmatifida* (Cochlospermaceae), and *Cylindropuntia alcahes* var. *gigantensis* (Cactaceae).

Main faunal species of the site include 141 species of vertebrates (91 birds, 13 mammals, 34 herpetofauna, and 3 fish), 2 freshwater prawns, and 159 species of arachnids. Among reptiles and amphibians, *Trachemys scripta nebulosa* stands out as the only tortoise present on the peninsula; other amphibians and reptiles that are notable at the site include *Hyla regilla, Eumeces lagunensis, Thamnophis hammondii, Elgaria paucicarinata,* and *Masticophis lateralis*. The 91 birds at the site include 54 resident species, 36 migratory species, and one transient species. Sixty-five of the bird species at the site are terrestrial and 26 are waterbirds. Of these numbers, the following terrestrial species are noted: *Vireo bellii, Polioptila californica, Geothlypis beldingi, Tachybaptus dominicus, Accipiter striatus, A. cooperii, and Falco peregrinus* (nesting on cliffs of the site). Migratory waterfowl recorded at the site include *Anas strepera, Anas americana, Anas cyanoptera, Tachybaptus dominicus, Recurvirostra americana, Tringa flavipes,* and *Actitis macularia*. Resident waterfowl at the site include *Gallinula chloropus, Egretta thula, Egretta tricolor, Butorides virescens,* and *Podilymbus podiceps*. Thirteen species of mammals are reported at the site, highlighted by bats due to their abundance, and the presence of *Spermophilus atricapillus* (Baja California rock squirrel). Fourteen species of scorpions have been recorded at San Isidro-La Purísima, and 11 species at San José Oasis. Crustaceans *Macrobrachium hobbsi* and *M. olfersii* have been recorded at La Purísima Oasis, and at San Gregorio Oasis the crabs *Callinectes arcuatus* and *Litopenaeus vannamei* are present. At La Purísima Oasis, *Fundulus lima* is a notable native fish species present, as is *Awaous tajasica* and *Dormitator latifrons* for San Gregorio Oasis.

Ancient cave paintings exist in San Miguel de Comondu, with little else recorded about the details of these ancient inhabitants. Historically, extensive livestock farming and artisanal agriculture exists among orchards in the vicinity of the site, and these activities provide the key economic support necessary for farmers, who are also often involved in the raising of livestock and forestry activities. Other activities include cutting and sale of palm and reed, and the collection of dates. Other main threats to the ecology of the site include hurricanes, structural change to wildlife habitat, the introduction of exotic species (e.g., *Phragmites australis, Phoenix dactylifera, Pennisetum ciliare, Cenchrus ciliaris,* and *Cryptostegia grandiflora*), and the impacts of cattle trampling and defecation nearby, which leads to waterbody eutrophication.

OASIS SIERRA DE LA GIGANTA, BAJA CALIFORNIA SUR: OVERVIEW

Oasis Sierra de La Giganta (Figures 3.20 and 3.21) is characterized by sheer slopes on the west side of Sierra de la Giganta, descending to coastal alluvial plains. Along the coast, small clusters of mangroves consisting of *Avicennia germinans, Laguncularia racemosa,* and *Rhizophora mangle* are present. Notable species at the site also include bighorn sheep (Ovis canadensis) that is under special protection, endemic flora including mesquite (Prosopis palmeri), and Baja California leopard lizard (Gambelia copeii). The site is within the restricted marine area of Parque Nacional Bahía de Loreto. Tourism, forestry, agriculture, and water extraction for urban consumption are amongst the main land uses of the site. The site is located in the State of Baja California Sur, Municipality de Loreto, west of the City of Loreto, which has an approximate population of 20,000 (one of the largest human settlements in the southern half of the Baja California Peninsula). The site ranges in elevation from 0 m to 735 m above sea level, and is

FIGURE 3.20 Composite aerial image of Oasis Sierra de La Giganta. Designated February 02, 2008; Area: 41,181 ha; Coordinates: 25°51′N 111°23′W; Ramsar Site number: 1,793.

FIGURE 3.21 Map of Oasis Sierra de La Giganta. Designated February 02, 2008; Area: 41,181 ha; Coordinates: 25°51′N 111°23′W; Ramsar Site number: 1,793.

comprised of five oases: La Primer Agua Oasis, Ligüi Oasis, Tabor Oasis, Juncalito Oasis, and Nutrí Oasis. Along the coastal fringes of the site there are small aggregations of mangroves, including *Avicennia germinans*, *Laguncularia racemosa*, and *Rhizophora mangle*. These basins and their wetlands along the eastern slope of the Sierra La Giganta provide water recharge to the aquifer, which is very important for human exploitation along this part of the State of Baja California Sur, covering the water needs for the municipality of Loreto. Oases at the site provide critical habitat for the subspecies of bighorn sheep, *Ovis canadensis weemsi*, which is listed under special protection in NOM-059-SEMARNAT-2010. This species is of tremendous ecological value and great economic potential in the region and in Mexico as a whole (Rodríguez and Álvarez 1996), and in the last 50 years it has seen a 40% reduction in its original habitat. Another species of mammal present at the site is the American badger (Taxidea taxus), also listed in NOM-059-SEMARNAT-2010. Other imperiled species of note at the site are the golden eagle (Aquila chrysaetos), iron stick (Olneya tesota), Mexican yellowshow (Amoreuxia palmatifida), and barrel cactus (Ferocactus townsendianus var. townsendianus).

SITE DETAILS

The area is comprised of a number of riparian areas, the most representative being the oases, which consist of bodies of fresh water, such as pools and reservoirs, which are fed by springs and subsurface flows of water. Oases support hydrophytic and riparian vegetation, surrounded by xeric scrub and deciduous forest, which support mammals, migratory birds, and include endemic species of fish, reptiles, amphibians, and arachnids. Between mountains, hills, and valleys the oases are characterized by vegetation dominated by palm groves, tule, and reed beds, surrounded by xeric vegetation of fleshy stemmed scrub (Fouquieria diguetti, Jatropha cuneata, and Caesalpinia arenosa) and cardon forest/scrub. Plant species include ironwood or blue mesquite (Prosopis palmeri), an endemic species of central Baja California Sur and distributed mainly on the formations of Sierra La Giganta, like *Cylindropuntia alcahes* var. *burrageana* and *C. bigelovii* var. *ciribe*, both endemic chollas to this region of the state. Cardon fleshy stemmed scrub at the site includes Mexican giant cactus (Pachycereus pringlei). The mezquital type of association at the site is dominated by mesquite (Prosopis articulata), and includes other species such as *Lysiloma candidum, Cercidium praecox,* and *C. microphyllum.* Interspersed among trees at the site are herbaceous plants, such as *Maytenus phyllanthoides, Jatropha cinerea, Vallesia glabra, Bourreria sonorae, Lophocereus schotii, Ruellia californica, Bebbia juncea, Matelea* spp., and genus *Krameria.* Halophytic vegetation at the site includes *Salicornia subterminalis, Allenrolfea occidentalis, Suaeda torreyana, Atriplex barclayana, Batis maritima,* and *Sesuvium portulacastrum.* In the vicinity of the site are approximately 302 endemic taxa, distributed among the slopes of eastern parts of the mountain, their connected small alluvial plains, and the associated islands. Within oases, there are 44 species of vertebrates: in the higher elevations, bighorn sheep (Ovis canadensis), puma (Felis concolor), and golden eagle (Aquila chrysaetos); in lower elevations, mule deer (Odocoileus hemionus), *Urocyon cinereoargenteus argenteu speninsularis, Lynx rufus, Procyon lotor, Antrozous pallidus minor,* and *Pipistrellus hesperus.* Rodents of the site include *Chaetodipus rudinoris, Dipodomys merriami platycephalus, Peromyscus eva, P. maniculatus coolidgei,* and *Ammospermophilus leucurus.* Herpetofauna at the site include *Gambelia copeii, Hyla regilla, Eumeces lagunensis, Thamnophis hammondii,* and *Masticophis lateralis.* Bird species of the site include *Vireo bellii, Polioptila californica* (both threatened), *Tachybaptus dominicus, Accipiter striatus,* and *A. cooperii.* The endemic hummingbird, *Hylocharis xantusii,* is also present at the site.

The site contains a number of petroglyphs, and paintings, within the caves of the area (Las Cuevas Pintas), as well as a number of ancient lithic work areas. All of the site is within a restricted zone of the Marina of the Bahía de Loreto National Park (established July 19, 1996). Further inland, tourism, forestry, and agriculture imposes pressure on water resources of the area. Urban development of the "Loreto Bay" subdivision puts additional pressures on water resources of the area, with its associated villas, commercial areas, marina, and commercial areas, not solely restricted to the municipality of Loreto, but also throughout the coastal area, from Nopoló to Ensenada Blanca.

Conservation of the site is a result of the presence of protected areas at Bahía de Loreto National Park, which has an officially approved management program that considers the ecological needs of the area. Among the activities carried out by the Park are:

1. Establishing site 'clean up' campaigns among beaches and seabeds, involving institutions, government agencies, and civic groups or organizations;
2. Developing strategies for the management and conservation of sea turtles;
3. Establishing surveillance operations in coordination with the municipal, state, and federal governments; and
4. Conducting outreach activities in coordination with the federal delegations of SEMARNAP and PROFEPA (Federal Attorney for Environmental Protection) on the rules of use and applicable legal provisions of the park.

Similarly, there is the Management Unit for the conservation of Loreto Wildlife, whose purpose is the conservation and management of the bighorn sheep population. Other efforts led by the Bahía Loreto National Park include long-term programmed activities, such as:

1. Establishing agreements with academic institutions for the protection of sites of paleonto-logical value;
2. Managing the creation of a biological station that has the equipment essential to carry out basic studies of oceanography, biology, and climatology;
3. Exploring/encouraging studies of the impacts of fishing gear, and the use of more sustain-able technologies, focused on the fishery.

PARQUE NACIONAL BAHÍA DE LORETO, BAJA CALIFORNIA SUR: OVERVIEW

The Coronados, Danzante, Montserrat, Catalana, and Del Carmen Islands form this archipelago, Parque Nacional Bahía de Loreto (Figures 3.22 and 3.23), off the eastern shore of Baja California. The diversity of marine mammals among the islands is higher than anywhere else in Mexico, with frequent sightings of blue whale (Balaenoptera musculus), fin whale (B. physalus), humpback whale (Megaptera novaeangliae), as well as orca and dolphins. *Spondylus calcifer, Pinctada mazatlanica,* and the *Purpura patula* (wide mouth rocksnail) are all protected mussel species found at the site, while the jumbo squid (Dosidicus gigas) comes to lay its eggs within the site's waters. The islands have an arid climate, rocky slopes, and desert vegetation, including seven endemic species, as well as small inlets with mangrove forests. Seaweed species (Sargassum spp.) are found to a 5 m depth, while *Amphiroa* spp. and others are resident at 20 m to 30 m depths. Among the islands, fishing is the main economic activity, and breaches in fishing laws and sustainable practices are a fundamental ecological threat to the site. Tourism for this area and site is important to visitors, not only to enjoy natural landscapes but also to provide a means to experiencing the historic ruins of missions and an ancient whale grease processing plant.

SITE DETAILS

Located in the area known as Bahía de Loreto, the Bahía de Loreto National Park is located in the Gulf of California, northwestern Mexico, off the coast of the municipality of Loreto, in the central eastern portion of the State of Baja California Sur. The total number of inhabitants in the municipality of Loreto is approximately 19,000. Although the municipality of Loreto is not within the Park, it is considered as part of its area of influence. The Loreto port also has an international airport. The islands and islets of the region occupy about 11.9% of the Park and the rest of the area is entirely marine.

The Bahía de Loreto National Park consists of a diversity of coastal environments, including rocky and sandy bottom seabeds, beaches, ravines, submarine canyons, and marine terraces. These environmental conditions foster high biological diversity, with insular environments that have high endemism of plants, insects, reptiles, and mammals. The park is inclusive of five islands, all of which contain 40% (1,385 species) of the total species of the Gulf of California. Of these species, 89 are in protection status by the NOM-059-SEMARNAT-2001 and/or -2010, and in the CITES.

Because of drainage conditions on the islands, there are no bodies of fresh water upon them, except for two springs: Agua Chica and Agua Grande, which are located in the east-central portion of Isla del Carmen. There are, similarly, no rivers on any of the islands. The relative scarcity of underground deposits of fresh water is due to low porosity and permeability to facilitate infiltration of surface water.

The island's mangroves include four species: black mangrove (Avicennia germinans), white man-grove (Laguncularia racemosa), red mangrove (Rhizophora mangle), and sweet mangrove (Maytenus phyllanthoides). Sargasso communities are represented by *Sargassum herporthizum, S. johnstonii, S. lapazeanum, S. macdougalii,* and *S. sinicola* to no more than 5 m deep. The rhodoliths at the site are represented by *Amphiroa beauvoisii, A. misakiensis, A. rigida, A. vanbosseae, A. valonioides, Corallina vancouveriensis, Jania adhaerens, Heteroderma gibbsii, Hydrolithon decipiens, H. farino-sum, Lithophyllum imitans, L. margaritae,* and *Porolithon,* to depths of 20 m to 30 m.

FIGURE 3.22 Composite aerial image of Parque Nacional Bahía de Loreto. Designated February 02, 2004; Area: 206,581 ha; Coordinates: 25°49′N 111°08′W; Ramsar Site number: 1,358.

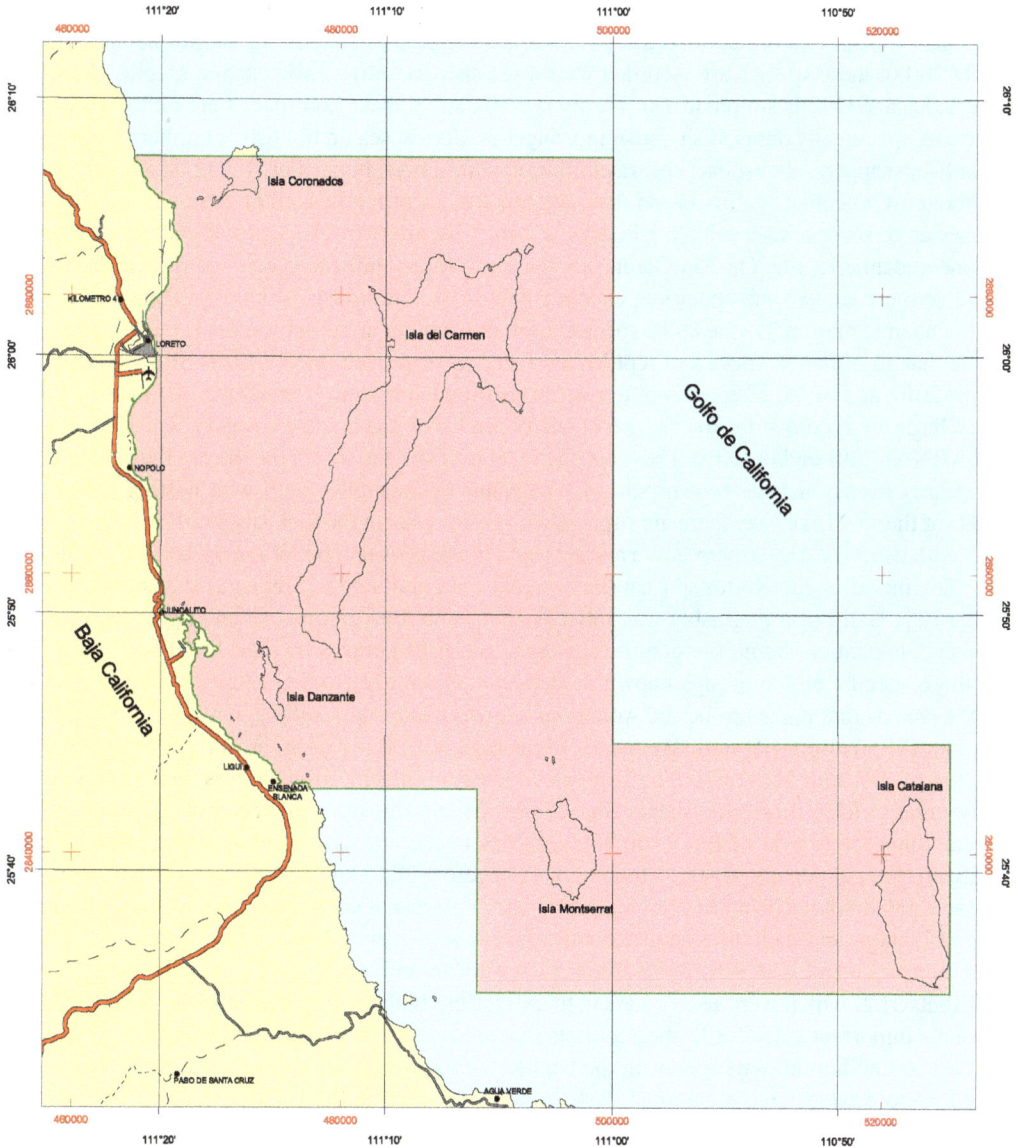

FIGURE 3.23 Map of Parque Nacional Bahía de Loreto. Designated February 02, 2004; Area: 206,581 ha; Coordinates: 25°49′N 111°08′W; Ramsar Site number: 1,358.

Bahía de Loreto National Park supports 262 species of higher plants, of which 120 reside within the coastal zone. About 100 species of phytoplankton have been studied in the marine waters of the Park, made up of diatoms, dinoflagellates, and silicoflagellates. The marine flora of the Park are represented by 161 species of macroalgae, of which 52 are endemic to the Gulf of California: 73% red algae, 16% green algae, and 11% brown algae. The terrestrial flora of the islands of the National Park (i.e., the 262 species) are of xeric scrub or fleshy stemmed scrub types, seven of which are endemic. Some cacti stand out at this site, such as the giant barrel cactus (Echinocactus platyacanthus), from Santa Catalina Island, with some almost 4 m high and one meter in diameter. On the beaches, vegetation that is adapted to dune conditions include blue palo verde (Cercidium floridum peninsulare) and iron stick (Olneya tesota). Palm trees, salt pine, and a variety of weeds have also been introduced to some of the islets of the site.

Marine fauna of the site include fan corals, spiny clams, sea urchins, starfish, tubeworms, and hermit crabs. Some marine fauna are economically important, such as chocolate clams, oysters, octopus, squid, giant sea cucumber, sea fans, black coral, crabs, sponges, starfish, and holothurians. There are more than 400 species of fish recorded in the central region of the Gulf of California and approximately 260 of these species are recorded within the site (including two endemic species: Axoclinus nigricaudus and Girella simplicidens). The most abundant of these 260 species are on the rocky reefs of the site, specifically damselfish, mulegino, ángel de Cortés, sea urchin fish or tamborillo, groupers, parrot fish, snappers, chubs, bacocos, rayadillos, cochitos, rays, butterflies, young ladies, old women, and botetes. On Santa Catalina Island there are ten species of reptiles, all of them endemic, either at the species or subspecies level. On Isla del Carmen there are two subspecies of endemic mammals, and one endemic reptile. On Isla Coronados there are three endemic species of mammals. On Isla Danzante there are two sub-species of endemic mammals, and on Isla Montserrat there are two species of endemic mammals, one at the species level and another at the subspecies level. It is important to note that 15 of the 50 species of reptiles are listed as imperiled in NOM-059-SEMARNAT-2001 and/or -2010, and of the 12 species of terrestrial mammals present, 11 are listed. A total of 235 species of birds are recorded for the site, of which 19 are listed under some category within NOM-059-SEMARNAT-2001 and/or -2010. The most abundant bird species within the site are the yellow-footed gull (Larus livens) and the brown pelican (Pelecanus occidentalis), both with nesting sites on the islands of the site. Likewise, there are ospreys with their nests on the high steep cliff areas of the site, along with herons, curlews, petrels, cormorants, and boobies. A number of introduced mammals exist at the site, including rat (Rattus sp.), cat (Felis catus), cottontail rabbit (Silvilagus sp.), and goats.

The Baja California Peninsula was inhabited by numerous groups of hunter-gatherers prior to Spanish colonization. Some missionaries categorized these peoples by their linguistic differences, into three specific ethnic groups known as Pericúes, Guaycuras, and Cochimíes. The Loreto area and the islands that make up the site were populated by Guaycuras, although it is possible to find at least four different pre-Hispanic languages in the region. In this region, as in the rest of the peninsula, the relationship between the indigenous people and the environment was based upon the use of resources to feed, dress, and shelter themselves without transforming the environment to a large degree. Upon the arrival of the Jesuit Missionaries in the late 17th Century, there were approximately 40,000 indigenous people in the region (Baegert 1989). Thus, the site has a rich history, with extensive paleontological, archaeological, and historical sites of value for scientific research, such as lithic workshops and shell piles on the islands of Del Carmen and Montserrat, including the remains of missions such as Ligüí. Areas of whale exploitation are evident from remnant areas that exist on Isla Montserrat, which were used as areas to extract the blubber from hunted whales. All of these sites have important value for hosting tourists that are interested in cultural heritage.

There are no human settlements on the islands. Until the coastal area in front of the municipality of Loreto was declared a National Park, human activities had caused the gradual deterioration of some habitat in the area, with associated decreases in certain commercial marine species that at some point in the past were much more abundant, as well as population declines and disappearances of certain endemic species on the islands. The Bay of Loreto had a burgeoning mother of pearl exploitation period (1876–1911), as well as a period when the salt from Isla del Carmen was extracted, a period when whales were heavily exploited. Presently, the main anthropogenic impacts are from motorized vehicles, which damage fragile ecosystems, such as on dunes and in wetlands; noise from boaters that disturb sea lion pups; and the overexploitation of fishing resources and illegal (unpermitted) fishing, all of which produce major impacts on the ecology of the site.

On July 19, 1996, a Protected Natural Area was declared at the site, officially declared as a National Marine Park (DOF 1996), called Bahía de Loreto. The Management Program for this area is one of the main instruments available to the Protected Natural Area, for its proper use and conservation/restoration. Among some of the activities that have been implemented at the site for environmental education purposes is working with civic organizations, fishing companies, and tourist service providers. Tourist and recreational activities at the site include trips to the islands of the Park and along the coast with kayaks, sailboats, recreational yachts, jet skis, and cruise ships. During these recreational activities, there are many opportunities for visitors to learn more about the natural history of the area. Diving, camping, and hiking continue to be popular activities as well at this site.

OASIS DE LA SIERRA EL PILAR, BAJA CALIFORNIA SUR, MEXICO: OVERVIEW

Located on the western slope of Sierra del Mechudo, Oasis de la Sierra El Pilar (Figures 3.24 and 3.25) is comprised of numerous oases that contribute to the hydrologic and biological functions of the landscape, supporting unique faunal species such as the peninsular clingfish (Gobiesox juniperoserrai) and the killifish (Fundulus lima), both considered endangered. This oasis complex represents very fragile ecosystems that are heavily influenced by extreme draught and human activities, such as agriculture and livestock ranching. Among the larger threats to the site are the presence of invasive fish species (Tilapia spp., Poecilia reticulata, Xiphophorus helleri, and X. maculatus), and invasive plants (Cryptostegia grandiflora); hydroelectric power stations in the watershed; and the lack of active management in the face of illegal hunting and extensive presence of livestock.

The oases of the Sierra El Pilar, located in the south-central portion of the State of Baja California Sur, Mexico, on the western slope of the Sierra el Mechudo, are part of the great mountain range known as La Giganta, and within the municipalities of Comondú and La Paz. Altitude of the site ranges approximately between 900 m above sea level and 100 m above sea level. The wetlands of La Sierra el Pilar are comprised of Rancho San Lucas; Mission San Luis Gonzaga; Rancho Las Cuevas of the San Luis-Las Bramonas Basin; the oases of San Pedro de la Presa; San Basilio; Rancho Merecuaco; Paso Iritú of the Santa Rita Basin; and the oases El Caracol and Arroyo Las Pocitas. Among all of the oases, riparian vegetation includes *Washingtonia robusta*, which is ubiquitous across the Baja California Peninsula. These wetlands support important ecological functions related to hydrology and biology, are key habitat for endemic fish, and serve as natural corridors and refuges for flora and fauna, such as migratory birds.

SITE DETAILS

These sub-basins on the western flank of the Sierra de La Giganta are extremely important for recharging the Santo Domingo Valley aquifer, which is a key agricultural area for the state (INEGI 1995). The region has developed hydraulic works for the storage and diversion of water, such as the El Higuajil Dam in the San Luis-Las Bramotas Basin, and also within the Santa Rita Basin (Flores 1998). The main wetlands (oases) of the Los Oasis de La Sierra El Pilar site occur in the San Luis-Las Bramonas Basin, the Santa Rita Basin, and the El Pilar-Las Pocitas Basin. The water within these basins flows along streams that are heavily dependent upon precipitation during the rainy season. The bodies of fresh water that are associated with the wetlands of the site, such as pools and reservoirs, are fed by springs and other subsurface-associated waters. Between mountains, hills, and valleys hydrogeomorphology drive the biota of the oases, which are dominated by mesic vegetation, where palm groves, tule, and reed beds are notable, surrounded by xeric vegetation of fleshy stemmed and cardon scrub.

The main species of flora within the wetlands of the site are *Bursera microphylla*, *Jatropha cinerea*, *Leucaena macrophylla macrophylla*, *Phoenix dactylifera*, *Phragmites australis*, *Prosopis articulata*, *Salix sitchensis*, *Typha domingensis*, *Urochloa mutica*, and *Washingtonia robusta*. Xeric flora at the site are dominated by *Bursera hindsiana*, *B. laxiflora*, *B. microphylla*, *Cercidium peninsulare*, *Encelia farinosa*, *Esenbeckia hartmanii*, *Euphorbia misera*, *Fouquieria columnaris*, *F. peninsularis*, *F. splendens*, *Franseria magdalenae*, *Jatropha cinerea*, *Larrea tridentata*, *Lycium brevipes*, *Machaerocereus eruca*, *M. gummosus*, *Mammillaria peninsularis*, *Opuntia comonduensis*, *O. cholla*, *O. clavellina*, *Pachycereus pringlei*, *Pereskiopsis porteri*, and *Stenocereus thurberi*.

The main faunal species of the site include *Fundulus lima* and *Gobiesox juniperoserrai* (Ruiz-Campos et al. 2002); endemic reptiles and amphibians, such as *Chilomeniscus stramineus*, *Cnemidophorus maximus*, *Ctenosaura hemilopha*, *Coluber aurigulus*, *Bogertophis rosaliae*, *Eridiphas slevini*, *Eumeces lagunensis*, *Gambelia wislizenii copeii*, *Gerrhonotus paucicarinatus*, *Petrosaurus thalassinus*, *Phyllodactylus unctus*, *Tantilla planiceps*, *Thamnophis digueti*, *T. elegans*, and *Urosaurus nigricaudus*; endemic birds such as *Geothlypis beldingi*, *Hylocharis xantusii*; and the endemic rock squirrel (Spermophilus atricapillus). Key threatened species at the site are *Fundulus lima*, *Gobiesox juniperoserrai*, *Geothlypis beldingi*, *Hylocharis xantusii*, and *Icterus cucullatus*. Other fauna of the site include many reptiles and amphibians, such as *Bipes biporus*, *Coluber flagellum*, *C. lateralis*, *Crotalus enyo*, *C. mitchelii*,

FIGURE 3.24 Composite aerial image of Oasis de la Sierra El Pilar. Designated February 02, 2008; Area: 180,803 ha; Coordinates: 24°44′N 110°55′W; Ramsar Site number: 1,794.

FIGURE 3.25 Map of Oasis de la Sierra El Pilar. Designated February 02, 2008; Area: 180,803 ha; Coordinates: 24°44′N 110°55′W; Ramsar Site number: 1,794.

C. ruber, C. viridis, Phyllodactylus xanti, Pseudacris regilla, Scaphiopus couchii, Thamnophis hammondii, Trachemys scripta, Urosaurus microscutatus, and *Uta thalassina*; resident birds such as *Auriparus flaviceps, Calypte costae, Carpodacus mexicanus, Centurus uropygialis, Phainopepla nitens, Vermivora celata,* and *Zenaida asiatica*; and of which many are migratory birds, such as *Charadrius wilsonia beldingi, Fregata magnificens rothschildi, Oceanodroma tethys tethys, Phaethon rubricauda rothschildi, Sula dactylatra californica, S. leucogaster brewsteri, Wilsonia pusilla*; and the mammals *Ammospermophilus leucurus, Bassariscus astutus, Chaetodipus baileyi, C. spinatus.* Aquatic invertebrates at the site include shrimp-like crustacean species such as *Macrobrachium americanum, M. hobbsi, M. michoacanus, M. olfersii,* and *M. tenellum* (Hernandez et al. 2007). The exotic aquatic fauna at the site include *Tilapia zillii, Poecilia reticulata, Xiphophorus helleri,* and *Xiphophorus maculatus* (Ruiz-Campos et al. 2002).

The wetlands of La Sierra El Pilar are embedded within agriculture and livestock areas that are important to the people of the area. Because these activities are water-limited, the springs, sub-surface waters, and hydraulic works such as dams provide the inhabitants, property owners, and ranchers with the water necessary for domestic activities, orchards, farm animals, and row crop cultivation. The vast predominance of land tenure at the site is from individual land owners. Many of the ranches established along the streams and oases of the area implement extensive cattle ranching. Significant immigration of people to the site stems from the Jesuit Mission of San Luis Gonzaga, which has both historical significance to the people of the area as well as cultural significance.

Adverse factors affecting the ecology of the site include the presence of exotic species, which definitively affect the ecological integrity of wetlands, such as from the invasive fish *Tilapia zillii*, *Poecillia reticulata*, *Xiphophorus helleri*, and *X. maculatus* (Ruiz-Campos et al. 2002) and invasive plants, such as *Cryptostegia grandiflora* (Arriaga et al. 1998). The construction of hydraulic works to take advantage of the water from springs and subsurface waters of the area produces negative impacts on the ecological characteristics of the site, namely by depriving wetland areas of sufficient water quantity and affecting water quality. These human designed waterworks reportedly were implemented without sufficient impact studies to determine the full potential for affecting the aquatic and riparian fauna and flora of the site. Other adverse impacts on the site include extraction of biotic resources without sufficient planning, management, or controls, and specific actions such as poaching (Arriaga et al. 1998). Another negative impact on the site is the establishment and promotion of buffelgrass for provisioning the extensive livestock in the region. This forage species poses a threat to native wetland plant species, by supplanting them.

There are research activities at the site that are being conducted by The Biological Research Center of the Northwest, located in the city of La Paz, which focuses on studies of biodiversity and conservation sciences. Only minor and infrequent visitors are noted at the site, which has relatively few recreational or tourism activities.

BALANDRA, BAJA CALIFORNIA SUR, MEXICO: OVERVIEW

Balandra (Figures 3.26 and 3.27) is a coastal wetland in the Gulf of California that supports the largest area of mangroves in La Paz Bay, and includes three of the four mangrove species found in Mexico: *Rhizophora mangle*, *Avicennia germinans*, and *Laguncularia racemosa*. The site is a major area for migratory and resident birds, such as *Pelecanus occidentalis*, *Pandion haliaetus*, *Fregata magnificens*, and *Egretta caerulea*. Notable reptile species at the site are *Dipsosaurus dorsalis* and *Callisaurus draconoides*, which are threatened. Marine mammal species richness at the site is the highest in the world, including whale species *Balaenoptera physalus*, *Balaenoptera edeni*, *Megaptera novaeangliae*, and *Eschrichtius robustus*. There are records of seven of the 11 known species of baleen whales and 20 of the 68 known species of toothed whales at the site. The wetlands of this area are typical of coastal ecosystems such as these, which provide coastline stabilization, protection against storms, and retention of sediments and nutrients. The site is a recreational center for local visitors, who use the site mainly for artisanal harvesting of bivalves and fish.

SITE DETAILS

The site is located within the Bay of La Paz, 27 km north of the city of La Paz, capital of the State of Baja California Sur, Mexico, and resides entirely at or very near sea level. The bay of the site is approximately 53 ha, with a lagoon of approximately 30 ha, and a mangrove dominated plant community of 22 ha. The proximity of the mangroves to seagrass beds or coral or rocky reefs, such as the Isla Gaviota meadow and the reefs located between Merito and Balandra, facilitate trophic transfer and habitat utilization by fish and invertebrates, providing organic material and energy to the marine species of the area, many of which spend their juvenile stages in the area, thus supporting a thriving ecosystem with lobster, shrimp, crabs, oysters, clams, mullets, catfish, and snappers. After their stay in the mangroves, juvenile fish move out to the meadows of sea grasses that grow in deeper water, eventually reaching rocky reefs and coral.

The eastern region of the Baja California Peninsula is one of the most arid regions that has a substantial presence of mangroves, in Mexico. Mangroves in arid regions such as Balandra are only found in the Red Sea, the Persian Gulf, and the Gulf of California in Mexico. The site has one of the few uncontaminated mangroves representative of the world's remaining arid regions.

FIGURE 3.26 Composite aerial image of Balandra. Designated February 02, 2008; Area: 449 ha; Coordinates: 24°19′N 110°20′W; Ramsar Site number: 1,767.

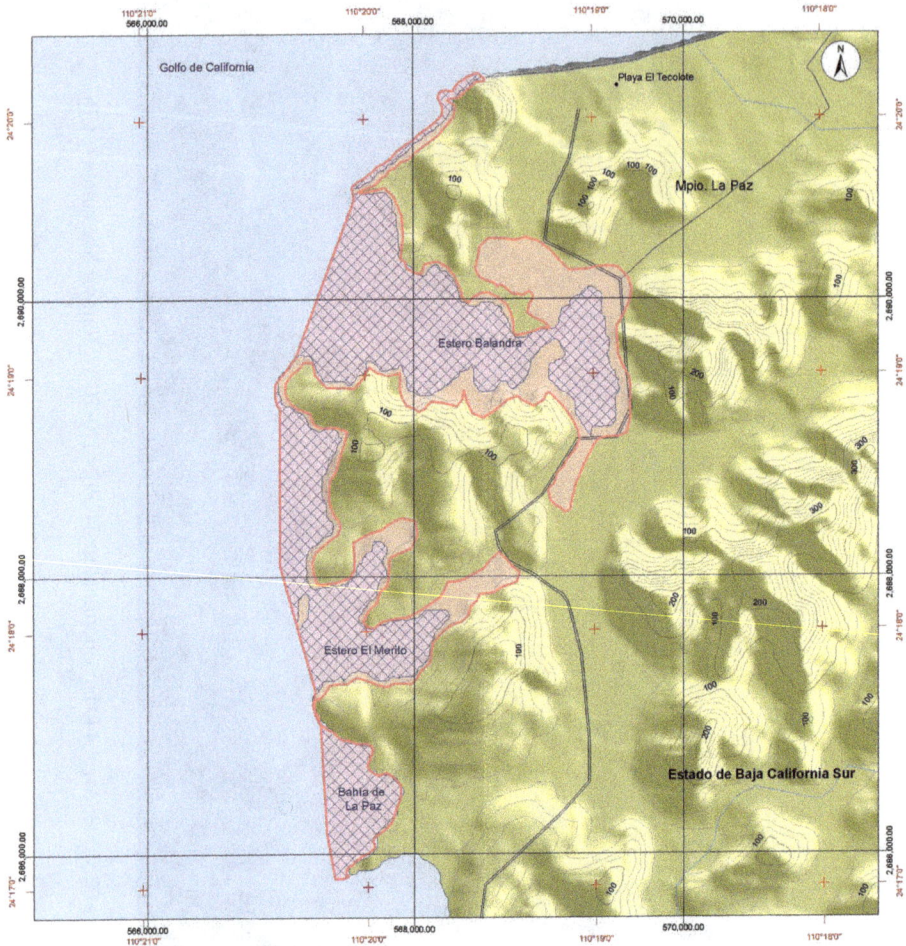

FIGURE 3.27 Map of Balandra. Designated February 02, 2008; Area: 449 ha; Coordinates: 24°19′N 110°20′W; Ramsar Site number: 1,767.

On the peninsular coast of the Gulf of California, unlike the continental coast (Sonora, Sinaloa, and Nayarit), mangrove ecosystems are typically small and discontinuous patches of vegetational growth. The site is also notable because of the 13 species of macroalgae that are found within the mangroves, of which seven are red algae, five green algae, and one a brown alga (Huerta-Múzquiz and Mendoza-González 1985), such as *Caulerpa sertulariodes*, *Spyridia filamentosa*, and *Polysiphonia simplex*. These algae contribute approximately 30% to 60% of the ecosystem's primary productivity.

Within the site a number of species of snapper, such as *Lutjanus argentiventris*, *L. coloured*, *L. novemfasciatus*, *L. aratus*, and *Hoplopagrus guentherii* use the mangroves and associated areas of seagrass for recruitment, in juvenile stages. The site functions as the last refuge for a number of aquatic species, as they enter the open ocean, and the heterogeneity of habitat types at the site offers a variety of refuge areas for organisms within sea grass, mangrove, and rocky and coral reef areas.

The Caleta-Laguna de Balandra is at the margin of the site, east of the Valley of La Paz; it consists of lithic tuff strata, of the Comondú Formation, which is in lateral contact in the eastern areas with a granitic batholith, whose maximum height is at approximately 1,250 m above sea level. This mountain area drops sharply towards the Bay of La Paz, creating 100 m valleys that split the Comondú Formation from east to west, which creates various stream beds that drain significant volumes of water into the bay during the rainy season. The western limit of the Valley of La Paz is represented mainly by the Comundú Formation that creates the last foothills of the Sierra de la Giganta (Padilla et al. 1985). The soil of the region is volcanic in origin, which has left extensive regions covered by these lava flows, fragmented material, and some destroyed volcanic cones. The hills of the area reach a maximum elevation of approximately 60 m, and are composed primarily of volcanic rock (Hausback 1984). Some of the surrounding hills are composed of unconsolidated sediments, which are predominantly covered by xerophytic vegetation; these ancient hills are what were once islands, and are now substantially elevated along with the remainder of the present peninsula (Sirkin 1985). Floodplains in the vicinity of the site are located immediately adjacent to the mangrove swamps of the site and extend inland towards the hills.

Notable fauna of the site include spiny clams (Scyllarides princeps), sea urchins (Strongylocentrotus sp.), starfish (Astropecten armatus), tube worms (Spionidae), and crabs (Maiopsis panamensis, and Stenocionops ovata). A number of species at the site are of economic importance, such as chocolate clam (Megapitaria squalida), frilled Venus (Chione undatella), rugose pen shell (Pinna rugosa), giant lion's paw (Nodipecten suibnodosus), and oysters (Crassostrea corteziensis, Spondylus princeps unicolor, and Crassostrea gigas); cephalopods (Abraliopsis affinis and Onychoteuthis banksii); and squid (Dosidicus gigas and Loligo opalescens). Compared to other lagoons within the Bay of La Paz, Balandra has a greater number of fish species, which could be explained by the oceanic influences upon the area, and the relatively greater heterogeneity of substrates. There are also a large variety of seabirds at the site, including pelicans and gulls (Fregata magnificens, Larus livens, L. californicus, and L. heermanni). Resident bird species include white heron (Egretta thula), yellow-crowned night heron (Nyctanassa violacea), brown pelican (Pelecanus occidentalis), osprey (Pandion haliaetus), the magnificent frigatebird (Fregata magnificens), little blue heron (Egretta caerulea), and gulls (Larus livens, L. californicus, L. heermanni). Observations at the site include 4 species of amphibians including Couch's spadefoot toad (Scaphiopus couchii) and Pacific tree frog (Hyla regilla), and 37 species of reptiles including the desert iguana (Dipsosaurus dorsalis), while the most abundant turns out to be the western zebra-tailed lizard (Callisaurus draconoides), listed as threatened under NOM-059-SEMARNAT-2010. Other species of reptiles found in the area are *Coleonyx switaki, C. variegatus, Sceloporus magister, Leptotyphlops humilis, Lichanura trivirgata, Chilomeniscus cinctus, Elaphe rosaliae, Eridiphas slevini, Hypsiglena torquata, Masticophis flagellum, M. lateral, Pituophis melanoleucus, Phyllorhynchus decurtatus, Salvadora hexalepis, Sonora semiannulata,* and *Trimorphodon biscutatus.*

Examples of the diverse and rich marine mammal populations of the area include the fin whale (Balaenoptera physalus), Eden's whale (Balaenoptera edeni), humpback whale (Megaptera novaeangliae), gray whale (Eschrichtius robustus), blue whale (Balaenoptera musculus), sei whale (Balaenoptera borealis), minke whale (Balaenoptera acutorostrata), sperm whale (Physeter macrocephalus), common bottlenose dolphin (Turciops truncatus), common dolphin (Delphinus delphis and Delphinus capensis), and orca (Orcinus orca). Other mammals of note at the site are the California sea lion (Zalophus californianus).

Due to the site's proximity to the city of La Paz, Balandra has always been an important place for the community, which has given it economic, scenic, scientific, educational, and recreational value for the people of the area. The Baja California Peninsula was inhabited before the Spanish colonization by numerous groups of hunter-gatherers who, to facilitate their evangelization during the Colony, were divided by missionaries according to their linguistic differences in three large ethnic groups known as Pericúes, Guaycuras, and Cochimíes. The area of La Paz was inhabited by Pericúes and in this region, as in the rest of the peninsula, the relationship that humans had with nature was tightly bound with the basic needs of people, i.e., food, clothing, and shelter. Nevertheless, there was no overexploitation and the resources were sustainable for the approximate 40,000 indigenous people of the region, until the 17th Century when the Jesuit Missionaries arrived. Since the late 19th Century, fishers and their families arrived at Balandra aboard sailboats made of poplar or metal, staying for several weeks to fish for a number of popular species, including chub, horse mackerel, mulatto snapper, grouper, and mullet. During the winter, dives were carried out for several hours in different parts of the area to capture mother of pearl. In the Bay of La Paz, shark was caught using fish bait, with their fins being consumed and any stored meat salted to prevent spoilage.

In the middle of the 20th century, roads and highways created additional routes for visitors, towards Pichilingue, places that before were only accessible by sea, increasing the influx of visitors.

Over time, the importance of fishing in the area decreased and the area became a favorite beach for people of the La Paz community. Presently, fishing in Balandra is not a formal practice, and it is only done for self-consumption; clam, crabs, and octopus are the main species harvested. Due to the arid environment, La Paz has few parks or green areas for recreation, increasing the relevance of the site's beaches for the recreation and leisure of its nearby populations, together with Isla Espíritu Santo, which is a favorite commercial fishing spot for locals. Local, national, and international visitors make their way to a number of beaches in the area, including those at La Paz, El Coromuel, La Concha, Costa Baja, El Caimancito, El Tesoro, Pichilingue, Balandra, and El Tecolote. Of all these beaches, only Balandra offers tourists the experience of enjoying a pristine beach, with a landscape minimally modified by real estate development or other major infrastructure.

The Balandra area is located within the Marine Priority Area No. 10, "Baja Island Complex Southern California" of the priority regions for conservation determined by the National Commission for the Knowledge and Use of Biodiversity. Due to its importance for the citizens of La Paz, Balandra has served as an educational area for schools, governmental personnel, and a variety of other organizations as a place of teaching about the ecology, biology, and conservation of the area. Since 2007, the group "Colectivo Balandra," comprised of citizens, organizations, and researchers has collaboratively worked to maintain the social values of the site, and the site itself.

HUMEDALES MOGOTE-ENSENADA LA PAZ, BAJA CALIFORNIA SUR, MEXICO: OVERVIEW

Humedales Mogote-Ensenada de La Paz (Figures 3.28 and 3.29) is a coastal lagoon separated from La Paz Bay by a sandy barrier (El Mogote). The mangroves of Ensenada de La Paz create small lagoons that are important nesting habitat for many bird species, including *Ardea herodias, Bubulcus ibis, Egretta rufescens, E. thula, E. tricolor, E. caerulea, Nyctanassa violacea, Nycticorax nictycorax, Eudocimus albus, Butorides striatus, Rallus limicola* (endemic), *Charadrius wilsonia*, and *Sternula antillarum*, with the majority of these species under special legal protection. 37% of the bird species at the site are migratory, with more than 20,000 migratory shorebirds utilizing the site for days to weeks for foraging and resting during the winter season. The most common mammal at the site is the California sea lion (Zalophus californianus), along with other noticeable mammal species, such as raccoon (Procyon lotor), coyote (Canis latrans), and gray fox (Urocyon cinereoargenteus). There are 390 recorded fish species at the site, including 14 species of sharks. The main human activities at the site are agriculture, livestock, a variety of industrial uses, and tourism. The site is designated as a Site of Regional Importance in the Western Hemisphere Shorebird Reserve Network (WHSRN, 2006).

The mangroves of the La Paz cove are important nesting habitat for wading birds, such as the great blue heron (Ardea herodias), cattle egret (Bubulcus ibis), reddish egret (Egretta rufescens), little blue heron (E. caerulea), yellow-crowned night heron (Nyctanassa violacea), black-crowned night heron (Nycticorax nictycorax), American white ibis (Eudocimus albus), striated heron (Butorides striatus), Virginia rail (Rallus limicola) [endemic], Wilson's plover (Charadrius wilsonia), and least tern (Sternula antillarum) [Becerril and Carmona 1997].

Noted species of special protection that utilize mangrove habitat at the site are the great blue heron (Ardea herodias), reddish heron (Egretta rufescens), Heermann's gull (Larus heermanni), yellow-footed gull (Larus livens), and elegant tern (Sterna elegans). Migratory birds in the winter season include the western sandpiper (Calidris mauri), semipaleated tildillo (Charadrius semipalmatus), picocurvo (*Numenius phaeopus*), and marbled godwit (Limosa fedoa).

SITE DETAILS

Vegetation of coastal dunes at the site is characterized by the presence of *Abronia maritima, Croton californicus, Amaranthus watsonii, Sporobolus virginicus, Maytenus phyllanthoides, Salicornia bigelovii, Monantochloe littoralis, Batis maritima, Suaeda californica, Atriplex barclayana, Lycium brevipes, Monantochloe littoralis, Batis maritima*, and *Sesuvium verrucosum*. The mangroves of

FIGURE 3.28 Composite aerial image of Humedales Mogote-Ensenada de La Paz. Designated February 02, 2008; Area: 9,184 ha; Coordinates: 24°09′N 110°21′W; Ramsar Site number: 1,816.

FIGURE 3.29 Map of Humedales Mogote-Ensenada de La Paz. Designated February 02, 2008; Area: 9,184 ha; Coordinates: 24°09′N 110°21′W; Ramsar Site number: 1,816.

the site are characterized by the presence of the red mangrove (Rhizophora mangle), the white mangrove (Laguncularia racemosa), and the most abundant, black mangrove (Avicennia germinans). On the dunes themselves are other typical scrubland species, dominated by white elephant tree (Jatropha cinerea), red elephant tree (J. cuneata), elephant tree (Bursera microphylla), palo Adán (Fouquieria diguetii), and desert plum (Cyrtocarpa edulis). Vegetation in the immediate vicinity of the coastline includes Mexican giant cactus (Pachycereus pringlei), mesquite (Prosopis articulata), desert plum (Cyrtocarpa edulis), blue palo brea (Cercidium floridum), cholla (Opuntia cholla), red elephant tree (Jatropha cuneata), *Machaerocereus gummosus*, chamizo (Ruellia peninsularis), ironwood (Olneya tesota), and red elephant tree (Bursera hindsiana). Macroalgae of the site include 128 species: 67 species from the Rhodophyta, 30 from the Phaeophyta, and 31 from the Chlorophyta.

Marine mammals of the site include California sea lion (Zalophus californianus), Guadalupe fur seal (Arctocephalus townsendi), common seal (Phoca vitulina), and northern sea elephant (Mirounga angustirostris) [Urbán-Ramírez et al. 1997]. The most common terrestrial mammals at the site are the raccoon (Procyon lotor), coyote (Canis latrans), and gray fox (Urocyon cinereoargenteus). Bird species of the site include 322 species of 18 orders and 52 families. *Sternula antillarum* is a particularly important migratory bird at the site, which utilizes the mangroves of the site for nesting, and is cataloged in NOM-059-SEMARNAT-2010 as a species of special protection. Other equally important species inhabiting the mangroves of the site are *Ardea herodias*, *Egretta rufescens*, *Larus heermanni*, *Larus livens*, *Sterna elegans*, *Zenaida asiatica*, *Z. macroura*, *Columbina passerina*, *Auriparus flaviceps*, and *Dendroica petechia*. Reptiles and amphibians of the site's mangroves include *Bipes biporus*, *Phyllodactylus unctus*, *Dipsosaurus dorsalis*, *Callisaurus draconoides*, *Sceloporus orcutti*, *Sceloporus zosteromus*, *Uta stansburiana*, *Urosaurus nigricaudus*, *Cnemidophorus tigres*, *Ctenosaura hemilopha*, *Leptotyphlops humilis*, *Masticophis flagellum*, *Phyllorhynchus decurtatus*, *Pituophis vertebralis*, *Salvadora hexalepis*, *Trimorphodon biscutatus*, *Eridiphas slevini*, *Bogertophis rosaliae*, *Lampropeltis getula*, *Chilomeniscus stramineus*, *Hypsiglena torquata*, *Crotalus enyo*, *Crotalus mitchellii*, and *Crotalus ruber*. Fish in the Bay of La Paz include 390 species of 251 genera and 106 families (Abitia-Cardenas et al. 1994). Up to 14 species of sharks have been recorded in the Bay (Arellano et al. 1991, Armenta et al. 2004). A total of 120 species of bivalves, 173 species of gastropods, eight species of cephalopods, three species of polyplacophorans, and two species of scaphopods have also been noted at the site. There are two groups of macrocrustaceans at the site, including stomatopods and decapods, with a total of 13 species.

A variety of archeological artifacts have been discovered at the site, including arrowhead projectiles, spears, and stone flakes with cutting edges. Within the vicinity of the mangroves at the site, there are shell fragments that are thought to be associated with a number of residents of the past. There are a number of signs that nomadic groups of Guaycuras were present at the site in the past. The wetlands and mangroves of the La Paz cove region have been of interest to the indigenous people of the area, specifically the Guaycuras and Pericúes. With the arrival of the Spanish, the cove was customarily used to shelter ships from inclement weather; a historical site of the mission attributed to Hernán Cortés, who on May 3, 1567 arrived at the bay named Villa de la Santa Cruz (currently Las Cruces), near the Bay of La Paz, was also founded. Although this colony did not prosper, in 1596 Sebastián Vizcaíno arrived at the site to establish what is the present Mission of Nuestra Señora de La Paz, in the cove of La Paz.

Presently, community members of the city of La Paz come to the mangroves of the site to gather amongst themselves and also extract mangrove oysters (Ostrea palmula). The mangrove waters also serve as a source of live bait, which the fishers of the area utilize. In the past, some thought it was best to drain the wetlands to make them more usable and accessible to humans, such as the area of Pichilingue, which was drained and filled in the 1970s. Currently, mangroves are the subject of scientific research, including studies by local institutions such as the Biological Research Center of the Northwest, the Autonomous University of Baja Southern California, and the Interdisciplinary Center for Marine Sciences/IPN, ongoing since the late 1990s.

At the bottom of the cove and in the vicinity of the wetlands of the site, between El Centenario and Chametla, agricultural activities occur. Residential, infrastructural, industrial, and recreational uses also exist in the coastal zone of the site. Recreational use and fishing is also common in the water bodies of the site. In the surrounding areas, mismanagement of solid waste, and illegal extraction of species (including endemic flora and fauna due to unregulated tourism) have been noted (Arriaga et al. 1998). A number of volunteer groups in the area work to educate and provide outreach, specifically aimed at sustainable use of the site, beach cleanup, and other resource monitoring activities. Most of the tourism activity at the site is focused on the aquatic environment, and the coastal zone, primarily related to spa and beach visits, and involving the anchorage of private vessels that come to the area for sailing and other water sports.

PARQUE NACIONAL CABO PULMO, BAJA CALIFORNIA SUR: OVERVIEW

Parque Nacional Cabo Pulmo (Figures 3.30 and 3.31) is one of the only coral reefs found in the eastern Pacific, and the only coral reef in the Gulf of California. Among the noteworthy fauna at the site are five endangered marine turtle species (Caretta caretta, Chelonia mydas, Dermochelys coriacea, Eretmochelys imbricata, and Lepidochelys olivacea), and six cetacean species (Balaenoptera

FIGURE 3.30 Composite aerial image of Parque Nacional Cabo Pulmo. Designated February 02, 2008; Area: 7,100 ha; Coordinates: 23°27'N 109°25'W; Ramsar Site number: 1,778.

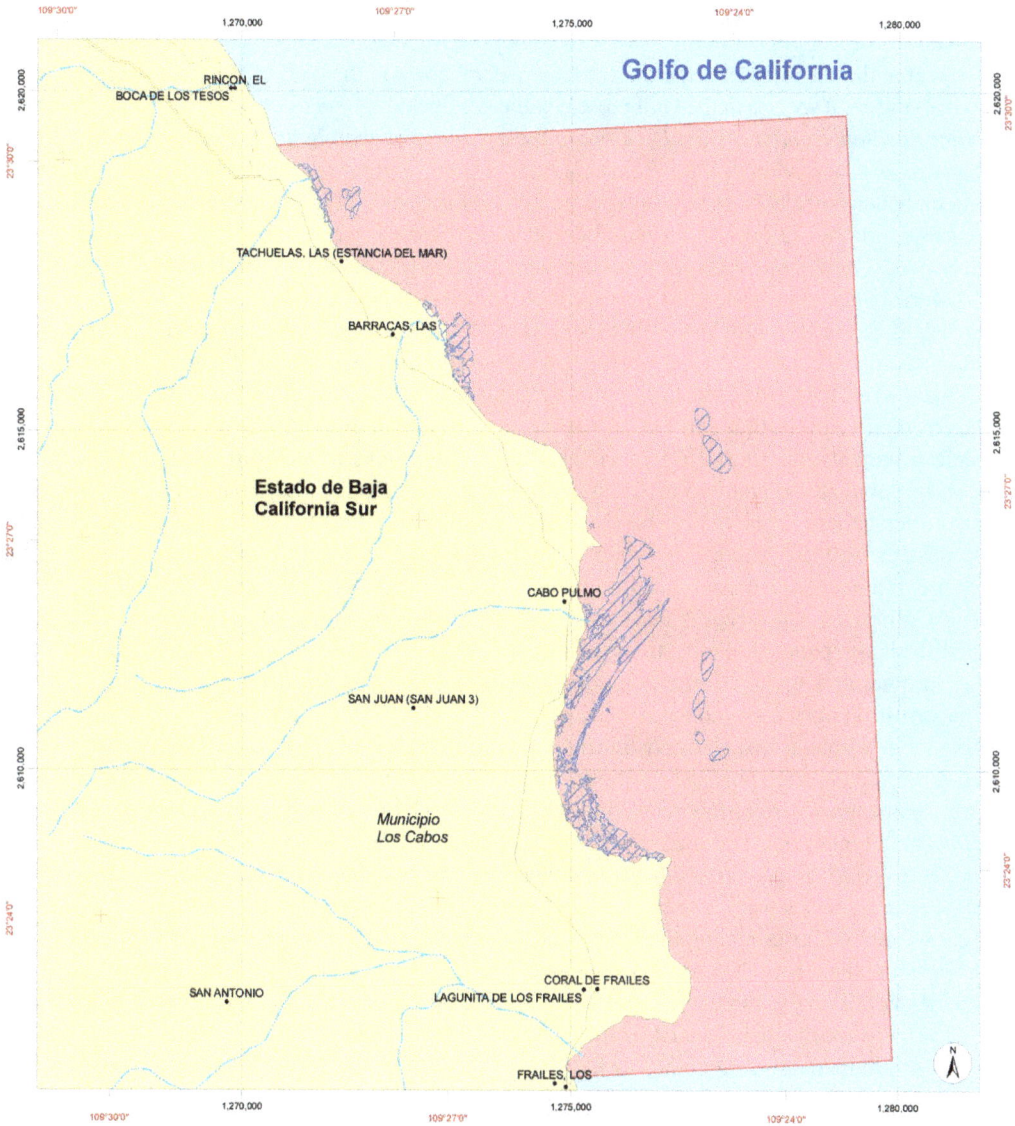

FIGURE 3.31 Map of Parque Nacional Cabo Pulmo. Designated February 02, 2008; Area: 7,100 ha; Coordinates: 23°27′N 109°25′W; Ramsar Site number: 1,778.

edeni, Balaenoptera physalus, Megaptera novaeangliae, Stenella longirostris, Steno bredanensis, and Tursiops truncatus), all under special protection. The site is home to 11 of the 14 species of hermatypic corals. There are 226 reef fish species observed at the site, and the site also supports numerous bird species. Nearly all of the site is marine, and the only land portion comprises the beaches within the Terrestrial Federal Marine Zone. Adverse factors at the site include sport fishing, nautical traffic, and pollution that damages corals and other species.

SITE DETAILS

The site is in a coastal area of the Baja California Peninsula, where semi-desert landscapes abound between the waters of the Pacific Ocean and the Sea of Cortez, and yet provides a region of elevated primary productivity that supports the biological diversity of marine species in the area. The coral reef present in the Bay of Cabo Pulmo constitutes one of the few reef areas in the Eastern Pacific, and the only reef in the Gulf of California or Sea of Cortez. The site is within the confluence of the

Panamanian, Californian, and Indo-Pacific biogeographical regions, and thus the biological diversity found in this region is the highest found along the Mexican Pacific coastline (Kerstitch 1989).

The reef at the site represents the most extensive coral cover in the Gulf of California, with 11 of the 14 hermatypic coral species in the Gulf present at the site, including *Pocillopora verrucosa*, *Pocillopora capitata*, *Pocillopora damicornis*, *Pocillopora meandrina*, *Pavona gigantea*, *Pavona clivosa*, *Porites panamensis*, *Psammocora stellata*, *Psammocora brighami*, *Fungia curvata*, and *Madracis pharensis*. Within the ichthyological community there are 226 reef species (Villarreal 1988), of the total 875 reef species listed for the Gulf of California (Finley et al. al. 1996). The molluscs of the reef include *Conus brunneus* and *Conus princeps*, commonly known as cones; scorpion snail (Murex elenensis); Chinese snail (Muricanthus princeps); kiosque rock shell (Thais kiosquiformis), and pearl oyster (Pinctada mazatlanica). Sea turtles, such as the olive ridley (Lepidochelys olivacea) and leatherback (Dermochelys coriacea) use the site for nesting, as well as hawksbills (Eretmochelys imbricata) and green sea turtles (Chelonia mydas), for feeding, breeding, or as a resting area during migration.

The bays of Cabo Pulmo and Los Frailes form alluvial valleys composed of granite clasts and volcanic fragments; an area of dunes arises within those areas to an approximate height of 5 m with an approximate width of 15 m. Cabo Frailes, which separates the two bays, forms a wide ridge between them; these sandbars extend out from the coast and are exposed at low tide in the near-coast areas as well. In general, the water in the bays is clear; the site is within two basins, within which are Trinidad Stream and San José Stream, which ultimately discharge their waters into the Gulf of California. The streams that drain both basins are intermittent and form well-integrated braided drainage patterns. Four streams flow into Cabo Pulmo, and five streams flow into Los Frailes, providing flow only during the rainy season (occasionally, in July-September).

The terrestrial plant community at the site is made up of fleshy stemmed scrub, with the following species: white elephant tree (Jatropha cinerea), ocotillo (Fouquieria sp.), and *Machaerocereus gummosus*. The coastal zone's physical environment is heterogenous in this area, comprised of stony beach and a dune of approximately 160 m by 25 m. There is a beach present at the site, with patches of strawberry bush and mesquite. The main floral species present at the site are *Sargassum*, *Gracilaria spinigera*, *Halymenia californica*, *Halymenia templetonii*, *Hypnea johnstonii*, and *Hypnea cervicornis*.

Main faunal species present at the site are those characteristic of the Gulf of California, with the fish, corals, and molluscs among the better studied, relative to the rest of the extant fauna at the site. With respect to the rest of the faunal communities, the data are scarce, although it has been noted that the species richness of some communities of cryptofauna, especially of polychaetes, is notably high. Apart from the species that depend for their survival on the reef, and the ecological processes that occur within it, there are some other faunal species that are temporary visitors that make use of the site for feeding, reproduction, and resting during migration. The species that frequent the site for feeding include genera *Dasyatis* (manta rays), *Caranx* (jack fish), *Kyphosus* (chubs), and *Mugil* (mullet). Migratory marine species also use the site, with species present under some category of protection by Mexican Law, such as turtles that periodically visit some site beaches, either for spawning or feeding, as follows: *Dermochelys coriacea* (leatherback sea turtle), *Caretta caretta* (loggerhead sea turtle), *Lepidochelys olivacea* (olive ridley sea turtle), *Chelonia mydas* (green sea turtle), and *Eretmochelys imbricata* (hawksbill sea turtle). Marine birds present at the site are common throughout the gulf, including the yellow-footed gull (Larus livens), the least tern (Sternula antillarum), the elegant tern (Sterna elegans), the brown pelican (Pelecanus occidentalis), herons (Ardea herodias and Casmerodius albus), as well as godwits, sandpipers, and curlews (Limosa fedoa, Numenius phaeopus, and Numenius americanus). You will also see a small colony of California sea lions (Zalophus californianus) at the site. Near the coast you can also observe other marine mammals that transit within and near the site, such as common bottlenose dolphin (Tursiops truncatus), the spinner dolphin (Stenella longirostris), and rough-toothed dolphin (Steno bredanensis). During winter you may observe the humpback whale (Megaptera novaeangliae), fin whale (Balaenoptera physalus), and Eden's whale (Balaenoptera edeni) at the site.

There are many paleontological remains of reef fauna at the site, and in Bahía Cabo Pulmo there is an archaeological site recorded by Massey (1955), which was used by bands of hunter-gatherer nomads belonging to the Pericúes group. Current land use includes recreational beach use, and much of the area is used for artisanal fishing (long line only, with daily limits for local family consumption only). Cattle ranching is traditional in the area and is practiced extensively. The population of those in the vicinity of Cabo Pulmo have shifted their economic focus from activities related to fishing, to tourism and the associated restaurant services, boat rental, and dive guiding that comes with that tourism industry.

A number of negative impacts to the reefs at the site include pollution related to boating, and the impact of tourism; solid waste on beaches and in the sea; and wastewater that is increasingly associated with human populations that exist and those that are increasing in the area from new development. The site was declared a National Marine Park in 1995. The Cabo Pulmo National Park Conservation and Management Plan was published in December 2006, which is a legal instrument that determines the activities that are allowed in the area, and those that are restricted within the interior of the protected natural area. Periodic scientific monitoring occurs at the site to measure the health of the reef, and to assess the abundance and diversity of sea turtles, by the Tortuguero Group of the Californias.

With the collaboration of some civil associations, such as Friends for the Conservation of Cabo Pulmo (ACCP), information on the natural resources of the site, both in Cabo Pulmo and the surrounding communities and cities, is shared broadly to increase knowledge of the site and its ecosystem functions. Norcross Corporation, ACCP, and the Autonomous University of Baja California Sur work to bring compelling examples of resource conservation to the classroom, as well as shared learning about the importance of protecting such areas for future generations continues between both the people of local communities, and visitors.

SISTEMA RIPARIO DE LA CUENCA AND ESTERO DE SAN JOSÉ DEL CABO, BAJA CALIFORNIA SUR: OVERVIEW

One of the main characteristics of Sistema Ripario de la Cuenca and Estero de San José del Cabo (Figures 3.32 and 3.33) is the presence of the San Jose Oasis and San Jose Estuary, which epitomize oceanic-to-desert interfaces, as important ecological transitional areas found in the Baja California Peninsula. The characteristic vegetation of this estuarine ecosystem is formed by typical oasis species, such as palms, and aquatic species. Floral species found in the riparian ecosystems of this site are *Washingtonia robusta* and *Erythea brandegeei*, both endemics from Baja California; *Populus brandegeei* var. *glabra*, an endemic of Sierra La Laguna; as well as *Prunus serotina, Ilex brandegeana, Heteromeles arbutifolia,* and *Salíx lasiolepis.* The site also plays an important role for migratory species, as it is the last resting stop for aquatic bird species migrating to areas in the south of Mexico, Central America, and South America. A total of 217 species of waterfowl have been observed and noted at the site, of which 97 are migratory, and 19 are considered at risk, such as the *Sternula antillarum browni.* Because of the role it plays for birds, this estuary has been recognized as an Important Bird Area (National Audubon Society). The water table at this oasis in the desert is close to the surface, and its soils are predominantly saline, often with a well-developed salt crust.

SITE DETAILS

The San José del Cabo basin and estuary's riparian ecosystems are located in southern Baja California Sur, Mexico. Biogeographically it belongs to the Cape Region and politically to the Municipality of Los Cabos. The closest and largest city within the basin is San Jose del Cabo, which is also located in the San José Basin, delimited by the watersheds of the La Laguna and La Trinidad mountain ranges, which with their intermittent surface runoff feed the main hydrology that forms San José Stream. The estuary supports a diversity of aquatic plant associations, including submersed vegetation, riparian vegetation, and cultivation-associated vegetation. The border between the estuary and the seawater of the Gulf of California consists of a thin sandbar that allows minimal marine intrusion. Thus, the characteristic vegetation of the estuary is formed by typical species of oases, such as palm groves, reeds, and other associated aquatic species of such areas.

A total of 217 species have been recorded at the site, 97 of which are migratory, and 19 of which are in some risk category, such as the California least tern (Sternula antillarum browni). Due to the

FIGURE 3.32 Composite aerial image of Sistema Ripario de la Cuenca and Estero de San José del Cabo. Designated February 02, 2008; Area: 124,219 ha; Coordinates: 23°03'N 109°41'W; Ramsar Site number: 1,827.

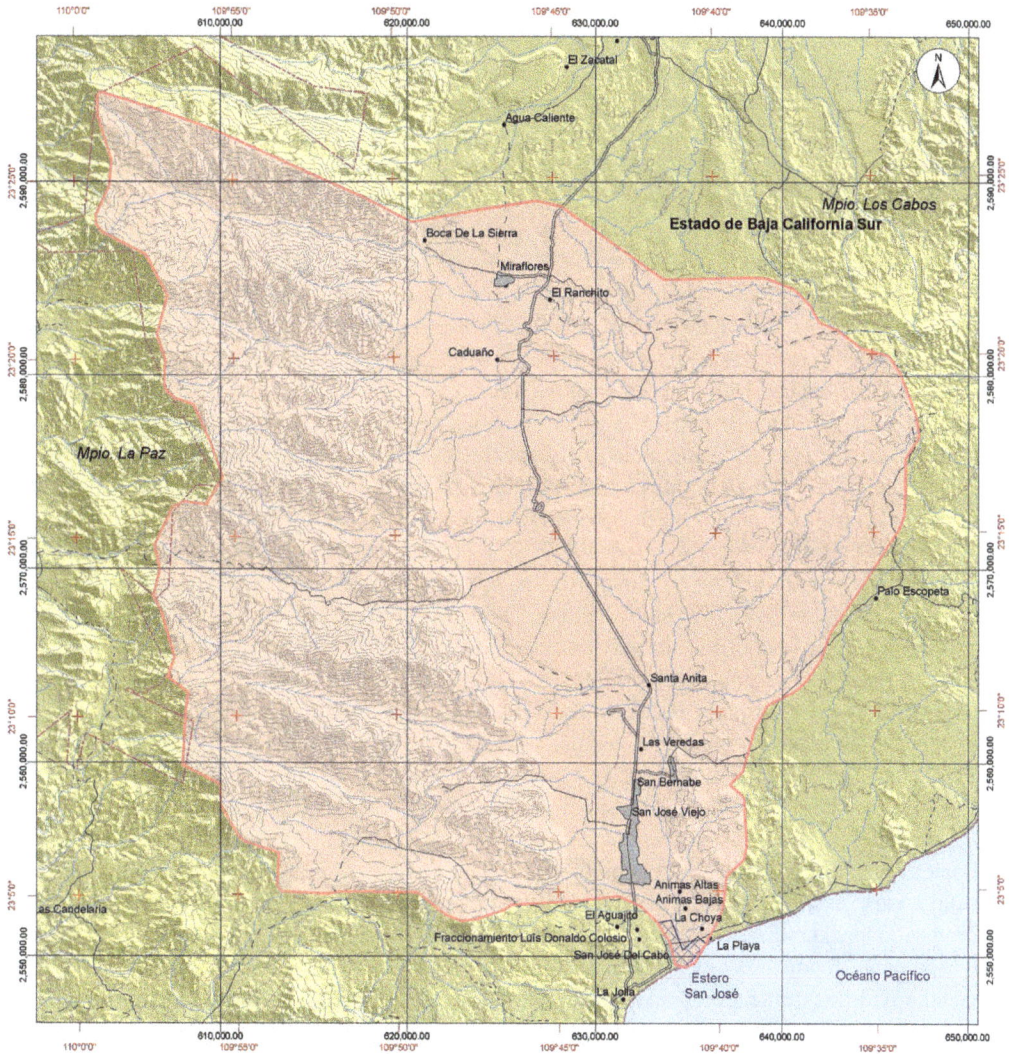

FIGURE 3.33 Map of Sistema Ripario de la Cuenca and Estero de San José del Cabo. Designated February 02, 2008; Area: 124,219 ha; Coordinates: 23°03′N 109°41′W; Ramsar Site number: 1,827.

importance of the avifauna of the estuary, the site has been recognized as an Area of Importance for the Conservation of Birds. Within the basin, there are artificial wetlands created by the construction of dams, such as Boca de la Sierra and Caduaño, which are relatively small. However, these water control structures have become key to providing both the local communities with water, as well as for maintaining important water supplies to maintain the ecology of the site, specifically for migratory birds.

The vegetation of the basin is represented mainly by deciduous forest, fleshy stemmed scrub, and oak forest, and to a lesser extent palm groves and gallery or riparian forest. On the periphery of the estuary is tule, reedbeds, and inland palm groves dominated by *Washingtonia robusta* (Mexican fan palm), shrubs, and herbaceous strata. Smaller areas of forest at the site consist of primarily guamúchil (Madras thorn) and mezquital (mesquite), the first dominated by abundant *Pithecellobium dulce* trees, and the second dominated by *Prosopis articulata*. The variety of environmental conditions at the site favors a great diversity of vertebrate fauna, from native amphibian species such as the *Bufo punctatus* and the *Pseudacris* frogs, which are distributed among the different types of vegetation of the basin, to Couch's spadefoot toad (Scaphiopus couchii), which is a common inhabitant of desert areas, and is found preferentially among fleshy stemmed scrubs of the site (Álvarez et al. 1988). Common Mexican tree frog (Smilisca baudinii) and bullfrog (Rana catesbeiana) are introduced species, the latter being a relatively scarce species. For reptiles at the site, oases are the sites of greatest species richness.

Of the 218 species in the San José Basin, 96% are found in oasis areas (Boca de la Sierra and Estero San José), since these bodies of water are most attractive to resident and migratory birds (Rodríguez-Estrella et al. 1997, Rubio et al. 1997). The estuary or oasis of San José del Cabo maintains a community of terrestrial and aquatic birds that are almost exclusively within the southern tip of the Peninsula.

In addition to the ecological characteristics of the plant communities, San José Basin's numerous floral species provide human utility; a total of 197 plant species of human utility are recorded, which represent about 18% of the flora reported for this region. The families with the largest number of 'useful' species are Compositae, Leguminosae, and Cactaceae; predominant human uses include food and medicine. In this region, there are 153 species endemic to the Peninsula, which represents about 14% of the total flora recorded, with the fleshy stemmed scrub community having the largest number of endemics. It is also important to note that within the oases of the basin, the presence of rubber vine (Cryptostegia grandiflora), an extremely aggressive invasive species that, without proper control, can affect water quality.

Main faunal species at the site are the 217 species of birds, 15 of which are protected and 4 of which are endemic, such as Xantus's hummingbird (Hylocharis xantusi), Belding's yellowthroat (Geothlypis beldingi), a passerine bird dependent on reeds and water bodies, which is in danger of extinction (Rodríguez-Estrella et al. 1997). The seven endemic mammal species of San José del Cabo basin mainly belong to genus *Rodentia*: three subspecies to the family Muridae, one Geomyidae, and the other of the Heteromyidae. There is also a species of the family Cervidae (mule deer) and a species of bat of the Vespertilionidae family at the site. The bat *Myotis peninsularis* is endemic to the Cape Region, while mule deer (Odocoileus hemionus peninsulae) are endemic to the Baja California Peninsula. *Chaetodipus dalquesti* (dalquesti mouse) is a species endemic to the lower parts of the area's mountain range, and is under special protection.

A total of 21 species of reptiles endemic to the Baja California Peninsula have been recorded at the site (Grismer 2002). Species under special protection per NOM-059-SEMARNAT-2010 are the San Lucan alligator lizard (Elgaria paucicarinata), worm lizard (Bipes biporus), spotted sandbow (Chilomeniscus stramineus), cat's eye nocturnal snake (Hypsiglena torquata), San Diego striped gecko (Coleonyx variegatus), San Lucan gecko (Phyllodactylus unctus), cape gecko (Phyllodactylus xanti), Sonoran woodpecker (Ctenosaura hemilopha), Baja California stone lizard (Petrosaurus thalassinus), cape scaly lizard (Sceloporus licki), Hunsaker scaly lizard (Sceloporus hunsaker), speckled rattlesnake (Crotalus mitchellii), and red diamond rattlesnake (Crotalus ruber). Fauna categorized as threatened at the site include common king snake (Lampropeltis getula), chuckwalla (Sauromalus obesus), western zebra-tailed lizard (Callisaurus draconoides), as well as the endemic Baja California nocturnal culebra (Eridiphas slevini), Baja California squeaky snake (Masticophis aurigulus), black-tailed tree lizard (Urosaurus nigricaudus), cachora (Eumeces lagunensis), and the Baja California rattlesnake (Crotalus enyo). Other mammals under some risk status include *Notiosorex crawfordi* (shrew), *Taxidea taxus* (American badger), and *Choeronycteris Mexicana* (trumpeted bat).

The socio-economic functions of the San José Basin, and in particular to the riparian ecosystems, the oases, and the estuary of the site are varied but all relate to the water cycle and the recharge of the aquifer. The riparian ecosystems and the oases are also sites of great attraction for eco-tourism activities; the sites highlighted for these activities are Boca de la Sierra, Miraflores, Capuano, El Cajón Waterfall, and San Miguelito Waterfall. With regard to cultural and historical values, the Jesuit Mission of San José del Cabo, established in 1730, is the historic center of the city. Among other cultural values in the basin are the rancherías that are established along the streams, which preserve the ancient traditions of the first Spanish settlers to the region. Within the estuary there is recreational use, and within the land area near the coast there are tourist accommodations, with surrounding areas primarily used for agriculture. Major tourism impacts are noted at the site due to where a marina has been developed. In the other areas of the basin there is substantial agricultural activity and forestry. The site with the greatest pressure and adverse environmental factors is the San José Estuary, since contaminated water is discharged into this body of water, in addition to the introduction of exotic fish such as tilapia. However, the most important threat to the ecology of the site is the development of large-scale tourism, which involves major marina construction and activities.

El Estero de San José del Cabo was declared a State Ecological Reserve in 1994, under the category of 'Subject to Ecological Conservation.' The decree specifies that the execution of public or private works within the area defined as the core zone of the State Ecological Reserve will not be authorized or allowed, except those strictly necessary for the recovery, conservation, and scientific research. However, the pressures of tourism and associated development in the area are substantial. On the northwestern end of the basin, on the Sierra La Laguna, is a Biosphere Reserve of the same name, where conservation and ecological restoration programs are ongoing. The Centro de Investigaciones Biológicas del Noroeste, S. C. (CIBNOR), the Autonomous University of Baja California Sur (UABCS), and the Interdisciplinary Center for Marine Sciences have conducted a variety of ecological and environmental studies at the site (Arriaga 1997, Breceda et al. 2007). For the State Reserve of the Estero de San José, the UABCS carried out a management program and Breceda et al. (2006) has conducted research on the effects of hurricanes on the vegetation of the estuary.

REFERENCES

Abitia-Cardenas, L.A., Rodriguez-Romero, J., Galván-Magaña, F., Cruz-Agüero, J., and H. Chávez-Ramos. 1994. Lista sitemática de la ictiofauna de la Bahía de La Paz, Baja California Sur, México. *Ciencias Marinas.* 20(2): 159–181.

Aguirre, A., Contreras, B., de la Cueva, H., Gonzalez, S., Martinez Rios, L., Martinez, V., Montes, C., Palacios, E., Paz Esparza, R., Salazar, M., and J. Serrano. 1999. *Opinión técnica sobre los proyectos turísticos "Cabo San Quintín" y "Bay Shores", en Bahía San Quintín, Baja California.* 32pp.

Álvarez, S., Gallina, P., González, A., and A. Ortega. 1988. Herpetofauna (pp. 167–184). In: L. Arriaga and A. Ortega (Eds.) *La Sierra de la Laguna de Baja California Sur.* Centro de Investigaciones Biológicas del Noroesunate. B.C.S. Publicación No. 1, La Paz, B.C.S.

Armenta-Martínez, L., Sánchez V., L., and C. Juárez O. 2004. Composición y distribución de larvas de peces en la bahía de La Paz (Golfo de California) durante epocas climáticas extremas (Verano 2001 - Invierno 2002). Res. XIII Reunión Nacional de la Sociedad Mexicana de Planctología, A. C. y VI Reunión Internacional de Planctología. Nuevo Vallarta, Nayarit, México, del 25 al 28 de abril.

Anderson, D.W. 1983. The seabirds (pp. 246–264 and pp. 474–481). In: T.J. Case and M.L. Cody (Eds.) *Island Biogeography in the Sea of Cortéz.* University of California Press, Berkeley, California. 508pp.

Arellano, P.L., Geraldo, H., Godoy, G., H.J., Juarez, O., C.C., Reyes Del C., S., Soto L., W., and C. Tovar F. 1991. *La comunidad ictioplanctónica en la boca de la ensenada de La Paz, B.C.S.* (verano-invierno). Res. II Congr. Nal. Ictiol. I–17.

Arriaga Cabrera, L., Aguilar Sierra, V., Alcocer Durán, J., Jiménez Rosenberg, R., Muñoz López, E., and E. Vázquez Domínguez. 1998. *Regiones hidrológicas prioritarias: fichas técnicas y mapa (escala 1:4,000,000).* Comisión Nacional para el Conocimiento y Uso de la Biodiversidad, México.

Arriaga Cabrera, L., Vázquez Domínguez, E., González Cano, J., Jiménez Rosenberg, R., Muñoz López, E., and V. Aguilar Sierra. 1998. *Lista de áreas prioritarias marinas de México. Regiones marinas prioritarias de México.* Comisión Nacional para el Conocimiento y uso de la Biodiversidad, México.

Baegert, J.J. 1989. *Noticia de la Península Americana de California.* (2nd. ed). En Español, Gobierno del Estado de B.C.S., La Paz.

Brandstein, K. 1998. *Women's perceptions regarding their health and their families in a remote village in México.* Thesis Master's Degree in Social Work and Master's Degree in Public Health. San Diego State University. San Diego California, USA. 121pp.

Breceda, A., Maya, Y., Castoreña, L., Martínez, G., Wurl, J., Miranda, R., and A. Valdez. 2007. *Manejo Integral de la Cuenca Hidrológica Forestal de San José del Cabo.* B.C.S. Informe Técnico CONAFOR-CONACYT (CO1–5671).

Castellanos, V.A. and J.G. Llinas. 1991. Aves migratorias: Patosy gansos (pp. 231–246). In: A. Ortega and L. Arriaga (Eds.) *La reserva de la biosfera del Vizcaíno en la Península de Baja California.* Centro de Investigaciones Biológicas del Noroeste de Baja California Sur A.C., México.

CITES (Convention on International Trade in Endangered Species of Wild Fauna and Flora). 2005. *Text of the convention (Appendices I, II, and III).* United Nations Environment Program. Geneva, Switzerland. 48pp.

Cody, M.L. and E. Velarde. 2002. The landbirds (pp. 41–54). In: Case, T. J., M. L. Cody, and E. Ezcurra (Eds.) *A New Island Biogeography of the Sea of Cortés.* Oxford University Press. 669pp.

Findley, L.T., Torre, J., Nava, J.M., van der Heiden, A.M., and P.A. Hastings. 1996. *Preliminary ichthyofaunal analysis from a macrofaunal database on the Gulf of California, México.* In: Proceedings of 76th Annual Meeting (p. 138). American Society of Ichthyologists and Herpetologists, 13–19 June, 1996. New Orleans.

Flores, E. 1998. *Geosudcalifornia. Geografía, agua y ciclones.* Universidad Autónoma de Baja California Sur. 277pp.

Gobierno del Estado de Baja California. 2006. *Programa de Ordenamiento Ecológico de la Región de San Quintín, Baja California (POESQ)*. Elaborado por la Facultad de Ciencias Marinas, Universidad Autónoma de Baja California, Mexico.

Grismer, L. 1999. Checklist of the amphibians and reptiles on islands in the Gulf of California, Mexico. *Bulletin of the Southern California Academy of Science*. 98(2): 45–56.

Grismer, L.. 2002. Amphibians and reptiles of Baja California including its Pacific Islands and the islands in the Sea of Cortés. University of California Press. Berkeley. 399pp.

Hausback, B.P. 1984.   Cenozoic volcanic and tectonic evolution of Baja California, Mexico.   Geology of Baja California Peninsula, In Pacific Section Society of Economic Paleontologists and Mineralogists. Special Paper 39, pp. 219-236.

Hernandez, L., Murugan, G., Ruiz-Fields, G., and A.M. Maeda-Martinez. 2007. Freshwater shrimp of the genus Macrobrachium (Decapada: Palaemonidae) from the Baja California Peninsula, Mexico. *Journal of Crustacean Biology*. 27: 351–369.

Howell, S.N.G. 2001. Regional distribution of the breeding avifauna of the Baja California Peninsula. In: R. Erickson and S. N. G. Howell (Eds.) *Birds of the Baja California Peninsula: Status, Distribution, and Taxonomy*. American Birding Association, Monographs in Field Ornithology.3: 10–22.

Huerta-Muzquiz, L. and A.C. Mendoza-González. 1985. Algas marinas de la parte sur de la Bahía de La Paz, Baja California Sur. Phytologia. 59(1): 35–54.

INEGI (Instituto Nacional de Geografía y Estadística). 1995. *Síntesis geográfica del estado de Baja California Sur. México*. 52pp.

Kerstitch, A.. 1989. Sea of Cortez marine invertebrates: a guide for the Pacific coast, Mexico to Ecuador. Sea Challengers. 120pp.

Martínez-Fragoso, J. 1992. *Bahía San Quintín: Un diagnóstico para suprotección*. Informe Técnico preparado con apoyo de Pronatura, Pro Esteros y CICESE. Ensenada, B. C. México. 119pp.

Massey, B.W. and E. Palacios. 1994. Avifauna of the wetlands of Baja California, Mexico: Current status. *Studies in Avian Biology*. 15:45–57.

Mellink., E. 2002. El límite Sur de la Región Mediterránea de Baja California, con base en sus tetrápodos endémicos. *Acta Zoologica Mexicana*. 85:11–23.

Page, W.G., Palacios, E., Alfaro, L., González, S., Stenzel, L.E., and M. Jungers. 1997. Numbers of wintering shorebirds in coastal wetlands of Baja California, Mexico. *Journal of Field Ornitholgist*. 68(4): 562–574.

Rebman, J., Gibson, J., Unitt, P., and E. Ezcurra. 1998. *Expedition to Sierra San Francisco and Sierra Guadalupe*, Baja Southern California, Mexico. Report. October-November 1997. 38pp.

Rodríguez-Estrella, R., Rubio, L., and E. Pineda. 1997. Los oasis como parches atractivos para las aves terrestres residentes e invernantes (pp. 157–186). In: L. Arriaga and R. Rodríguez-Estrella (Eds.) *Los Oasis de la Península de Baja California*. Centro de Investigaciones Biológicas del Noroeste. B.C.S. Publicación No. 13, La Paz, B.C.S.

Ruiz-Campos, G., Castro-Aguirre, J.L., Contreras-Balderas, S., Lozano-Vilano, M.L., González-Acosta, A.F., and S. Sanchez-González. 2002. An annotated distributional checklist of the freshwater fishes from Baja California Sur, México. *Reviews in Fish Biology and Fisheries*. 12: 143–155.

Serrano-González, J. 1998. *Dictamen que se emite sobre la situación actual que presentan los terrenos que serán afectados por obras de infraestructura a futuro en Punta Mazo y Punta Azufre de la Bahía de San Quintín, Baja California*. Informe no publicado presentado ante la Directora del Centro INAH en Baja California. 5pp.

Tershy, B.R. and D. Breese. 1997. The birds of San Pedro Mártir Island, Gulf of California, Mexico. *Western Birds*. 28:96–107.

Torreblanca, R.E. 2003. *Diagnóstico de las pesquerías comerciales de la región de Bahía de los Angeles en el año 2003*. Tesis de Licenciatura en Oceanología. Universidad Autónoma de Baja California. Ensenada, Baja California. 123pp.

Velarde, G.M.E. 1989. *Conducta y ecología de la reproducción de la gaviota parda (Larus heermanni) en Isla Rasa, Baja California*. Tesis doctoral, Facultad de Ciencias, UNAM.

Velarde, G.M.E. 1989. *Conducta y ecología de la reproducción de la gaviotaparda (Larus heermanni) en Isla Rasa, Baja California*. Tesis Doctoral. Facultad de Ciencias UNAM. 129pp.

Velarde, E. and D.W. Anderson. 1994. Conservation and management of seabird islands in the Gulf of California. Setbacks and successes. In: Nettleship, D.N., J. Burger and M. Gachfeld (Eds.) *Seabirds on Islands: Threats, case studies, and Action Plans (Birdlife Conservation Series No. 1)*. BirdLife International, Cambridge. 1:229–243.

Velarde, E. and E. Ezcurra. 2002. Breeding dynamics of Heermann's Gulls (pp. 313–325). In: Case, T., M.Cody and E. Ezcurra (Eds.) *A New Island Biogeography of the Sea of Cortés*. Oxford University Press. 669pp.

Villalobos-Hiriart, J.L., Nates-Rodriguez, J.C., Diaz-Barriga, A.C., Valle-Martínez, M.D., Flores-Hernández, P., Lira-Fernández, E., and P. Schmidtsdorf-Valencia. 1989. *Crustáceos estomatópodos y decápodos intermareales de las islas del Golfo de California*. México. Listados Faunísticos de México. Instituto de Biología, UNAM. México, D.F.

4 Mexico
States of Sonora and Sinaloa

The 16 Ramsar wetlands along the western coast of northern mainland Mexico, in the states of Sonora and Sinaloa, provide key wetland ecosystem services for the continental areas of Mexico, as well as for the migratory organisms along the Pacific Flyway and the 'blue corridors' of oceanic organisms. These 16 coastal wetlands along the northern coastal region of mainland Mexico (Figure 4.1) are:

- Sistema de Humedales Remanentes del Delta del Río Colorado
- Humedales de Bahía Adair
- Humedales de Bahía San Jorge
- Canal del Infiernillo and Esteros del Territorio Comcaác
- Humedales de la Laguna la Cruz
- Isla San Pedro Mártir
- Estero el Soldado
- Complejo Lagunar Bahía Guásimas-Estero Lobos
- Humedales de Yavaros-Moroncarit
- Lagunas de Santa María-Topolobampo-Ohuira
- Sistema Lagunar San Ignacio-Navachiste-Macapule
- Laguna Playa Colorada-Santa María la Reforma
- Ensenada de Pabellones
- Sistema Lagunar Ceuta
- Playa Tortuguera el Verde Camacho
- Laguna Huizache-Caimanero

SISTEMA DE HUMEDALES REMANENTES DEL DELTA DEL RÍO COLORADO, BAJA CALIFORNIA AND SONORA: OVERVIEW

Sistema de Humedales Remanentes del Delta del Río Colorado (Figures 4.2 and 4.3) includes the remnants of the Colorado River Delta, and it represents ideal habitat for many migratory and resident species within the desert area of Northwestern Mexico. The site forms part of the Pacific Flyway for migratory waterfowl along their journey from Canada or the U.S. to the southern continent, including species like *Dendroica coronata*, *Tachycineta bicolor*, and *Vermivora celata*. Other bird species use this area as breeding and nesting grounds, such as *Charadrius vociferus* and *Himantopus mexicanus*. The site offers a number of regional-scale ecosystem services, such as desert aquifer recharge, and flood prevention and attenuation. The vicinity around the site is predominantly agricultural.

SITE DETAILS

The site's existing wetland boundaries generally follow the boundaries of natural accumulation and flow of water in an area that was originally the Colorado River Delta, prior to the construction of the Morelos Dam in 1950. Owing to the enormous transformation of the delta during the 20th century, and the concomitant disappearance of natural surface water flows, including the interruption of the natural flow of the Colorado River, cessation of most (but not all) flooding events in the area has occurred. Wetlands present in the area now are often isolated within the larger landscape matrix of remnant topographic and soil conditions, which still contribute to the much-reduced accumulation of water and

DOI: 10.1201/9781003046394-4

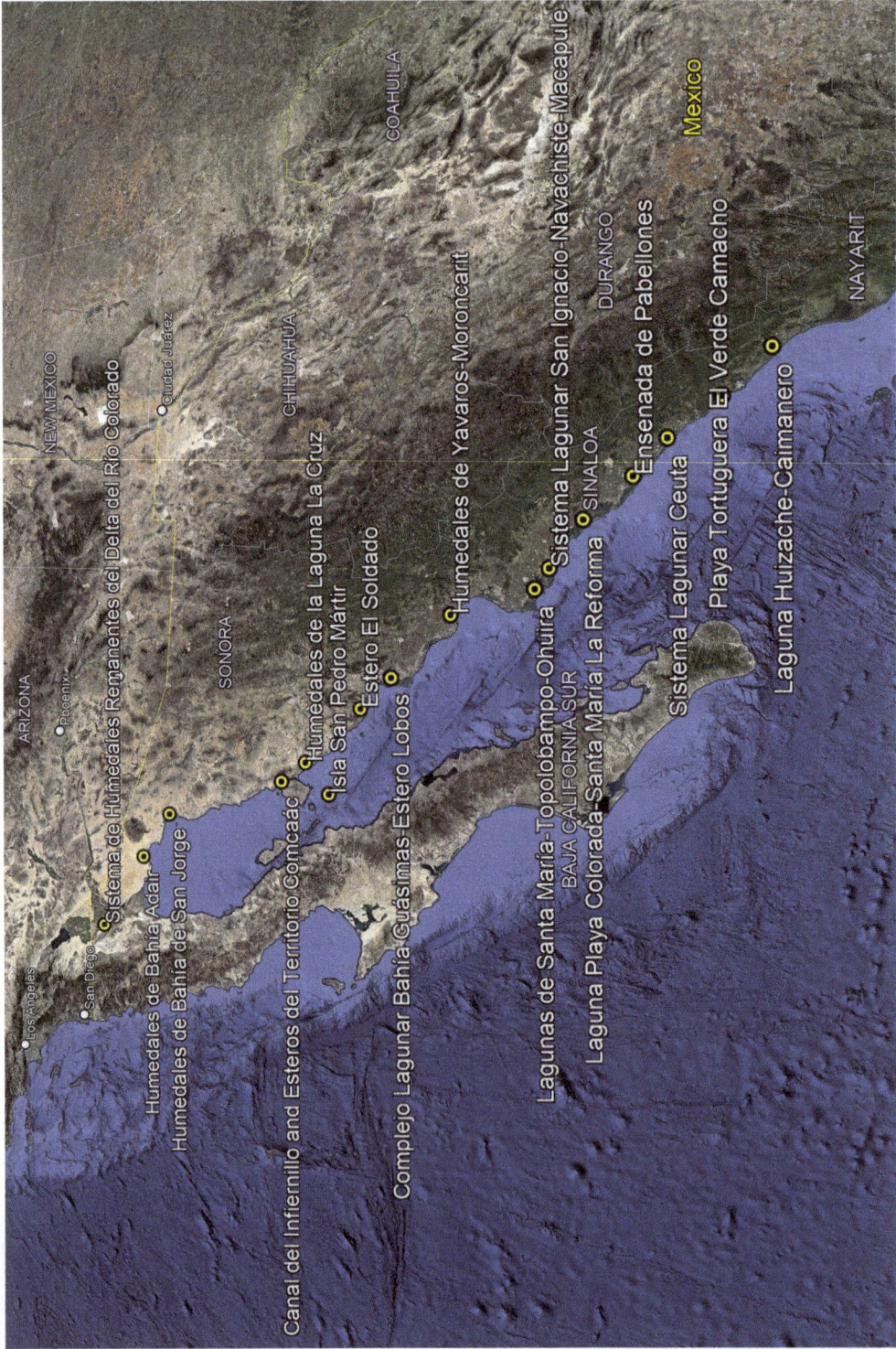

FIGURE 4.1   Referenced Ramsar wetlands generally shown along the western coast of northern mainland Mexico, in the states of Sonora and Sinaloa.

FIGURE 4.2 Composite aerial image of Sistema de Humedales Remanentes del Delta del Río Colorado. Designated February 2, 2008; Area: 127,614 ha; Coordinates: 32°19′N 115°16′W; Ramsar Site number: 1,822.

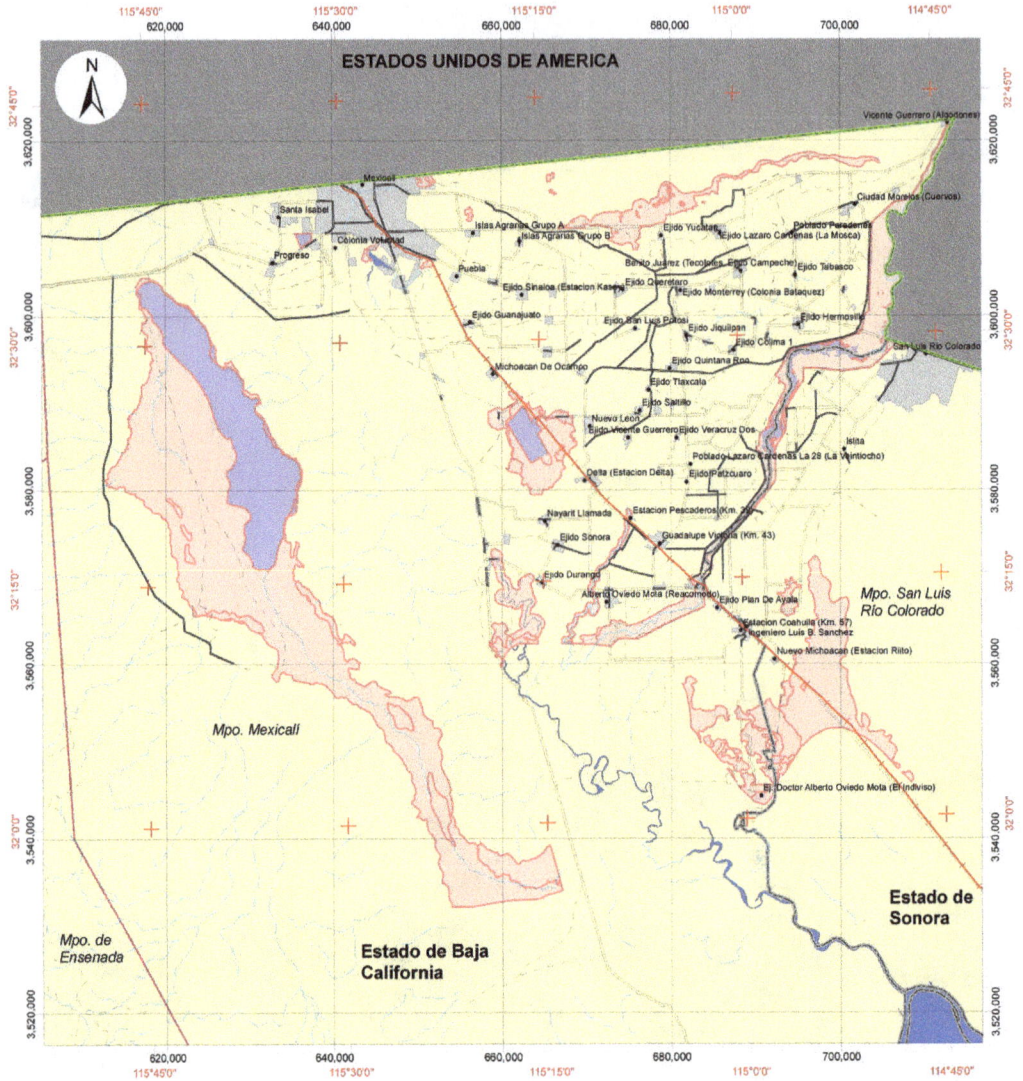

FIGURE 4.3 Map of Sistema de Humedales Remanentes del Delta del Río Colorado. Designated February 2, 2008; Area: 127,614 ha; Coordinates: 32°19′N 115°16′W; Ramsar Site number: 1,822.

interception of runoff that is part of landscape-level ecosystem processes. Such isolated wetlands are associated with the larger wetland area of the site, periodically, and are dependent upon the tenuous biological and physical connections in the region. In the eastern portion of the site, the presence of wetlands is limited mainly by the natural boundary of the Colorado River Delta and floodplains. Landscape modifications have also occurred in the Mexicali Irrigation District (the agricultural valley), which have considerably altered the natural flow of water throughout this dissected wetland landscape (e.g., canals built for irrigation). Most of the wetlands that make up the site are located to the northwest, within what is now referred to as the Agricultural Valley of Mexicali, an area surrounding the city of Mexicali within the lower basin of the Colorado River, i.e., the Colorado River Delta.

The site is at an average elevation of 3 m above sea level, with some areas of the site as high as 10 m to 20 m above sea level, especially in the northern areas of the watershed. All of the wetlands of the area (i.e., remnant wetlands) are of great importance to the ecology of the site because they represent refuge areas for migratory and other resident species within the otherwise agriculturally dominated landscape. Because the wetland plant communities of the site are located within areas of greater soil moisture, vegetation and water bodies of the site together provide the unique habitat conditions required for a number of key migratory and native species. In some places, the presence of beavers has even been reported, as well as mammals that were heretofore considered extirpated in this region of the country. Hinojosa-Huerta et al. 2004 report, solely for the Mesa de Andrade wetlands (located in the northwestern part of the site), approximately 100 bird species that benefit from these types of vegetational conditions. Of these approximately 100 bird species, at least 13 species are protected by statute (both Mexico, United States, and California); the protected species include Yuma Ridgeway's rail, willow flycatcher, and the black rail These same species and approximately 250 more have been recorded along all of the remnant wetlands of the Colorado River, demonstrating the importance of cross-border cooperation and conservation efforts that exist along the migratory route of birds throughout the countries of North America. Although intensively farmed, the site, which resides in this dominantly agricultural valley, provides key ecological functions that mitigate the loss of the vast areas of wetlands that once formed the Colorado River Delta.

The hydrology of the Mesa de Andrade wetlands is supplemented and indeed maintained by infiltration of water from the All American Canal, in California, where the canal crosses a sandy zone of the Mesa de Andrade wetlands (Cortez-Lara and García-Acevedo 2000). The phenomenon of water upwelling in the Andrade wetlands is due to subsurface water pressure in this area (Calleros et al. 1991), with this subsurface water flowing along a south/south-west direction, moving the infiltrated water from the All American Canal toward the Valley of Mexicali, where the water is extracted with a series of wells to the south of the Mesa de Andrade wetlands and used for agricultural purposes (Cortez-Lara and García-Acevedo 2000). Wetlands directly associated with the riparian corridor of the Colorado River Delta and its flood zones are fed by water from the Morelos Dam, but which is less than what is necessary to maintain a fully functional ecosystem of these characteristics. Some of the wetlands of the area are also fed or supplemented by irrigation water and agricultural return water, paradoxically, such that ecologically important wetlands of the area are now dependent upon the region's agriculturally engineered hydrology, to maintain their existence. In the southern portion of the site, species such as the totoaba (Totoaba macdonaldi) and even the highly endangered vaquita (Phocoena sinus) are dependent upon the ecological functions of the contributing river and delta/wetland's water and nutrients, as well as are numerous economically important fish species. The site contains listed (in the United States and California) species, including three in danger of extinction, two threatened, and five in special protection, with (Mexico) listed species including at least one in danger of extinction, one threatened, and four in special protection status. In addition, considering the degree of deterioration that the Colorado River wetlands have undergone in this region, any wetland in this region, no matter how small, should be considered critically important due to its rarity and contribution to habitat for the above-listed species, as well as others that support ecosystem functions in any way.

An important bird found in the Mesa de Andrade wetlands is the California black rail (Laterallus jamaicensis coturniculus), which is listed as endangered in Mexico. Other threatened vertebrate species at the site, protected by the Official Mexican Standard NOM-059-SEMARNAT-2010 (DOF 2010), are Castor canadensis (also included in the CITES), Chaetodipus arenarius, Peromyscus maniculatus, Callisaurus draconoides, Coleonyx variegatus, Crotaphytus collaris, Sauromalus obesus, and Uta stansburiana.

Greater than 350 species of birds are supported by the site's wetlands, either temporarily or permanently. Examples of bird species frequenting the site are migratory neotropical land birds, such as the yellow-rumped warbler (Dendroica coronata), the tree swallow (Tachycineta bicolor), and the orange-crowned warbler (Vermivora celata), as well as birds of prey such as the burrowing owl (Athene cunicularia), the hen hawk (Circus cyaneus), and the bird hawk (Accipiter striatus). Migratory waterfowl are present at the site, with ducks being the most common, such as blue-winged teal (Spatula discors), shoveler duck (Anas clypeata), cinnamon teal (Anas cyanoptera), and mallard duck (Anas platyrhynchos). Common wetland shorebirds at the site include long-billed sandpiper (Limnodromus scolopaceus), Wilson's phalarope (Phalaropus tricolor), least sandpiper

(Calidris minutilla), and American avocet (Recurvirostra americana), as well as two species of shorebirds that nest in the Mesa de Andrade wetlands: killdeer (Charadrius vociferus) and black-necked stilt (Himantopus mexicanus) [Hinojosa et al. 2004].

Due to migratory-seasonal uses of the site, certain species occur at the site at key stages of their lives, while some species use the area solely for resting during their migratory passage along the Pacific Flyway, including *Dendroica coronata, Tachycineta bicolor,* and *Vermivora celata.* Other species use the site for reproduction or nesting, such as *Charadrius vociferus* and *Himantopus mexicanus.*

Observations of Hinojosa-Huerta et al. (2004) confirm that wetlands at the site regularly support a clapper density of 21 per 100 ha of marsh, which represents a population estimate of approximately 172 individuals, maintaining this second largest population in Mexico of the subspecies *Rallus longirostris yumanensis,* Yuma Ridgeway's rail (Hinojosa-Huerta et al. 2004). The Mesa de Andrade wetlands supports a small population of the rare California black rail (Laterallus jamaicensis coturniculus), of which only approximately 50 pairs exist in Mexico (Hinojosa-Huerta et al. 2004).

HUMEDALES DE BAHÍA ADAIR, SONORA: OVERVIEW

Humedales de Bahía Adair (Figures 4.4 and 4.5) extends along 76 linear km of coastline, from Punta Borrascoso to Estero La Cholla, and includes estuaries, salt flats, and pools. The site is comprised of three habitat types: estuaries, artesian wells, and salt marshes within the Gran Desierto de Altar, one of the most arid and extreme deserts of North America. The site supports 12 faunal species found under special protection in the Official Mexican Standard NOM-059-SEMARNAT-2001 and/or -2010 (DOF 2002, 2010), such as the endemic and endangered desert pupfish (Cyprinodon macularius), and species in the CITES, such as marine turtles *Caretta caretta, Chelonia mydas, Dermochelys coriacea,* and *Lepidochelys olivacea.* Three fish species that are endemic to the northern Gulf of California (Gillichthys seta, Anchoa mundeoloides, and Leuresthes sardina), as well as two endemic floral species, *Distichlis palmeri* and *Suaeda puertopenascoa.* The main hydrologic ecosystem service of the site is recharge of the Sonoyta-Puerto Peñasco Aquifer, which is of prehistoric formation. The main land uses at the site are eco-tourism and real estate development in the coastal zone, as well as salt extraction, conservation, scientific research, environmental education, subsistence fishing, and oyster mariculture. A portion of the site is found in the Alto Golfo and Delta del Río Colorado Biosphere Reserve, and an adjoining area of El Pinacate and el Gran Desierto de Altar Biosphere Reserve, which follows the IUCN management categories of "Strict Nature Reserve," "Wilderness Area," and "Managed Resource Protected Area."

SITE DETAILS

The wetlands of Bahía Adair range from those areas that reside at the lowest level of low tide, following the coastline, to 20 m above the maximum high tide. Most of the Bahía Adair Wetlands are embedded in the Upper Gulf of California and Colorado River Delta Biosphere Reserve, within that buffer zone, and adjoin the El Pinacate and Gran Desierto de Altar Biosphere Reserve. The Bahía Adair Wetlands are located on the northwest coast of the State of Sonora, covering part of the municipalities of San Luis Río Colorado and Puerto Peñasco. The coastal lengths of the site contain a number of estuaries, namely El Borrascoso, Las Lisas, San Judas, Los Paredones, Cerro Prieto, and La Cholla. There are no permanent inhabitants within the site, only a non-resident fishing cooperative. The closest city to the site is Puerto Peñasco, with a population of approximately 63,000. The next largest nearby municipality is San Luis Río Colorado, with approximately 193,000 inhabitants, which lies inland and approximately 89 km to the northwest.

Tidal flow within this estuarine ecosystem is the driver for a variety of wetland habitat conditions, including channels, salt marshes, mud flats, and hypersaline salt flats. Thus, marshes are covered by halophytic shrubby vegetation, such as *Allenrolfea occidentalis, Batis maritima, Distichlis palmeri, Frankenia salina, Monanthochloe littoralis, Arthrocnemum subterminale, Suaeda esteroa,*

FIGURE 4.4 Composite aerial image of Humedales de Bahía Adair. Designated February 2, 2009; Area: 42,430 ha; Coordinates: 31°35′N 113°53′W; Ramsar Site number 1,460.

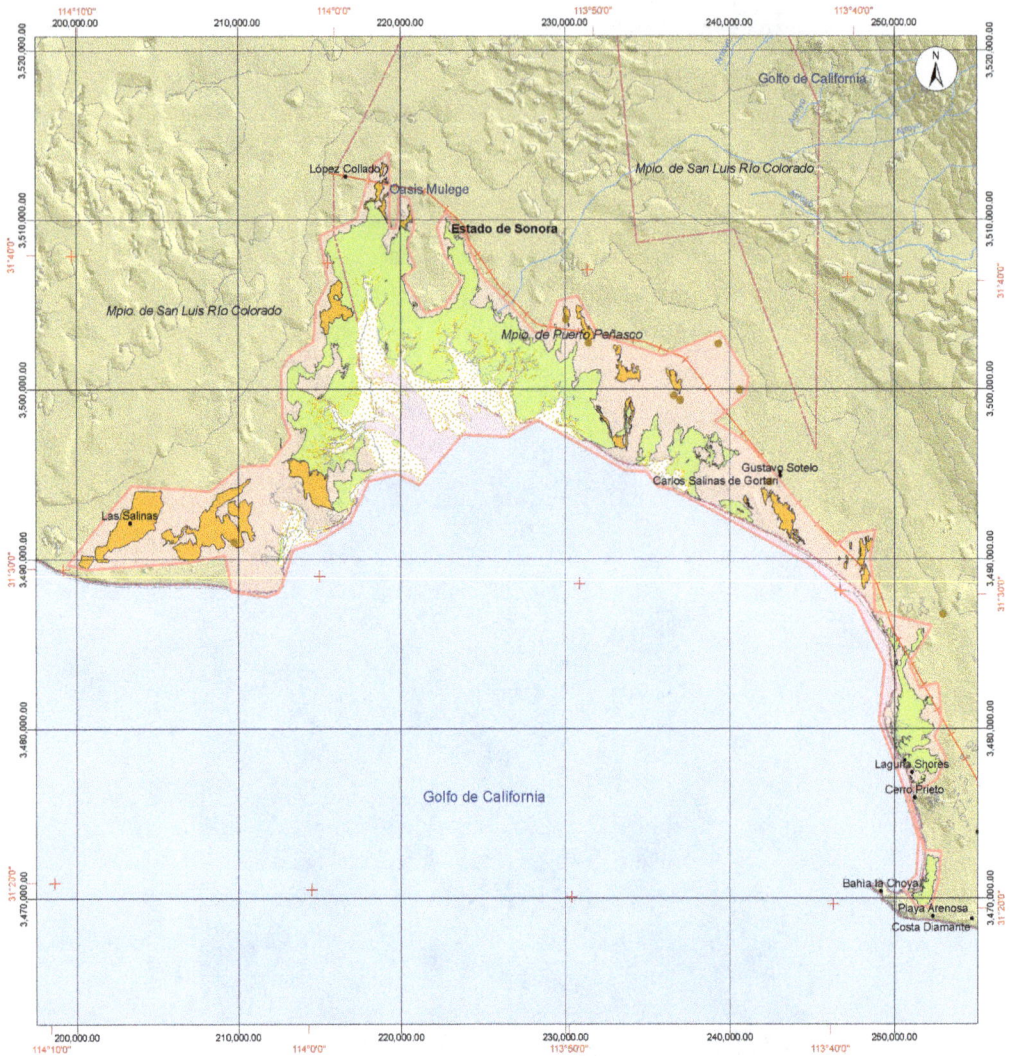

FIGURE 4.5 Map of Humedales de Bahía Adair. Designated February 2, 2009; Area: 42,430 ha; Coordinates: 31°35′N 113°53′W; Ramsar Site number 1,460.

and the endemic *Suaeda puertopenascoa* (Felger 2000). This distinctive plant community is the basis of a high primary productivity that sustains a diverse trophic web at the site. Channels and marshes at the site also serve as spawning and feeding grounds for commercial fish species (Glenn et al. 2006). These estuaries maintain connectivity between terrestrial and aquatic ecosystems that mediate nutrient and species flow, melding attributes of both ecosystem types and resulting in rich nutrient environments. The estuaries of the site serve as nesting, resting, and feeding habitat for resident and migratory birds, as part of the Pacific migration corridor, or Flyway (Brusca et al. 2006). Marsh areas at the site tend to be surrounded by unconsolidated dunes and hypersaline mud flats, of marine and evaporative origin (Ezcurra 1984). These muddy flats may have thick salt crusts and lack vegetation (Ezcurra et al. 1987). Artesian springs, also referred to as wells, are found on the shores of hypersaline mud flats. Wells are ecosystems of high ecological importance because they are the last available habitat for species adapted to the particular conditions of the site, and because they provide fresh water for mammals and birds in this otherwise extremely arid region of North America (CONANP 2007); the Gran Desierto is too arid even for most cacti to survive. Each of the habitats that comprise the Bahía Adair Wetlands support peculiar flora and fauna and are characterized by a high degree of specialization, due to the conditions at the site's sea-land interface. Twelve species of animals have been recorded in the Bahía Adair Wetlands that are under some category of protection in the Official Mexican Standard (DOF 2002, 2010) and/or listed by the International Convention on endangered Species of Fauna and Wild Flora (CITES 2005). There are also five species of endemic plants and animals within the estuaries of the site.

The Adair Bay wetlands are notable for their support of the following organisms:

- 86 species of birds including reddish egret (Egretta rufescens), Heermann's gull (Larus heermanni), yellow-footed gull (Larus livens), short-eared owl (Asio flammeus), peregrine falcon (Falco peregrinus), Yuma Ridgeway's rail (Rallus longirostris yumanensis), piping plover (Charadrius melodus), white-headed eagle (Haliaeetus leucocephalus), savannah sparrow (Passerculus sandwichensis), and California black rail (Laterallus jamaicensis coturniculus);
- herpetofauna such as chuckwalla (Sauromalus obesus), western zebra-tailed lizard (Callisaurus draconoides), flat-tailed horned lizard or flat-tailed false chameleon (Phrynosoma mcallii);
- desert pup (Cyprinodon macularius), which is endemic to the region and is in danger of extinction; and
- threatened and protected sea turtles in Mexico, such as loggerhead sea turtle (Caretta caretta), Pacific green sea turtle (Chelonia mydas), Atlantic green or white turtle (Chelonia mydas), leatherback sea turtle (Dermochelys coriacea), and olive ridley sea turtle (Lepidochelys olivacea).

Channels and marshes serve as nursery areas for larval and juvenile stages of fish and invertebrates (Carrera and de la Fuente 2003). For example, in the estuaries of the area there is an important recruitment of early stages of penaeid shrimp (Litopenaeus stylirostris); for this species, maximum concentration of mature females is in the coastal zone outside the estuaries along the Sonora coast (Calderón-Aguilera et al. 2003). Estero La Choya is also one of the most important breeding sites for the warrior swimming crab, *Callinectes bellicosus*; it is estimated that the site supports the production of 375 tons of this species, per year (25% of the fishing production of Puerto Peñasco).

The climate of the region is of the dry-desert type (Contreras 1993). The temperature that occurs ranges from 8° C in winter to 42° C in summer (Valdés-Casillas et al. 1999). The region has two seasons, the mid-latitude winter season from November to May, and the subtropical summer season from June to October (Mosiño and García 1974). Extraordinary climatic anomalies occur in this region, such as the El Niño Southern Oscillation (Valdés-Casillas et al. 1999), which can dramatically influence the water cycle for the site and region. The rainy season at the site occurs both in summer and winter, but these are very sporadic. The number of rainy days per year is approximately five (Lavin and Organista 1988). It typically rains at the site during summer months, with an average annual precipitation of 73 mm and high evaporation rates (0.9 m/year; Álvarez-Borrego and Galindo-Bect 1974).

At the landscape level, these coastal wetlands serve as a link between the Sonoran Desert and the marine ecosystem of the Gulf of California. For example, herpetofauna and terrestrial mammals, such as coyote (Canis latrans) subsidize their diets extensively with marine resources. Desert dust can also provide nutrients to marsh vegetation (Okin et al. 2005). Comparatively, these estuaries export energy in the form of organic matter, mainly in the form of detritus derived from halophilic plants to the marine environment, out to the open ocean ecosystem.

The Adair Bay wetlands are of historical and archaeological importance, as evidenced by findings of inhabitants of the greater desert region for more than 37,000 years (Hartmann 2002). Of main importance for the pre-Hispanic cultures at the site, and in the region, were the wells and water holes, as they used them as a source of water in this very dry environment. The more recent cultures in the area were the Areneros and Pinacateños. The Pinacateños were still inhabiting the region when it was visited by Padre Kino in 1698 (Hartmann 2002, Felger and Broyles 2007). Currently the area has cultural and religious importance for the Tohono O'odham ethnic group, whose members live in the cities of Sonoyta, Puerto Peñasco, and in Arizona, USA. This ethnic group considers wells and water holes as particularly sacred areas, associated with different important rituals.

The estuaries to the south of the area, La Cholla and Cerro Prieto, are presently very attractive for tourism due to their proximity to the city of Puerto Peñasco. In these areas aquatic-recreational activities occur regularly, and the manual capture of molluscs, eco-tourism or low-impact tourism, environmental education, sport fishing, flora and fauna observation, boat tours for tourist purposes, and all-terrain vehicle rides are quite common. Most of the tourists who visit the area are North Americans. Development of roads in the area, including the Coastal Highway connects the Gulf of Santa Clara with Puerto Peñasco, increasing tourist access to this area, among other visitors.

HUMEDALES DE BAHÍA SAN JORGE, SONORA: OVERVIEW

Humedales de Bahía de San Jorge (Figures 4.6 and 4.7) includes eight types of habitats that connect terrestrial ecosystems with one of the most productive marine ecosystems of the world, the Gulf of California, embedded within an extremely arid part of North America. The site supports a tremendous diversity of flora and fauna, all of which are adapted to the extreme conditions of the region, such as halophytic plants, four endemic species of fish (Gillichthys seta, Anchoa mundeoloides, Colpichthys regis, and Leuresthes sardina), and two endemic plant species (Distichlis palmeri and Suaeda puerto-penascoa). The site is along the Pacific Flyway and serves as a nesting habitat for birds like Wilson's plover (Charadrius wilsonia), the least tern (Sternula antillarum browni), the American oystercatcher (Haematopus palliatus frazari), and the large-billed savannah sparrow (Passerculus sandwichensis rostratus). The channels and marshes of the site serve as refuge and feeding habitat for larval and juvenile stages of fish and invertebrates. The site also supports a resident population of bottlenose dolphin (Tursiops truncatus). The site's wetlands are important for flood mitigation and the prevention of coastal erosion. The major land uses in the vicinity of the site are aquaculture and fishing.

SITE DETAILS

The Bahía San Jorge Wetlands extend along 38 linear km of coastline, from the northern end of Estero Almejas to the southern end of Estero San Francisquito, including mud flats, marshes, salt flats, and coastal dunes. The Bahía San Jorge Wetlands range from the lowest level of low tide, following the coastline, to 20 m above maximum high tide, comprising the Federal Maritime Terrestrial Zone, encompassing coastal dunes and some salt flats that are within the terrestrial-wetland ecotone. At its northeast end, the site boundary coincides with the Puerto Peñasco-Caborca Highway. The site encompasses the La Salina and Almejas estuaries, as well as the San Francisquito estuary, as well as part of the municipalities of Puerto Peñasco and Caborca.

The San Jorge Bay Wetlands comprise several general habitat types, including estuaries, intertidal mud flats, salt flats, coastal dunes, sandy beaches, permanent shallow marine waters, and terrestrial-wetland ecotone (Valdes-Casillas et al. 1999). The marsh areas tend to be surrounded by unconsolidated dunes and soils with intermittent flooding of marine and evaporative origin, called salt pans; resulting salt flats can have thick salt crusts and lack vegetation (Ezcurra et al. 1987). Twenty-three species have been recorded in the San Jorge Bay Wetlands that are under some category of protection in NOM-059-SEMARNAT-2001 and/or -2010 and/or listed by the International Convention on endangered Species of Fauna and Wild Flora (CITES 2005). The site also supports seven endemic species of the Gulf of California.

Fauna at the site with special protections are the least tern (Sternula antillarum browni) [Mellink and Palacios 1993], reddish egret (Egretta rufescens), yellow-footed gull (Larus livens), peregrine

FIGURE 4.6 Composite aerial image of Humedales de Bahía de San Jorge. Designated February 2, 2010; Area: 12,198 ha; Coordinates: 31°06′00″N 113°04′11″W; Ramsar Site number 1,983.

FIGURE 4.7   Map of Humedales de Bahía de San Jorge. Designated February 2, 2010; Area: 12,198 ha; Coordinates: 31°06′00″N 113°04′11″W; Ramsar Site number 1,983.

falcon (Falco peregrinus), Heermann's gull (Larus heermanni), golden hawk (Buteo regalis), Harris' hawk (Parabuteo unicinctus), elegant tern (Sterna elegans), pink rattlesnake (Crotalus atrox), little horned viper (Crotalus cerastes) [Zuria-Jordan 1996], and the horse-killing lizard (Gambelia wislizenii) [Mellink and Palacios 1993, Zuria-Jordan 1996].

Fauna at the site that are threatened include the red-billed tropicbird (Phaethon aethereus), prairie falcon (Falco mexicanus), large-billed savannah sparrow (Passerculus sandwichensis rostratus), tortoise (Gopherus agassizii) [Lemos Espinal 2007], western zebra-tailed lizard (Callisaurus draconoides), and black storm petrel (Oceanodroma melania) [Mellink and Palacios 1993, Zuria-Jordan 1996]. The site also supports the endangered Mexican shearwater (Puffinus opisthomelas), the Colorado Desert fringe-toed lizard (Uma notata), and the myotis fisherman bat (Myotis vivesi) [Arriaga et al. 2000].

Immediately adjacent to the terrestrial-wetland ecotone are cactus species characteristic of the Sonoran Desert, including the senita cactus (Lophocereus schotii) [Morzaria-Luna et al. 2007]. The site also hosts a resident population of bottlenose dolphin (Tursiops truncatus) [Mellink and Orozco-Meyer 2006] in the off-shore areas, loggerhead turtle (Caretta caretta), the Pacific green sea turtle (Chelonia mydas), the Atlantic green sea turtle or white turtle (Chelonia mydas), the leatherback sea turtle (Dermochelys coriacea), and the olive ridley sea turtle (Lepidochelys olivacea) [Honan and Turk-Boyer 2001, CONANP 2007]. All species of sea turtles found at the site are protected in Mexico (SEMARNAT 2002) and in the world (CITES 2005).

In the estuaries of Bahía San Jorge there are four species of fish endemic to the Northern Gulf of California: the small sucker (Gillichthys seta) [Barlow 1961], the olive ridley (Anchoa munde-oloides), charal silverside (Colpichthys regis), and gulf grunion (Leuresthes sardina) [Zuria and Mellink 2005]. There are two endemic plant species at the site: saltgrass (Distichlis palmeri), which is endemic to the Gulf of California coast, and the Puerto Penasco suaeda (Suaeda puertopenascoa), which exists only in the estuaries of northern Sonora (Felger 2000).

As a whole, the site provides habitat for highly diverse biological communities, including at least 151 species of birds (Mellink and Palacios 1993, Zuria-Jordan 1996); 62 species of fish (Zuria-Jordan 1996, Zuria and Mellink 2005); 6 terrestrial mammals (Zuria-Jordan 1996, Arriaga et al. 2000); 1 marine mammal (Orozco-Meyer 2001); 9 reptiles; 54 vascular plants (Morzaria-Luna et al. 2007); 11 algae, 14 dinoflagellates and diatoms (Juárez Romero 2007); 33 molluscs; and 31 crustaceans (Villalobos-Hiriart et al. 1989). It is worth mentioning that relatively few comprehensive biological studies have been carried out in the area and that the actual number of species is probably much higher. Although in the entire Gulf of California there are only 50 endemic species of fish (Castro Aguirre et al. 2005), four of these endemic species are found in the San Jorge Bay Wetlands, as follows: the small sucker (Gillichthys seta) [Barlow 1961], the olive ridley (Anchoa mundeoloides), the charal silverside (Colpichthys regis), and the gulf grunion (Leuresthes sardina) [Zuria and Mellink 2005].

The tides of the site are of the semi-diurnal mixed type, with two high tides and two low tides of different amplitude; the maximum amplitude is approximately 10 m and they result in current velocities of 1.5 m*sec^{-1} to 3 m*sec^{-1} (Kasper-Zubillaga and Carranza-Edwards 2005). The marine zone off of San Jorge Bay is an upwelling zone that fosters phytoplankton blooms, contributing to the high productivity of the area (Parker 1964). The waters of the San Francisquito Estuary may be affected by the presence of agrochemicals from runoff, from the agricultural areas of the Caborca Valley (CIAD 2006).

The marshes of La Salina, Almejas, and San Francisquito present a low species richness; in these areas there is only halophytic vegetation, which rarely exceeds 50 cm to 60 cm in height (Felger 2000). Most of the halophytic species at the site are perennial and salt-excreting, including succulent and non-succulent species; non-perennial species include dwarf glasswort (Salicornia bigelovii), an annual species, and jauja or saldillos (Suaeda spp.), a short-lived perennial species (Morzaria-Luna et al. 2007). Most of these halophytic plants are widely distributed, and some are regional endemics. Suaeda puertopenascoa is known only to exist in the estuaries of northwestern Sonora, and saltgrass (Distichlis palmeri) is the only grass endemic to the Sonoran Desert (Felger 2000). The largest marsh areas of the site are found in the vicinity of Estero Almejas.

The Bahía San Jorge Wetlands have social and cultural importance dating from approximately 9000 BC or earlier, as evidenced by the inhabitants of the northern Sonora region since that time (Huckell 1996). The site is located within the territory formerly known as Chichimecatlalli, in the Ootam pre-Hispanic region, which encompasses Sonora and Arizona, based on cultural, history, and ceramic evidence. The Hohokam and Trincheras were also in this region (Braniff 2009); the Trincheras culture used species from the San Jorge Bay Wetlands as a source of food and raw materials (Mitchell and Foster 2000). Currently, the San Jorge Bay Wetlands are of great socioeconomic importance for local communities due to the site's role in fishing and aquaculture production. Occasionally, local residents visit the sandbars or beaches to camp and rest. The estuary area is rarely used for recreational activities or tourism.

CANAL DEL INFIERNILLO AND ESTEROS DEL TERRITORIO COMCAÁC, SONORA: OVERVIEW

A channel between Tiburón Island and the Sonora coast in northwestern Mexico, Canal del Infiernillo and Esteros del Territorio Comcaác (Figures 4.8 and 4.9) is characterized by the presence of seagrass beds, mangrove estuaries, seasonal creeks, and small coral reef patches. The seagrass

FIGURE 4.8 Composite aerial image of Canal del Infiernillo and Esteros del Territorio Comcaác; Designated November 27, 2009; Area: 29,700 ha; Coordinates: 29°10′N 112°14′W; Ramsar Site number: 1,891.

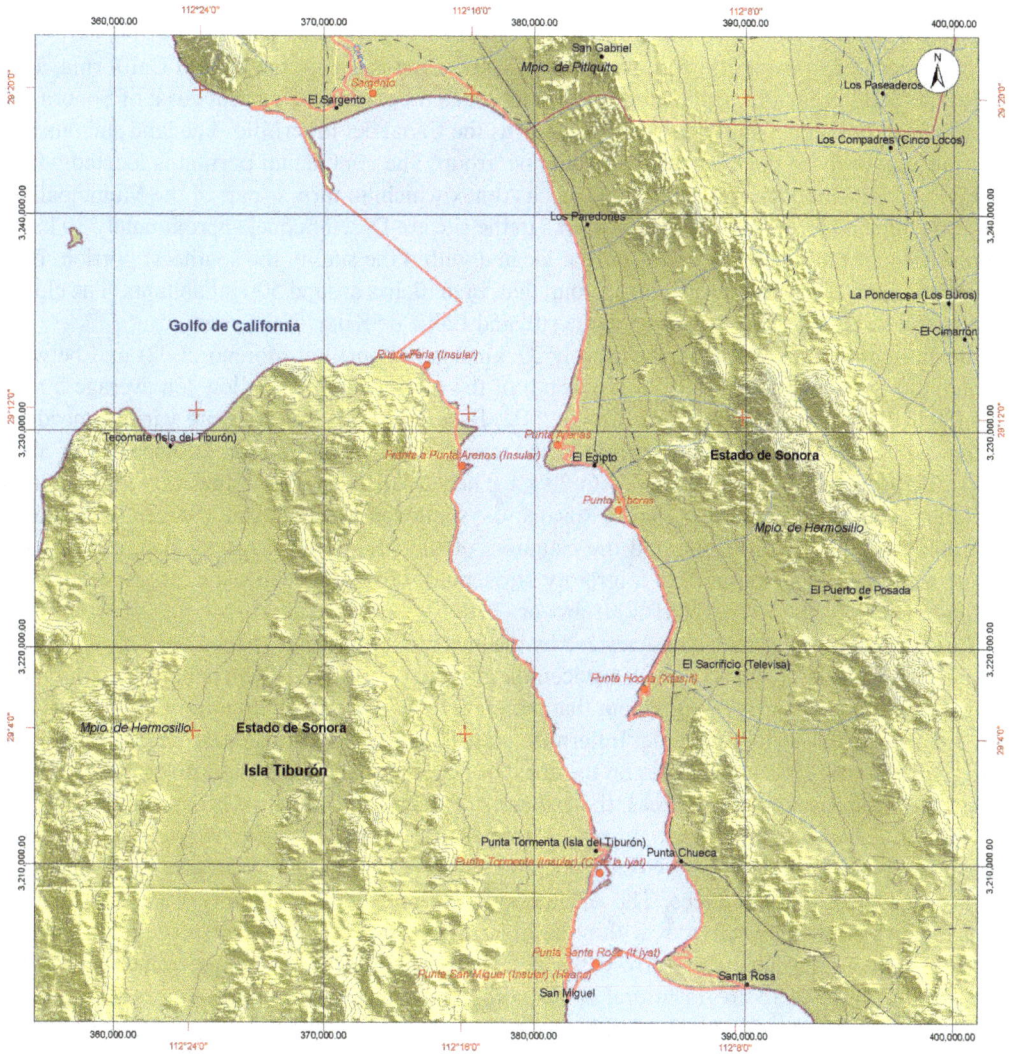

FIGURE 4.9   Map of Canal del Infiernillo and Esteros del Territorio Comcaác; Designated November 27, 2009; Area: 29,700 ha; Coordinates: 29°10′N 112°14′W; Ramsar Site number: 1,891.

beds are the largest concentration of annual marine grasses in the eastern Pacific, with its mangroves located at the northern limit of this vegetation type (i.e., coastal mangrove forests). These wetlands provide refuge, substrate, and food to several species that are the basis for commercial and artisanal fisheries. In addition, the site supports 81 endemic invertebrate species of the Gulf of California and several threatened species, such as mangroves (Avicennia germinans, Laguncularia racemosa, and Rhizophora mangle), totoaba (Totoaba macdonaldi), marine turtles (Eretmochelys imbricata, Caretta caretta, Dermochelys coriacea, Lepidochelys olivacea, and Chelonia mydas), and brant (Branta bernicla). For more than 2000 years, the ethnic group of Comcaác has inhabited this region as evidenced from their extensive ecologically-relevant traditional knowledge base, which they utilize to further their fisheries management practices. In contrast to several other regions in the Gulf, the seabeds at this site have been less disturbed by the nets of shrimp vessels, but they are currently threatened by overfishing and tourism.

SITE DETAILS

The site is located in the northwestern region of Mexico and is part of the Gulf of California, adjacent to the State of Sonora, and includes the strait located between the central coast of Sonora and Tiburón Island, within a maritime region known as the Canal del Infiernillo. The land surrounding the strait is entirely of the Comcaác (Seri) ethnic group. The continental portion is located within the Ejido de Desemboque and its Punta Chueca Annex which, in turn, is part of the Municipalities of Hermosillo and Pitiquito. The closest towns to the site are Desemboque, approximately 30 km to the northeast of the site, and Punta Chueca, located within the site, in the southeast portion. Both are communities of the Comcaác ethnic group that, in total, has around 500 inhabitants. The closest mestizo towns are Puerto Libertad, to the north, and Bahía de Kino, to the south.

The Infiernillo Channel is approximately 37 km long (Bourillón-Moreno 2002) and between 1.8 and 10 km wide (Basurto 2005). The depth of the marine zone is shallow (on average 5 m or 6 m) [Bourillón-Moreno 2002, Torre-Cosío 2002, Basurto 2005], and there are various meadows of *Zostera marina* and *Ruppia maritima* in the area, as well as more than 200 species of algae (Bourillón-Moreno 2002 and Torre-Cosío 2002). The site contains nine estuaries with mangrove vegetation, with the largest areas (Santa Rosa) at the south-southeast end and (Sargento) the north-northeast end. The marine zone and the estuaries of the site are a refuge for fish, crustaceans, marine mammals, and thousands of migratory birds, many of which are in some category of protection under NOM-059-SEMARNAT-2001 and/or -2010.

The estuaries of the site are characterized by the presence of the three mangroves on site, namely white (Laguncularia racemosa), black (Avicennia germinans), and red (Rhizophora mangle) mangrove. Felger and Moser (1987) point out that the mangroves play(ed) an important role in Seri culture, so much so that in the Canal del Infiernillo (marine zone) there are at least 72 sites of cultural value, with another 190 cultural sites on the coast of Isla Tiburón (Luque and Robles 2006).

As commonly occurs in arid areas, the estuaries of the site are negative because in the interior part the saline concentration (up to 40 ppt) is greater than in the marine zone (on average 35 ppt) [Martínez-Yrizar 2006 and Torre-Cosío 2002]. The waters of the channel have high productivity, and are rich in fishing resources. The members of the Comcaác community have the exclusive rights to carry out fishing activities within the Infiernillo Channel and waters surrounding Tiburón Island (Bourillón-Moreno 2002, Torre-Cosío 2002, Basurto 2005). On the continental shore of the Infiernillo Channel there are 14 littoral zones with a maximum width of 0.5 km to 8 km, where sedimentary dynamics of the coastal zone occur (Lancín 1985). The east coast of Isla Tiburón is completely sandy, and the mainland coast of the channel is also sandy, with the exceptions of Punta Ignacio, Punta Sargento, and Cabo Tepopa, which are rocky areas. At the mouths of the estuaries of the site the bottom sediments are composed of shell fragments and small stones, referred to locally as "tepetate." Within the channel there are also muddy areas with only a few small rocky reefs, which are actually large extensions of mussel beds known locally as "chorales," that serve as substrate areas for a tremendous diversity of animals (Torre-Cosío and Bourillón-Moreno 2000).

Mangroves and submerged vegetation are undoubtedly the most important types of plants in the Canal del Infiernillo since they are the basis of high marine productivity. Submerged vegetation, seagrass (Zostera marina, Ruppia maritima, and Halodule wrightii), and algae (e.g., species of the genera Caulerpa and Sargassum), along with the mangrove detrital cycles are the most important elements of productivity at the site. The main groups of fauna represented in the Infiernillo Channel and its catchment area are annelids (193 species), arthropods (319 species), molluscs (445 species), echinoderms (86 species), and chordates (546 species) [Torre-Cosío 2002, Macrofauna Gulf Database 2008]. Records in the area indicate that there are four species of whales, which are included in the IUCN Red List, as follows: humpback whale (Megaptera novaeangliae), Bryde's whale (Balaenoptera brydei), fin whale (Balaenoptera physalus), and gray whale (Eschrichtius robustus). Under special protection at the site are brant (Branta bernicla), reddish egret (Egretta rufescens), yellow-footed gull (Larus livens), brown pelican (Pelecanus occidentalis), California sea lion (Zalophus californianus), green sea turtle (Chelonia mydas), loggerhead turtle (Caretta caretta), hawksbill turtle (Eretmochelys imbricata), leatherback sea turtle (Dermochelys coriacea), olive ridley sea turtle (Lepidochelys olivacea), rugose pen shell (Pinna rugosa), and channel corvina (Cynoscion spp.).

The Infiernillo Channel, along with its coasts have been part of territory of the Comcaác ethnic group for many years (Bowen 2000, Felger and Moser 1985, Luque 2002). All the estuaries and many locales along the coast and in the marine zone have names in the native Comcaác language and, according to Luque and Robles (2006), many are sacred sites, including the areas of feeding and hibernation for the green sea turtle (Chelonia mydas). Archaeological remains at the site are very scarce because, until a few decades ago, the Comcaác were nomadic according to archaeological evidence that is available; nevertheless, the Comcaác are the only group that has constantly inhabited the area. According to Felger and Moser (1985) Comcaác fishers can locate, day or night, the areas with seagrasses in and around the site, which indicates the extensive knowledge, the rituals, and sacred meanings of many of the places due to their importance for fishing, bird refuge, or refuge for members of the Comcaác community, for example during turbulent or other weather threats. Currently, many of the fishing grounds that were used since pre-Hispanic times are still in use. The coastal fishery in the region focuses mainly on crab, scallops, and fish (e.g., croaker and mullet). Comcaác fishers are dedicated almost entirely to crab fishing, which is carried out with "Chesapeake" type traps, and nets for crab, the latter of which are illegal, but are beginning to be used more due to the need to have higher incomes from fishing, compounded by the increasing scarcity of the product caused by a greater number of foreign fishers.

The waters and coasts of the Infiernillo Channel are one of the most important sites at the national level for bivalve molluscs, crustaceans, and commercially important fish, with the main extraction activities occurring as artisanal fishing. The main developed fisheries in the Channel are the bivalves *Atrina* spp. and *Pinna rugosa*, and the warrior swimming crab (Callinectes bellicosus), producing around 70 mt and 3350 mt, respectively, each season (Basurto 2002). There is also fishing for rays (Dasyatis depterura and Rhinobatos spp.), mullet (Mugil spp.), and croaker (Cynoscion spp.) [Torre-Cosío 2002] at the site. Currently, the extraction of scallops (Atrina spp. and Pinna rugosa) is the major fishing activity (Basurto 2005, 2008). Coastal fishing is still the main source of income for the Comcaác communities (Luque and Doode 2007). In recent years, tourism has developed in an incipient way without having hotel infrastructure or services. Tourism is developing predominantly in the southern and central zone of the Channel del Infiernillo, near the town of Punta Chueca, which includes cultural tourism that is of interest to visitors, mostly foreigners (Desemboque and Punta Chueca), leading those visitors to learn more about the way of life of the Comcaác community, and buy handicrafts from local community artisans/vendors. In fact, the second greatest source of income within the site's communities is the sale of handicrafts, which is carried out mainly by women of the area.

HUMEDALES DE LA LAGUNA LA CRUZ, SONORA: OVERVIEW

As with a number of the sites along this western coast of North America, Humedales de la Laguna La Cruz (Figures 4.10 and 4.11) is part of the Pacific Migratory Route and an important rest area for migratory birds. The site provides food and refuge for a total of 154 species of birds, 84 of which are aquatic bird species, and of those, 9 species have populations at the site that are over 1% of the total regional population. In terms of fish, 96 species have been identified at the site, which is vital for some species during adverse climatic conditions and as a breeding area for a variety of fish, mollusc, and crustacean species. The site is particularly important for species of the Families Gobiidae, Atherinidae, Gerreidae, and the Engraulidae, during their larval and juvenile stages.

The site is characterized by its mangrove forests, which are unique because they are located within the northern limit of their distribution, lack fresh water inputs, and are within the Sonoran Desert area. As a result, species such as the *Frankenia* spp., which are adapted to the very saline conditions, are found at the site. This site also supports threatened species such as the green sea

FIGURE 4.10 Composite aerial image of Humedales de la Laguna La Cruz. Designated February 2, 2013; Area: 6,665 ha; Coordinates: 28°47′15″N, 111°52′53″W; Ramsar Site number: 2,154.

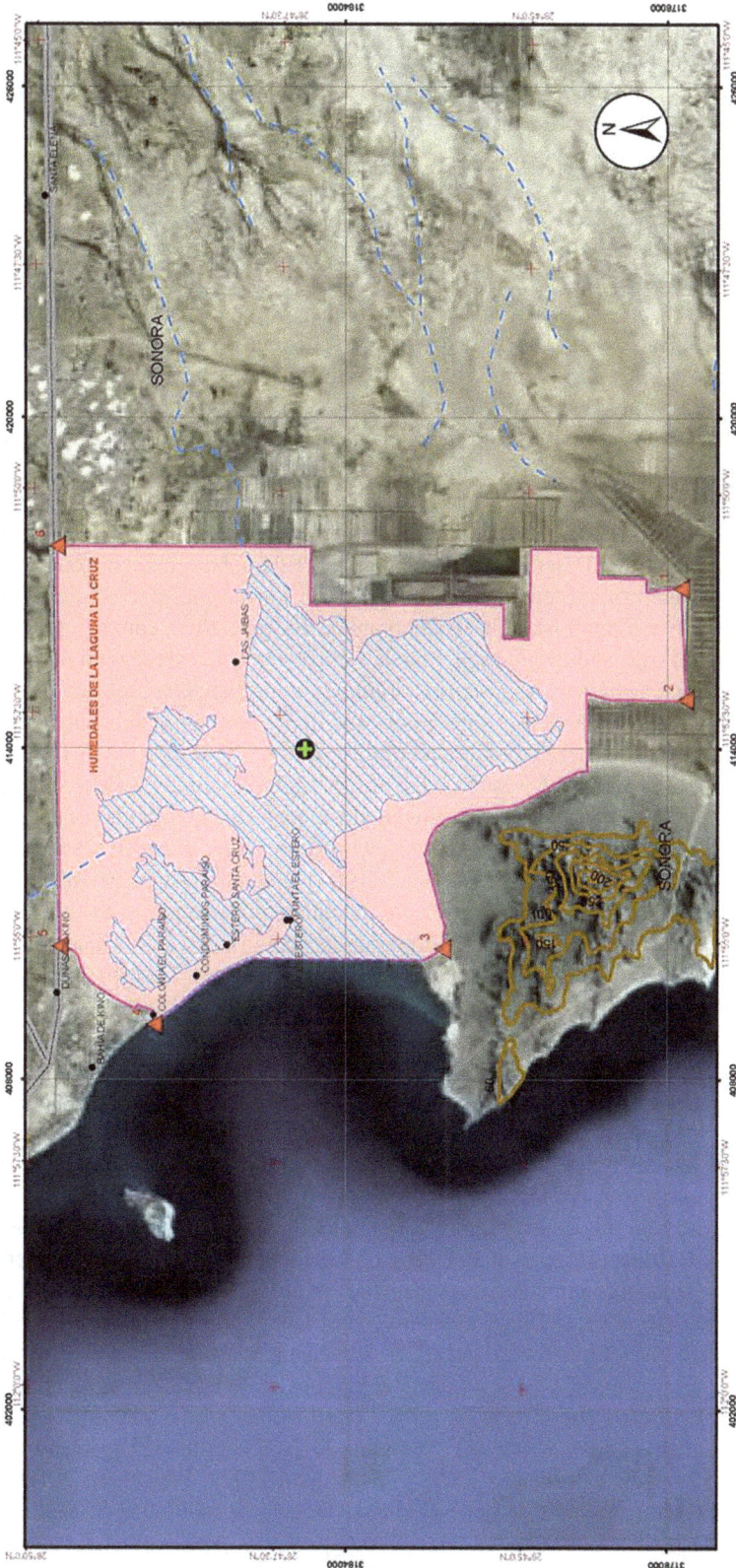

FIGURE 4.11 Map of Humedales de la Laguna La Cruz. Designated February 2, 2013; Area: 6,665 ha; Coordinates: 111°52′53″W 28°47′15″N; Ramsar Site number: 2,154.

turtle (Chelonia mydas) and the elegant tern (Sterna elegans), among others. Furthermore, the site is significant for local community members because it maintains the fisheries of the area, via its detrital processes, which is of economic importance to the people of the area. The site is also an important area for tourism and research. The main threats to the site are related to the shrimp farms located on its outskirts, and their effluent.

SITE DETAILS

Laguna La Cruz is located in the northwest of Mexico, on the east coast of the Gulf of California, in the State of Sonora, municipality of Hermosillo, just south of the fishing and tourist town of Kino Bay. Laguna La Cruz is composed of saline vegetation, mangroves, mud flats, and marshes, as well as permanent channels, and it is considered a negative estuary. Laguna La Cruz is the result of formation processes over the last 10,000 years (Moreno et al. 2005), historically the mouth of the Sonora River. Its current mouth is approximately 1.1 km wide and it is directed towards the west-southwest of Kino Bay. The site has an average depth of 1 m, with some channels to about 5 m in depth. In the estuary, there are semi-diurnal mixed tides with an amplitude of 1 m (Valdez-Holguin 1994) and at the center of the lagoon there are a variety of habitats that include extensive muddy and sandy plains inundated by tides (Grijalva-Chon et al. 1996). Throughout its channels are important areas of black mangrove (Avicennia germinans) and to a lesser extent red mangrove (Rhizophora mangle), on the perimeter. These mangrove areas are critical feeding areas and refugia for marine species and aquatic birds. Within the lagoon, there is a small island with Sonoran Desert vegetation, comprising *Pachycereus pringlei*, *Cylindropuntia* spp., and *Lycium* spp. Marshes of the site are covered with *Allenrolfea* spp., *Salicornia* spp., *Frankenia palmeri*, *Sporobolus virginicus*, and *Distichlis palmeri*. The site has a salt marsh area that is flooded only during spring tides and in the summer months, between May and September (Hannah 2008). In general, the site's mangrove community is intact, with the exception of disturbed or cut mangroves along the western edge of the lagoon. Laguna La Cruz supports larval, post-larval, and juvenile fish (99 species), including many of commercial importance (Grijalva-Chon et al. 1996). The lagoon is also a refuge for 84 waterfowl species, many of which are protected under Official Mexican Standard NOM-059-SEMARNAT-2010 (Fleishman 2011a, 2011b). Likewise, the lagoon is critical for aquifer recharge and forms a natural barrier that decreases the physical force of waves during large storm events. Laguna La Cruz maintains mangrove forests, which represent some of the northernmost limits of their distribution along the American Pacific region, and at the site are among just a few areas where mangroves are found within the Sonoran Desert. Unlike other mangrove areas, those of Laguna La Cruz depend directly on tides, given that they receive no hydrologic input from rivers, permanent streams, or heavy rain events.

Laguna La Cruz supports resting habitat for migratory waterfowl, especially for shorebirds, such as gulls and terns (Fleischner and Gates 2009). In addition, Laguna La Cruz creates a diverse landscape matrix of habitat types, such as marshes, channels, hyper-saline flats, and mud flats, which, combined, create a transitional area between upland terrestrial ecosystems that are one of the most productive areas in this desert region, with one of the most diverse marine ecosystems of the world, the Gulf of California (Gienn et al. 2006). Laguna La Cruz also supports wintering, feeding, and reproduction areas for many faunal species that are protected by national and international legislation, such as the elegant tern (Sterna elegans), least tern (Sternula antillarum), Heermann's gull (Larus heermanni), the yellow-footed gull (Larus livens), peregrine falcon (Falco peregrinus), Cooper's hawk (Accipiter cooperii), least bittern (Ixobrychus exilis), reddish egret (Egretta rufescens), brant (Branta bernicla), Yuma Ridgeway's rail (Rallus longirostris yumanensis), American oystercatcher (Haematopus palliatus frazari) [Wittman 2012]. There are 19 species of waterfowl identified as nesting species at Laguna La Cruz (Fleishman 2011a, Fleishman 2011b, Wittman 2012), among which several are under some level of risk category in NOM-059-SEMARNAT-2010, including *Rallus longirostris yumanensis*, *Haematopus palliatus frazari*, and *Ixobrychus exilis*. The endangered green sea turtle (Chelonia mydas) utilizes La Laguna La Cruz, as well as other species of sea turtles (Becerra 2012). The totoaba (Cynoscion macdonaldi) is a critically endangered marine fish at the site, as well as the sea cucumber, *Isostichopus fuscus* (Bruckner et al. 2003).

The site also supports 154 bird species, including 84 waterfowl species and ten species of herons (Fleishman 2011a, 2011b). The lagoon is located within an area at the northernmost nesting record of the roseate spoonbill (Platalea ajaja) [Fleishman and Blinick 2011]. La Laguna La Cruz, as a negative estuary, has very saline conditions, providing ideal habitat for halophytic plant species such as *Frankenia* spp. (Hannah 2008).

The site is an important breeding area for a multitude of species, both molluscs and crustaceans, ranging from microscopic species of plankton to macro-invertebrates; these larvae live in the estuary until they are ready to join the zooplankton community of the Gulf of California. The contribution of nutrients to the marine environment from Laguna La Cruz, and its role as a nursery area, contribute to the complex trophic web of the Eastern Region of the Great Islands of the Gulf of California, which supports 35 (36%) species of marine mammals. Thirty-one of these species are cetaceans of the suborders Odontoceti and Mysticeti, representing just over a third of the species of cetaceans of the world.

The first inhabitants of the Bahía de Kino region were the members of the indigenous community of the Comcaác or Seris, with archaeological vestiges in the area approximately 2,000 years old. These indigenous people inhabited the central areas of the current territory of the State of Sonora, mainly off the coast and on the islands of the Gulf of California, covering an extensive territory that was limited to the south by the Yaqui River and to the north by the Altar Desert (Sheridan 1999). The human populations of Bahía de Kino have always depended on the resources of the site, eventually spurring economic development through commercial and sport fishing (Moreno et al. 2005). There is moderate human use at the site, mainly represented by artisanal fishing activities. Families visit the estuary to camp and spend outdoor time as well, and the area naturally provides cultural and religious significance for the local people, which is evidenced in small sanctuary near the mouth of the estuary.

The Kino Bay Center for Cultural and Ecological Studies has been supported by Prescott College (in Arizona, USA), which provides youth leadership and community involvement in economic, social, and ecological development. Associated programs have been responsible for creating and maintaining an educational platform in community schools, and in collaboration with the other programs, to facilitate classes, field trips, workshops, community events, and youth events. Currently these programs reach many hundreds of young people from the ages of 6 to 17, during the school year. Some activities and projects support classes and practicums in elementary schools, including activities of ecology clubs and community cleanups (Robledo-Mejía et al. 2012).

ISLA SAN PEDRO MÁRTIR, SONORA: OVERVIEW

A small island of 127 ha off the coast of Sonora, regarded as one of the best-preserved islands in the Gulf of California, Isla San Pedro Mártir (Figures 4.12 and 4.13) is a favorite site for marine birds that hosts very large colonies of blue-footed boobies (Sula nebouxii), brown boobies (Sula leucogaster), brown pelican (Pelecanus occidentalis), and red-billed tropicbird (Phaethon aethereus). Two endemic lizards inhabit the site: *Uta palmeri* and *Cnemidophorus martyris*. There is also a colony of approximately 2,500 California sea lions (Zalophus californianus) at the site. The island is bordered by trenches more than 900 m deep, where the waters of the northern and southern parts of the Gulf meet and generate upwelling, which provides highly productive areas for sea life. Cliffs and steep slopes dominate the island, which has very little vegetation apart from a small forest of Mexican giant cactus (Pachycereus pringlei) and seasonal meadows of wild poppy (Sphaeralcea hainesii). Its large guano deposits were exploited until 1978, when the island was declared a protected area. Presently there are only temporary fishing camps on the island, and eco-tourism is still a minor activity. The altitudinal range of this small island is from 0 m to 305 m above sea level.

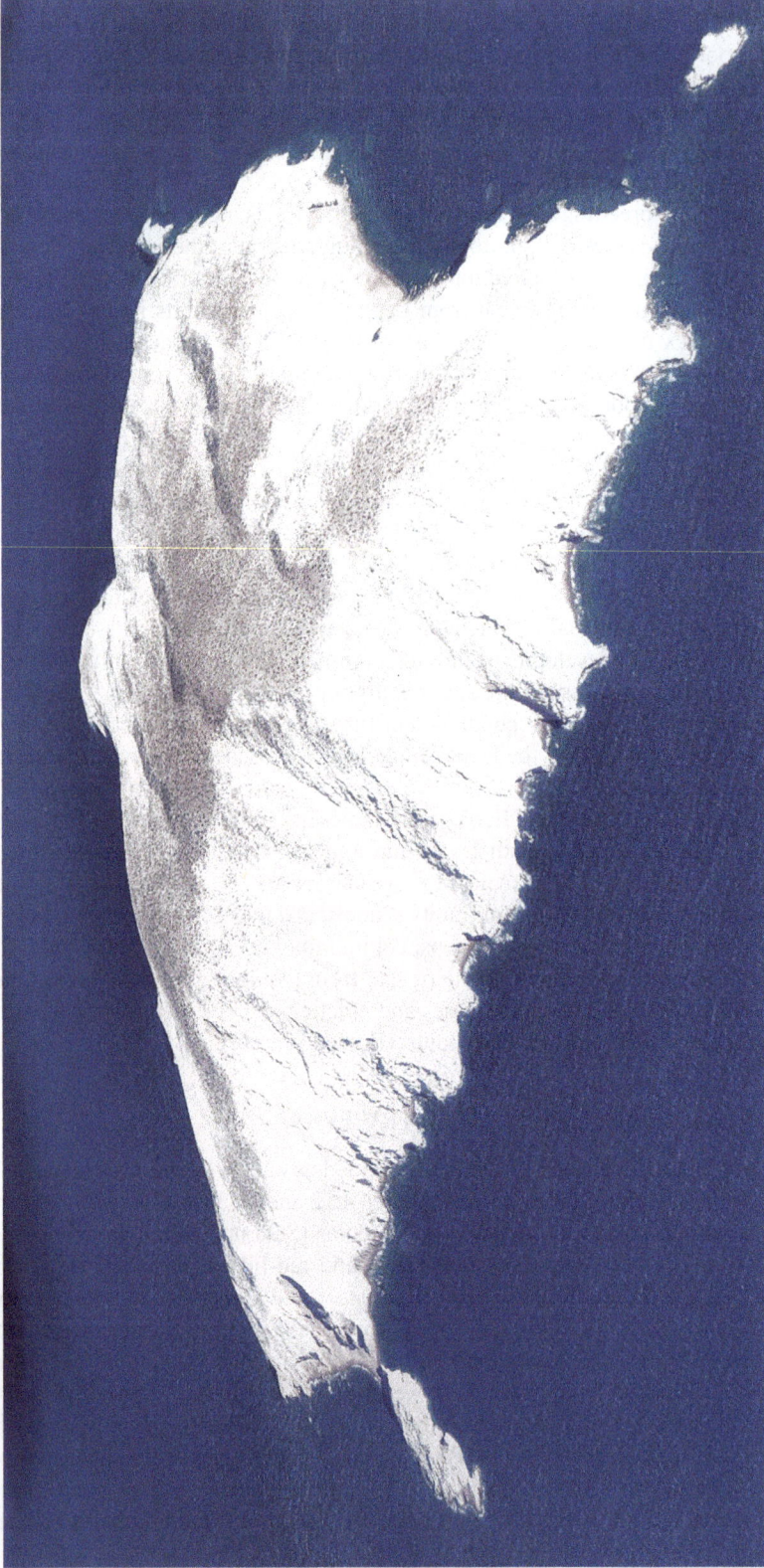

FIGURE 4.12 Composite aerial image of Isla San Pedro Mártir. Designated February 02, 2004; Area: 30,165 ha; Coordinates: 28°23'N 112°19'W; Ramsar Site number: 1,359.

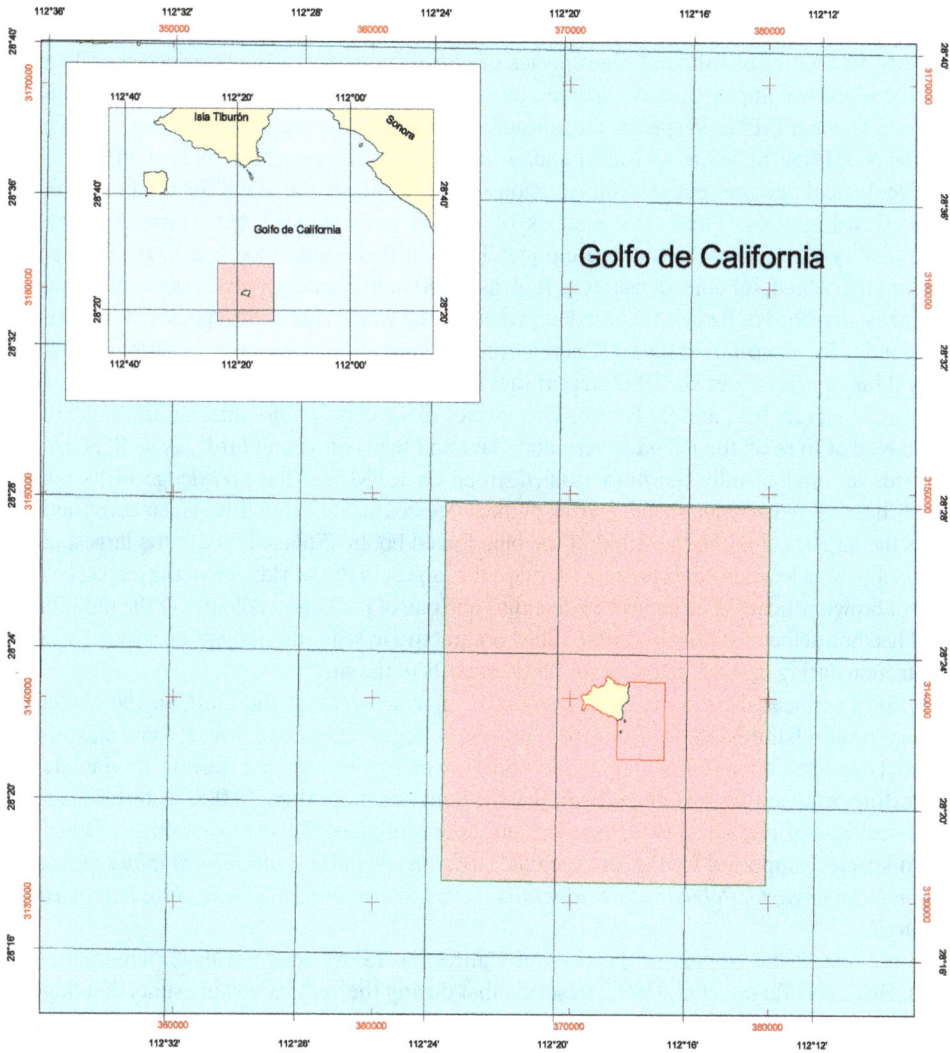

FIGURE 4.13 Map of Isla San Pedro Mártir. Designated February 02, 2004; Area: 30,165 ha; Coordinates: 28°23′N 112°19′W; Ramsar Site number: 1,359.

SITE DETAILS

The islands of the Gulf of California have been recognized worldwide as a unique ecosystem and they constitute one of the most intact archipelagos on the planet (Case and Cody 1983, Tershy et al. 1992). San Pedro Mártir Island (ISPM) can be considered one of the best-preserved sites within this great archipelago. It is the most oceanic island in the Gulf of California, since it is located more than 741 km from the coast of Sonora, a distance almost equal to the distance to the coast of Baja California. This makes the site difficult to access and therefore with a degree of human disturbance that is much less than the rest of the islands of the Gulf of California. The physical characteristics of the site place the island on the border between two distinct biogeographical areas, allowing for the observation of species with a clear tropical affinity, as well as species that dominate in temperate regions (Thompson et al. 1979, Brusca 1980).

Isla San Pedro Mártir is an extremely biologically rich site, with 27 terrestrial species of plants and 53 terrestrial bird species recorded. In the coastal marine areas of the site there are records of 36 species of seabirds, 68 species of fish, and nine species of marine mammals. The two species of lizards that inhabit the island accompany a snake species, related to rattlesnakes, which are endemic to this island. Of the total fauna of ISPM, 35 species are considered under some category of special protection, either within the NOM-059-SEMARNAT-2001 and/or -2010, by the IUCN, and/or by the CITES.

The geological age and geographic isolation of the ISPM set the stage for the evolution of two endemic lizard species. These two species of lizards have evolved behavioral, morphological, and life-history characteristics that are unique: The San Pedro side-blotched lizard (Uta palmeri), endemic to this island, inhabits densities as high as 2,200 individuals per ha (Wilcox 1980), one of the three highest densities of lizards recorded anywhere in the world. The other species of lizard endemic to the island is the whip-tailed lizard (Cnemidophorus martyris), of which very little is known about its natural history. Tershy et al. (1992) report that it feeds on the remains of fish and aerial insects, as well as beetle larvae, lice, and fly larvae. The species of western diamondback rattlesnake (Crotalus atrox atrox) that lives on the island is very abundant and feeds on young birds, as well as lizards.

Seabirds are undoubtedly the most studied group on ISPM and the knowledge of its population dynamics has allowed a better understanding of the processes that occur in this insular ecosystem. ISPM supports the largest colony in the world of the blue-footed booby (Sula nebouxii); the largest colony of brown booby (Sula leucogaster brewsteri), perhaps the largest in the world; one of the largest colonies in Mexico of brown pelicans (Pelecanus occidentalis); and one of the largest colonies of the red-billed trop-icbird (Phaethon aethereus) [Tershy 1998]. ISPM is attractive to seabirds, perhaps because of its isolation and protection during nesting, afforded by the sheer cliffs of the site.

The ISPM is located near the most productive marine areas of the Gulf, in the center of the migratory route of the Monterey sardine (Sardinops sagax caerulea), where there has also been recorded 17 species of seabirds and 10 shorebirds, which, although not nesting on the island, use it as a resting or a feeding site. In general, the site contains more than 20,000 waterbirds or 10,000 pairs of seabirds of one species or another, and approximately 500,000 shorebirds. Other impor-tant bird species supported by the site include *Larus heermanni*, *Synthliboramphus craveri*, *Sula nebouxii*, *S. leucogaste*, *Pelecanus occidentalis*, *Larus heermanni*, *L. livens*, *Sula leucogaster*, and *S. nebouxii*.

Like the rest of the islands in the Gulf of California, ISPM does not have perennial rivers or streams. However, Tershy et al. (1992) describe that during the very rare rain events that huge tribu-taries of water flow within the low areas of the topography of the island, which when falling into the sea form waterfalls of whitish water, washing away the accumulated guano and bird's nests in its path. The site is sparse in vegetation and contains 27 species of plants, with none of these species endemic to the island. The flora are dominated by a Mexican giant cactus (Pachycereus pringlei) that covers almost all of the high parts of the island. In the spring, extensive meadows of wild poppy (Sphaeralcea hainesii) grow at the site.

Main fauna species at the site include two endemic lizard species. Land birds include spar-rows (Emberizidae), doves, hummingbirds, and owls. Five species of land bird nest on the island, including peregrine falcon (Falco peregrinus), mourning dove (Zenaida macroura), common raven (Corvus corax), and mockingbird (Mimus polyglottos), which are all present throughout the region of the Sonoran Desert. The boobies, and to a lesser extent the brown pelican, are major elements of the trophic web on San Pedro Mártir. The coastal area that surrounds the island, made up of large boulders, is literally covered by one of the largest colonies of sea lions (Zalophus californianus) in the Gulf of California, whose population is estimated at around 2,500 individuals (Zavala-González 1990).

The site contains a series of stone walls that served as containment for the accumulation of guano, and remains of the walls of houses that were built by the guano extraction workers who lived on the island at the beginning of the 20th century. The archaeological value and degree of conservation of these areas are unexplored, however it is thought that the site has never been permanently inhabited by indigenous people. There are no permanent human settlements at the site, however there are temporary fishing areas along the rocky beach of Barra Baya for commercial fishing and temporary research camps (named Punta Cuervito). It is believed that the first Spaniards in the area visited the islands of the Gulf of California in 1539, under the command of Hernán Cortes and Francisco de Ulloa. However, the European explorations were very sporadic during subsequent years. Between 1720 and 1750 a small oyster pearl fishery was established in the region (Bahre 1983), which may have resulted in regular visits to Isla San Pedro Mártir. Between 1880 and 1950, the site was one of the three most important sites for sea lion hunting, from which their skin and oil were used (Nelson 1921, Bahre 1983). Thereafter, shark fishers began to use sea lions as bait, a practice that continues to this day. Sea lion hunters and other sporadic visitors to the site were thought to collect bird eggs. Guano mining activities on the island led to a complete modification of the island's surface, causing the loss of vegetational cover, and subsequent soil erosion (Tershy et al. 1992).

Presently, the surrounding area is economically supported by commercial fishing, sport fishing, and uncontrolled tourism. In adjacent marine waters, the use of bottom seine nets and illegal spearfishing has resulted in overfishing. Currently, there is a management program for all the Islands of the Gulf of California, which was originally published in 2000. This program includes San Pedro Mártir Island, under the protection of area zoning; on June 13, 2002 San Pedro Mártir Island was designated a Biosphere Reserve.

ESTERO EL SOLDADO, SONORA: OVERVIEW

Despite its relatively small size, Estero El Soldado (Figures 4.14 and 4.15) contains high biodiversity, and is thought of as a unique site among the estuaries in the Sea of Cortez for its relative health, and is yet quite representative of natural coastal wetlands of the Mexican Pacific Ocean. Three species of mangrove occur at the site, the black mangrove (Avicennia germinans), the red mangrove (Rhizophora mangle), and the white mangrove (Laguncularia racemosa). Approximately 408 species have been recorded at the site, including 121 species of invertebrates, 80 fish species, 75 bird species, 11 reptile species, 9 amphibian species, 9 mammal species, and 103 plant species. The estuary contributes to flood control during cyclones and severe storms, which generate large waves and surges. This estuary supports fisheries in the region, as well as research and educational activities.

The boundaries of the designated upland-marsh areas include El Soldado Hill and further upland approximately 500 m, toward San Francisco Bay, Sonora, which adjoins the site. The Estero El Soldado is located approximately 20 km northwest of the city of Guaymas, Sonora (approximate population 142,000), and less than 10 km southeast from the Bay of San Carlos, Sonora. The site is a complex of coastal lagoon, estuary, mangrove community, coastal dune, thorny scrub, and the littoral zone of the adjoining bay, beneath nearby Soldado Hill.

SITE DETAILS

The site supports some notable species, including the black brant (Branta bernicla nigricans), the western zebra-tailed lizard (Callisaurus draconoides inusitatus), the chuckwalla (Sauromalus obesus townsendi), the common side-blotched lizard (Uta stansburiana), Merriam's kangaroo rat (Dipodomys merriami), and the cactus mouse (Peromyscus eremicus). The above-listed lizard and terrestrial mammals are endemic. The site is home to 19 species with special protection:

- nine birds: the great blue heron (Ardea herodias), the reddish egret (Egretta rufescens), the basking falcon (Falco peregrinus), Heermann's gull (Larus heermanni), yellow-footed gull (Larus livens), wood stork (Mycteria americana), Virginia rail (Rallus limicola), least tern (Sternula antillarum), and elegant tern (Sterna elegans);
- an amphibian: the narrow-mouthed toad (Gastrophryne olivacea);
- a mollusc: Panama pearl oyster (Pinctada mazatlanica);

FIGURE 4.14 Composite aerial image of Estero El Soldado. Designated February 02, 2011; Area: 350 ha; Coordinates: 27°57′48″N 110°58′33″W; Ramsar Site number 1,982.

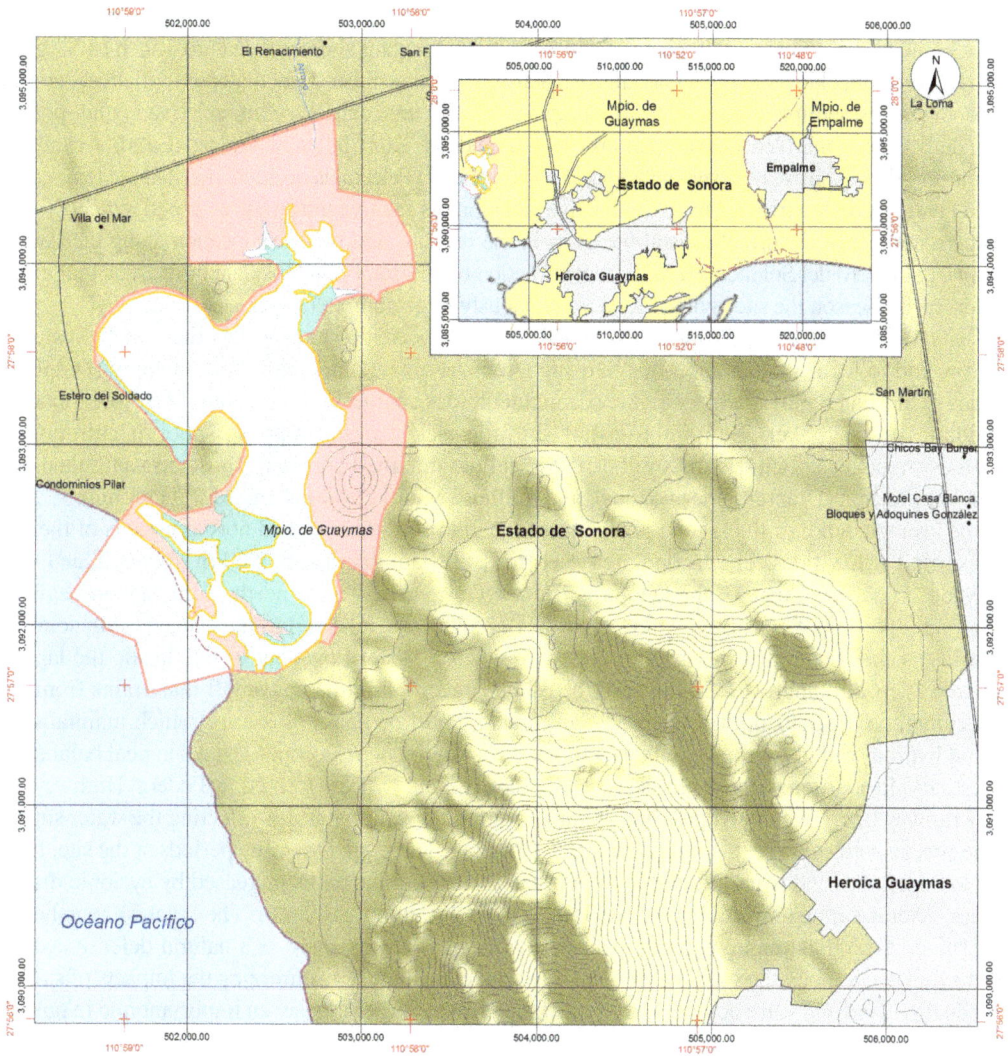

FIGURE 4.15 Map of Estero El Soldado. Designated February 02, 2011; Area: 350 ha; Coordinates: 27°57′48″N 110°58′33″W; Ramsar Site number 1,982.

- two land mammals: the desert bighorn sheep (Ovis canadensis mexicana) and the antelope jackrabbit (Lepus alleni), which is also endemic; and
- six floral species: three mangroves (Avicennia germinans, Laguncularia racemosa, and Rhizophora mangle), the Sonora guaiacum (Guaiacum coulteri), the ironwood (Olneya tesota), and the old man's head (Mammillaria thornberi).

Of the nine above bird species, two are near-threatened per the IUCN Red List: the reddish egret (Egretta rufescens) and Heermann's gull (Larus heermanni).

The estuary is temporary habitat for the growth of three species of crustaceans: two shrimp species: blue shrimp (Litopenaeus stylirostris) and yellowleg shrimp (Farfantepenaeus californiensis), and the arched swimming crab (Callinectes arcuatus). The shrimp in their post-larval stage and crabs in their megalopal stage enter the lagoon to feed and protect themselves, remaining there until their juvenile and/or adult stage, varying the dominance of each species according to the month in which they enter the lagoon.

The estuary is a place for many species of birds to seek refuge for feeding, resting, and nesting. The site is home to approximately 75 bird species, including abundant shorebird seabirds, such as Virginia rail (Rallus limicola), lapwing (Pluvialis squatarola), oystercatcher (Haematopus palliatus), curlew (Catoptrophorus semipalmatus), spotted sandpiper (Actitis macularia), whimbrel (Numenius phaeophus hudsonicus), marbled godwit (Limosa fedoa), and the short-billed sandpiper (Calidris canutus). Swimming-diving birds at the site include the brown pelican (Pelecanus occidentalis) and the long-eared shag (Phalacrocorax auritus). Due to its relative isolation, its relative intactness as a wetland ecosystem among a number of disturbed coastal wetlands in the regions, and the site's location along the Pacific Flyway, the Estero del Soldado exerts a great attraction on migratory and resident birds, such that in the winter–spring season the site is noted for a tremendously large number of species of waterfowl.

The estuary is also important habitat for dozens of species of marine fish that use the estuary to grow and to feed, such as the Mazatlán sun (Achirus mazatlanus), the short anchovy (Dolphina anchovy), northern gulf anchovy (Anchoa mundeoloides), large scale anchovy (Anchovia macrolepidota), ronco croaker (Bairdiella icistia), crevalle jack (Caranx hippos), snook (Centropomus sp.), the tongue cover (Citarichthys gilberti), the yellow-mouthed corvin (Cynoscion xanthulus), the Peruvian mojarra (Diapterus peruvianus), and mullets (Mugil cephalus and Mugil curema).

The average depth of the lagoon is 60 cm, so much of the southern and northern areas of the site are exposed at low tides. The lagoon area itself measures approximately 2.2 km in length, and 1.35 km wide. The depth range of the lagoon is from 1.6 m to 2 m. To the north of the lagoon begins a large rocky portion of coastline. A few streams, of pluvial origin, drain to the estuary, with some of them lost within the plains that lead to the lagoon, before water flows out to sea. Inside the lagoon there are islets of mangroves. To the east, the estuary receives significant runoff that drains from the Bacochibampo, El Soldado, La Ventana, San Martín, and Los Pajaritos Rivers, which maintain the natural hydrodynamics of the lagoon and whose conservation depends upon the ecological balance of the estuary. In addition to these physical factors, the artificial barriers formed by Federal Highway 15, and a railway line, have interrupted and shunted overland flow unnaturally, affecting the water supply to the site, as with some of the other estuaries of the area. There are two rainy periods at the site, from July to September and from December to February, with summer rains increased by cyclonic disturbances. Average annual rainfall at the site ranges from 200 mm to 400 mm. The brackish marshes of the estuary as well as the mangroves and other wooded wetlands serve as a natural defense against storms that occur along the coasts near Guaymas, and contribute to minimizing the impacts of storms by reducing wind and wave action. The 25 ha of mangroves at the site play an important role in purifying waters, and contributing organic matter for the detrital ecosystem services of the site.

Predominant vegetation cover at the site is scrub with various associations and ecotones, depending mainly on topography, soil type, proximity to the coast, and human impacts, with four general types of plant associations:

- submerged vegetation: dominated by seagrasses (Zostera marina and Ruppia maritima);
- coastal scrub: with the three types of mangroves described above;
- salty scrub: halophytic plants (small shrubs) predominate, located in flooded areas, adjacent to mangrove areas or deltas. The genera of plants most common of this type of vegetation are Salicornia and Batis; and
- coastal dune: includes widely distributed species that reproduce vegetatively and with high osmotic pressure such as Palmer's seaheath (Frankenia palmeri), four wing saltbush (Atriplex canescens), milky grass (Euphorbia misera), big galleta (Hilaria rigida), beach purslane (Sesuvium verrucosum), and romerito (Suaeda nigra).

There is evidence of the presence and use of this body of water and adjacent lands by ethnic groups such as the Yaquis, Seris, Low Pimas, and their ancestors (Bowen 1965, Thomson 1973, Findley 1976). Five sites known as conchales have been found, which are areas with large amounts of mollusc shells and fish bones, used as food by native indigenous populations. The importance of the shells is historical-cultural since they add to key aspects of the paleoecology of the estuary. The nomads, or Comcaác, have lived for 2,000 years on the central coast of the Sonora Desert, Isla Tiburón, Isla San Esteban, and other nearby locales, using the Gulf of California as a source of food and for seafaring transportation. The Guaímas were hunter-gatherers from the Comcaác group who traveled the coast from El Cochorit to San Carlos, in Bacochibampo and San José de Guaymas, obtaining their food (clams, oysters, crabs, fish, and other marine species) along the way of their journeys. After the Spanish conquests, the Comcaác were decimated, along with their territory, almost to the point of extermination at the beginning of the 20th century. Currently the Comcaác community is settled in Punta Chueca, Hermosillo, El Desemboque, and Pitiquito with about 1,000 inhabitants on Isla Tiburón, and part of the nearby mainland. The Infiernillo Channel and the coasts of Isla Tiburón have been officially recognized by presidential decree as an exclusive fishing zone of the Comcaác People and communities.

Hydrologic works, such as water projects, and urban growth near the site have had substantial impact due to activities, such as tourism and other impacts related to development. Notable research activities at the site include those by the Instituto Tecnológico y de Estudios Superiores de Monterrey, the University of Arizona, the Technological Institute of the Sea (ITMAR), the University of Sonora, the Center of Higher Studies of the State of Sonora (CESUES), the National College of Professional Technical Education (CONALEP), and many other institutions that in one way or another have used the site for educational purposes. Environmental education activities and visitor infrastructure exist at the site, with conservation/recreational programs focused on understanding and awareness of the concepts surrounding the sustainable use of the area.

COMPLEJO LAGUNAR BAHÍA GUÁSIMAS-ESTERO LOBOS, SONORA: OVERVIEW

A wetland ecosystem located along the northwest coast of Mexico, Complejo Lagunar Bahía Guásimas-Estero Lobos (Figures 4.16 and 4.17) has mangrove areas comprised of *Avicennia germinans*, *Laguncularia racemosa*, and *Rhizophora mangle*. An important hibernation area for aquatic migratory and coastal birds, the site supports 4% of the aquatic migratory bird populations and approximately 10% of the coastal birds observed in the northern Pacific coastal zone, annually. Among the bird species found at the site are *Egretta rufescens*, *Rallus longirostris*, *Rallus limicola*, *Sternula antillarum*, *Grus canadensis*, and *Branta bernicla*. Mammal species such as *Zalophus californianus californianus*, *Tursiops truncatus*, *Globicephala macrorhynchus*, *Delphinus delphis*, and *Myotis vivesi* are also observed at the site. The site's bays and estuaries are reproduction, nursery, and development habitat for species such as the blue shrimp (Penaeus stylirostris). Fishing (for shrimp and oyster), agriculture, hunting, and extensive livestock activities are the most common of human activities at the site.

SITE DETAILS

The site is located in the State of Sonora, comprising part of the municipalities of San Ignacio Río Muerto, Guaymas, and Empalme, and is located 27 km west of Ciudad Obregón (approximate population of 433,000) and 23 km to the east of Guaymas (approximate population of 113,000). Average altitude at the site is approximately 7 m, ranging from sea level to a maximum of 105 m. The Bahía Guásimas-Estero Lobos Lagoon Complex contain a multitude of lagoons that are distributed along the coastline, including Estero Tosalcahui, Estero el Colorado, Estero el Bosque, Estero Lobitos, Estero Lobos, Estero la Culebra, Estero la Piedrita, Estero el Escondido, Estero las Arenitas, Estero Guaycan, Estero la Pitahayita, Estero San Francisquito, Estero Luna, Estero Bairo, Estero el Siuti, Estero Camapochi, Estero los Algodones, Estero los Tecolotes, Estero las Cruces, Estero Mapo, and Estero Bachoco. These coastal lagoons are associated with fluvial deltaic ecosystems, produced by irregular sedimentation or surface subsidence that originates from the effect of compaction load, in

FIGURE 4.16 Composite aerial image of Complejo Lagunar Bahía Guásimas-Estero Lobos. Designated February 02, 2008; Area: 135,198 ha; Coordinates: 27°32′N 110°29′W. Ramsar Site number 1,790.

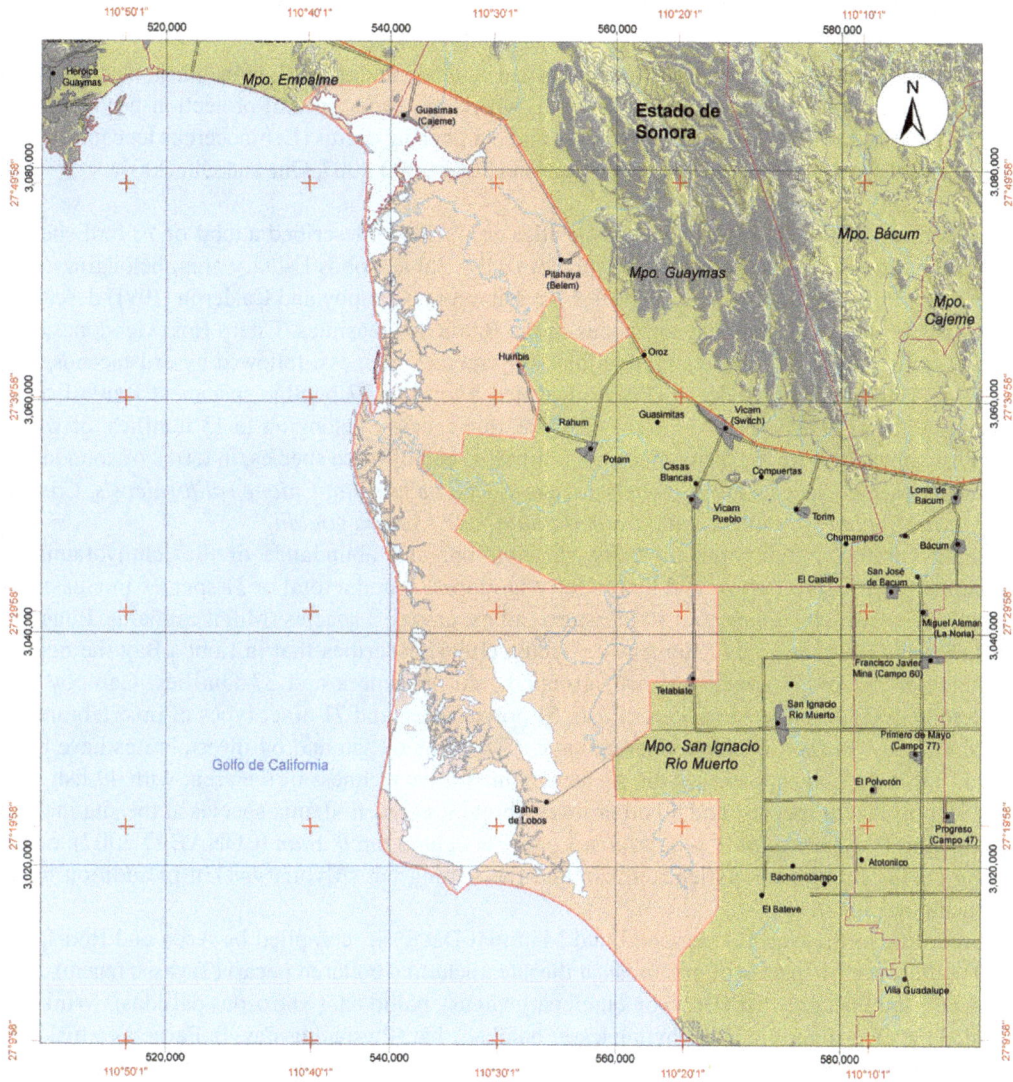

FIGURE 4.17 Map of Complejo Lagunar Bahía Guásimas-Estero Lobos. Designated February 02, 2008; Area: 135,198 ha; Coordinates: 27°32′N 110°29′W. Ramsar Site number 1,790.

addition to depressions formed by non-marine processes during reduced sea level. The site is dominated by emergent low-coastal vegetation and coastal floodplain areas, as well as mangroves.

Species of concern at this site, per NOM-059-SEMARNAT-2010, are (CONABIO 2007, Arreola 1995, Scott and Carbonell 1986):

- Birds: reddish egret (Egretta rufescens), mangrove rail (Rallus longirostris), Virginia rail (Rallus limicola), elegant tern (Sterna elegans), least tern (Sternula antillarum), sandhill crane (Grus canadensis), Heermann's gull (Larus heermanni) [all subject to special protection], and brant (Branta bernicla), considered threatened;
- Mammals: common bottlenose dolphin (Tursiops truncatus), long-finned pilot whale (Globicephala macrorhynchus), common short-nosed dolphin (Delphinus delphis), California sea lion (Zalophus californianus californianus), Sonoran woodrat (Neotoma phenax),

fisher-myotis-bat (Myotis vivesi) [all subject to special protection], and the Curaçao long-nosed bat (Leptonycteris curasoae), considered threatened; and

• Plants: black mangrove (Avicennia germinans), white mangrove (Laguncularia racemosa), and red mangrove (Rhizophora mangle), which are under special protection per NOM-059-SEMARNAT-2001 (Arreola 1994), and clambering cactus (Echinocereus leucanthus), a floral species subject to special protection (CONABIO 2007, Ohr and Ohr 2007).

Palacios and Mellink (1995) and Valdés-Casillas et al. (1997) described a total of 70 bird species at the site. Arreola (1995) described 56 species in the Bahía Lobos Estuary area, belonging to 35 genera and 15 families, of which 35 species are migratory. Campoy and Calderón (1991) described a total of 106 species of benthic organisms in the Bahía de Guásimas, Estero Los Algodones, and Bahía Lobos, where polychaetes predominated by species richness, followed by crustaceans, and molluscs. Enríquez and Calderón (1990) identified a total of 97 benthic species at Bahía Lobos. Audeves et al. (1997) described a total of 45 benthic species, belonging to 13 families, of which the most abundant were the Veneridae and Tellinidae, and of these species, in terms of abundance, the most observed species are: *Chione subrugosa, Chione compta, Chione californiensis, Cardita laticostata, Tellina straminea, Anadara perlabiata, and Lyonsia gouldii.*

Yepiz (1990), studying the diversity, distribution, and abundance of the ichthyofauna of Guásimas Bay, Estero Lobos, and Estero los Algodones, found a total of 31 species in Guásimas Bay, 47 in Estero Algodones, and 49 in Estero Lobos, among 3 species (Mugil cephalus, Eugerres axillaris, and Eucinostomus entomelas). Arreola (1995) describes that in Lobos Bay the nekton is characterized by 74 species of fish, grouped into 60 genera and 37 families. Campoy and Calderón (1993) describe 31 species of fish, 51 polychaetes, and 71 other types of invertebrates in Guásimas Bay. In the Los Algodones estuary, 47 species of fish and 64 invertebrates have been noted. Bahía de Lobos possesses the greatest comparative richness of the area, with 49 fish species, 96 polychaete species, and 86 other invertebrate species. Endemic species at the site include macroalgae *Chondracanthus squarrulosus and Eucheuma uncinatum* (CONABIO 2007), clambering cactus (Echinocereus leucanthus), and the fishing bat (Myotis vivesi), in addition to 12 ichthyofaunal species.

According to the North American Land Mammal Database, compiled by Arita and Rodríguez (2004), there are 24 species of mammals at the site, including collared pecari (Tayassu tajacu), coyote (Canis latrans), gray fox (Urocyon cinereoargenteus), pallid bat (Antrozous pallidus), wrinkled-bearded bat (Mormoops megalophylla), lesser baldback bat (Pteronotus davyi), Parnell's whiskered bat (Pteronotus parnellii), Mexican funnel-eared bat (Natalus stramineus), North American porcupine (Erethizon dorsatum), Botta's pocket gopher (Thomomys bottae), white-tailed deer (Odocoileus virginianus), Virginia opossum (Didelphis virginiana), northern desert shrew (Notiosorex crawfordi), antelope jackrabbit (Lepus alleni), desert rabbit (Sylvilagus audubonii), river otter (Lontra longicaudis), American hog-nosed skunk (Conepatus mesoleucus), skunk skipjack (Mephitis macroura), long-tailed weasel (Mustela frenata), American badger (Taxidea taxus), northern cacomixtle (Bassariscus astutus), white-nosed coati (Nasua narica), common raccoon (Procyon lotor), and lynx (Lynx rufus), in addition to a variety of marine mammals like the common bottlenose dolphin (Tursiops truncatus), long-finned pilot whale (Globicephala macrorhynchus), short-beaked common dolphin (Dolphinus delphis), and California sea lion (Zalophus californianus californianus) [CONABIO 2007]. Among the site's fish species are mullet (Mugil cephalus), crappie (Eugerres axillaris), and *Eucinostomus entomelas*, in addition to the following endemic species for the Lagoon Complex of Guásimas Bay-Lobos Estero and its surroundings: *Acanthemblemaria crockeri, Chaenopsis alepidota, Tomicodon boehlkei, and Stathmonotus sinucalifornici* (Yepiz 1990, Campoy and Calderon 1993, Arreola 1995, Castro et al. 2005).

The site is located in the Sonora Norte Hydrological Region, in the Río Yaqui Basin. The Yaqui River has an average annual volume of approximately 2,800 million cubic meters, which is used to irrigate a wide area of cultivated land, approximately 450 thousand ha, in the Yaqui Valley. In this watershed, the site's wetlands constitute an important physical regulator of regional microclimate, predominantly characterized by arid and semi-arid upland conditions (CAN 2003).

The site is within the territory of the Yaqui Yoreme Nation, and a large part of the wetland is within Mexico's fishing zone. The fishing fleet of San Ignacio Río Muerto, a municipality where Estero Lobos is located, is made up of eight fishing cooperatives, with 200 small vessels; the main fishing towns in the area are Bahía de Lobos and Los Médanos. One of the fishing cooperatives belongs to members of the Yaqui Yoreme Nation (Secretaría de Gobernación, 2004). Within the site, the Yaqui depend on artisanal fishing, exploiting various invertebrates and fish, but the main resource is shrimp and crab (Luque and Gómez 2007). The Yaqui belong to the group of Cahítas, who arrived in the territory of Sonora, coming from the Gila River in 1300 AD. In 1607, they had their first contact with the Spanish, and through conflict were predominantly deported to Yucatán with the remaining population of 3,000 protected by landowners in the region. Since 1911, they gradually returned to the State of Sonora, and in the 1930s the ethnic group was granted control of their lands, incorporating it into the national system that recognizes the legitimacy of Yaqui traditions (Olavarría 1995).

Current land use (including water use) at the site includes fishing, aquaculture, hunting, livestock, irrigated agriculture, a variety of urban uses, fishing, aquaculture (mainly shrimp and oyster), and salt exploitation. Bahía Lobos is subject to intense fishing exploitation and receives considerable amounts of agro-industrial and domestic waste from the catchment area of the Yaqui River Basin (Calderón and Campoy 1993). Another adverse impact on the site and area comes from the destruction of mangroves for the creation of shrimp farms (Cervantes 1994). Villegas et al. (1985) describe a high number of samples within the site for associated pesticides, such as lindane, aldrin, heptachlor, and DDT, and most frequently diazinon, dieldrin, and parathion. Additionally, a number of researchers are working at the site, based at local universities and with the US Fish and Wildlife Service. There is local hunting and recreational use, and a general absence of other organized tourism at the site.

HUMEDALES DE YAVAROS-MORONCARIT, SONORA: OVERVIEW

Humedales de Yavaros-Moroncarit (Figures 4.18 and 4.19) is a lagoon complex that is critical habitat for a number of species including yellow-footed gull (Larus livens), peregrine falcon (Falco peregrinus), and the black brant (Branta bernicla nigricans). There are an additional 66 species of birds at the site that are on priority conservation lists under the North American Wetlands Conservation Act (NAWCA) and the Neotropical Migratory Bird Conservation Act (NMBCA). Each year over 50,000 shorebirds visit the site's marshes, mud flats, and mangroves, including important wintering sites for 47,000 ducks, geese, and other waterfowl. Due to its high biodiversity and availability for human use, the entire region of the site, the southern coast of Sonora, is a marine priority area for Mexico. The gray whale (Eschrichtius robustus), killer whale (Orcinus orca), and sea lion (Zalophus californianus) are among marine mammals observed here. Adverse factors affecting the site include uncontrolled fishing and tourism activities. The most common land uses within the site include agriculture and ranching. Due to its variety of resource conditions and biological diversity, the southern coast of Sonora hosts 17 of the marine priority areas of Mexico, which were established in 1998.

SITE DETAILS

The site possesses biophysical conditions to support approximately 287 species, of which 42% are marine invertebrates, 7% are fish, 23% birds, 1.4% algae, 31% higher plants, and less than 1% are mammals (Valdés et al. 1994). The site provides important wintering areas for 47,000 ducks, geese, and other waterfowl, including Anas acuta, Spatula discolor, Anas americana, Sternula antillarum, Ardea herodias occidentalis, Sula nebouxii, Egretta rufescens, and Platalea ajaja (Valdes et al. 1994).

Within the site there is Yavaros Bay, a shallow lagoon with an average depth of two meters and a total volume of 154 million cubic meters (Gilmartin and Revelante 1978), and with a well-developed system of tidal channels, which go from the mouth (with depths of 10 m) to the center of the lagoon.

FIGURE 4.18 Composite aerial image of Humedales de Yavaros-Moroncarit. Designated February 02, 2010; Area: 13,627 ha; Coordinates: 26°43′39″N 109°31′00″W; Ramsar Site number: 1,984.

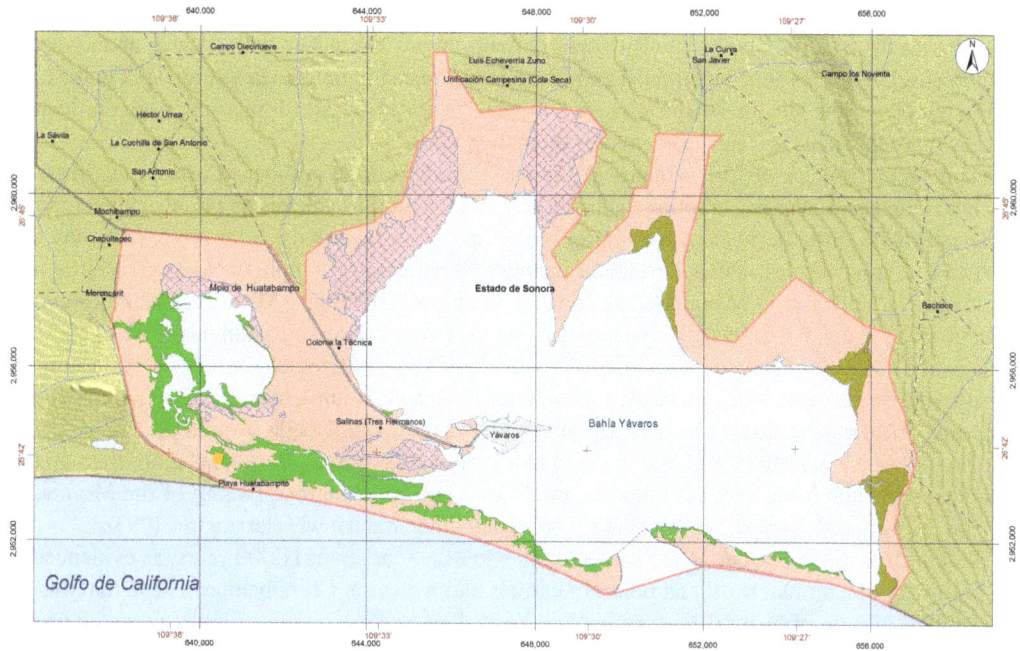

FIGURE 4.19 Map of Humedales de Yavaros-Moroncarit. Designated February 02, 2010; Area: 13,627 ha; Coordinates: 26°43′39″N 109°31′00″W; Ramsar Site number: 1,984.

This network of channels, together with the width and depth of the mouth, has allowed the permanent transit of larger vessels (shrimp and sardine boats) to the port of Yavaros. Yavaros Bay is within the Río Mayo Basin, with surface runoff that runs 294 km from its source in the Sierra Madre Occidental to its mouth in the Gulf of California (INEGI 1993). The climate of the site is dry, with the hottest months of July, August, and September reaching an average maximum temperature of 30 °C; coldest months are December, January, and February with minimum average temperature of 16.3 °C.

Terrestrial vegetation in the floodplain and lands adjacent to the coast is halophytic scrub, which represents 33% of the entire terrestrial area. Between halophytic vegetation and agricultural areas, especially around Yavaros Bay, there are areas of cacti that predominate the xeric scrub plant community. Inhabitants of Moroncarit use the local red mangrove as fuel.

Quite a few studies of marine fish and invertebrates have been conducted in the area (Gómez et al. 1974) as well as on feeding and sheltering habitat for shorebirds, and wintering ground for ducks, geese, and other waterfowl. Orcas (Orcinus orca) and sea lions (Zalophus orca) have been recorded in Yavaros Bay, and rare observations of individual jaguars (Panthera onca) have been periodically recorded close to the mangrove swamps, just several meters from the shoreline.

Fishing is an important element of the people in the community of Yavaros, focusing on harvesting shrimp, crab, clams, Maura pen shell, Chinese snail, tilapia, chihuil, sole, sierra, milkfish, shark, and manta, with a number of shrimp and sardine boats originating from Yavaros. Pond-based shrimp farming on land has increased in the area, due to the decrease in offshore catches, both for shrimp and sardines, leading to the modification of environmental conditions in some of the largest areas of the wetlands of the site, which are also critical areas for the natural shrimp fishery (SEPESCA 1987).

The main floral species of the site are halophyte scrub, characterized by the presence of red sand verbena (Abronia maritima), shrubby seablite (Suaeda fruticosa), dwarf saltbush (Atriplex barclayana), iodine bush (Allenrolfea occidentalis), and saltgrass (Distichlis spicata). Between the halophyte vegetation and the agricultural areas, especially around Yavaros Bay, cacti predominate, such as prickly pear cactus (Opuntia sp.), cholla (Opuntia cholla), and organ-pipe cactus (Pachycereus pecten-aboriginum and Stenocereus thurberi). In various areas of the site and in the channel that joins it with the bay, red mangrove (Rhizophora mangle) predominates, with a lesser abundance of black mangrove (Avicennia germinans) and white mangrove (Laguncularia racemosa). On the bank and in ditches of the site, there are salt pine (Tamarix sp.) and cattail (Typha sp.), each an invasive species that represents serious problems for the native flora.

Main faunal species of the wetlands of the site include a large number of arthropods (insects, arachnids, and crustaceans), amphibians (frogs and toads), reptiles (snakes and turtles), fish (marine and freshwater), birds (terrestrial and shore), small and large mammals, from mice, raccoons, jaguars, to the great whales. Other key fauna include desert tortoise (Gopherus agassizii), little toad (Gastrophryne olivacea), Mexican rose boa (Lichanura trivirgata), Mexican black kingsnake (Lampropeltis getula nigritus), Arizona coral snake (Micruroides euryxanthus), West Mexican coral snake (Micrurus distans), Mexican west coast rattlesnake (Crotalus basiliscus), gila monster (Heloderma suspectum), Mexican beaded lizard (Heloderma horridum), Mexican bat (Choeronycteris mexicana), field rat (Neotoma phexna), jaguar (Panthera onca), royal eagle (Aquila chrysaetos), great blue heron (Ardea herodias), and blue-winged chickadee (Anas acuta). Each year more than 50,000 individuals of shorebirds visit the marshes, mud flats, and the mangrove swamp of the Moroncarit Lagoon, as well as 47,000 wintering ducks, geese, and other waterfowl (Harrington 1993).

Humans have been present on the banks of the Río Mayo for about 11,000 years, as evidenced by the discovery of a human skull and bones of camels and wolves at Chinobampo. These remains are now at the Museum of Natural History in New York. There are also indications of nomadic groups who lived near the estuaries during the first millennium AD, based on the presence of shell piles and deposits of marine animal remains. Today, Yavaros, a community of around 4,000 people, is primarily focused on agriculture and fishing. Meanwhile, Moroncarit, with a population of 1,350, is located at the edge of the Estero Moroncarit and near the Valley of Mayo's agricultural fields. Approximately 73% of the population there also work in agriculture or fishing.

The region faces several environmental challenges, including a decline in fishing resources, conflicts over natural resource management between the tourism sector and unregulated hunting, and a notable decline in the mangrove species Avicennia germinans, likely due to the impacts of shrimp farms, changes in groundwater patterns, and the presence of a newly identified beetle that causes harm to the tree species. The communities of Huatabampo and Yavaros have expressed their opposition to hunting activities that harm fishing, the mangrove forest, and provide no benefit to the communities. These activities are causing an increase in the fragmentation and destruction of the wetland and mangrove ecosystem, along with modifications to the hydrological patterns of the site, as well as contamination from waste, wastewater, and sewage. There are also signs of overexploitation of game species and mortality of non-game species at the site. The Gulf of California's diverse habitats and biological diversity have suffered intense exploitation for 50 years, leading to the decline of rare species and the extinction of many commercially important fish species. Extensive agriculture, livestock, and aquaculture developments are causing desertification, and overexploitation of water is making the soil infertile for agriculture.

A multidisciplinary group has been focused on the environmental well-being of the Yavaros-Moroncarit wetland in recent years and has been tasked with conducting studies to conserve and better understand the ecological processes of the site. Key people from the communities of Huatabampo and Yavaros, as well as local officials, have expressed interest in reorganizing economic activities and protecting the wetland. Pronatura-Noroeste has offered to support and advise on the activities in the area with their extensive experience in conservation. In 2006, Pronatura-Noroeste started a project with the Mexican Fund for Nature Conservation and the local non-profit conservation organization Mangle Negro. This Protection Initiative and community development project has a strong communication component and has involved education and awareness events, conflict analysis workshops, and the development of a logo and slogan for the lagoon. A wealth of information and policy instruments are available to implement a comprehensive and sustainable community development scheme. Various forms of tourism take place in Moronacrit Lagoon, including hunting, kayaking, birdwatching, panga rides, and off-road adventure tourism. However, there is community opposition to hunting tourism, as it does not benefit the community. The lagoon is also visited by regional tourists during Holy Week for activities like hiking and camping.

LAGUNAS DE SANTA MARÍA-TOPOLOBAMPO-OHUIRA, SINALOA: OVERVIEW

Lagunas de Santa María-Topolobampo-Ohuira (Figures 4.20 and 4.21) is a complex of three coastal lagoons, with a total of eight islands: six in Ohuira Bay, one in Topolobampo Bay, and one in Santa María Bay. Mangroves at the site include *Rhizophora mangle, Laguncularia racemosa, Avicennia germinans*, and *Conocarpus erectus*. The site is home to 84% of the migratory waterfowl that make their way across Mexico during the winter. The site is subject to flooding and storms caused by tropical cyclones that regularly occur in the area. Among the potential factors at the site that could cause degradation in water quality, and the landscape as a whole, are the large amounts of waste-water discharges, especially agricultural runoff, to the coastal zone. Fishing is the most important human use in the area.

The site is located within the Area of Protection of Flora and Fauna "Islas del Golfo de California," a World Heritage Site and UNESCO Biosphere Reserve. The site is located in the northwestern region of Mexico, in the north of the State of Sinaloa, in the municipality of Ahome, which includes three coastal lagoons at Santa Maria, Topolobampo, and Ohuira. The closest city is Los Mochis (approximate population of 331,000), located 20 km east of Topolobampo. Altitude of the site ranges from 0 m to 35 m. Among the islands at the site are species of special concern, including *Egretta rufescens*, which nests in the region and is subject to protection by NOM-059-SEMARNAT-2001 and/or -2010. Within lagoon ecosystems, and in the adjacent seas, feeding and breeding areas for the olive ridley sea turtle (Lepidochelys olivacea), green sea turtle (Chelonia mydas), leatherback sea turtle (Dermochelys coriacea), and hawksbill sea turtle (Eretmochelys imbricata) are also all subject to special protection by NOM-059-SEMARNAT-2001 and/or -2010 and in the IUCN Red List. Flora of the islands, protected under NOM-059-SEMARNAT-2001 and/or -2010 as well as the IUCN Species Survival Commission, also include mangroves (Rhizophora mangle, Laguncularia racemosa, Avicennia germinans, and Conocarpus erectus), with special protection status ensuring their ecological functions during the phases of reproduction for numerous species of commercially important species, such as shrimp, oysters, and various fish species. Other floral species of interest at the site are the cacti species *Peniocereus marianus*; the site is also the southernmost distribution of the cactus *Lophocereus schotii*. Other important species that occur on the islands of the site are the Mexican yellow show (Amoreuxia palmatifida) and the Sonora guaiacum (Guaiacum coulteri), as well as other notably beautiful cacti: *Ferocactus townsendianus* var. *townsendianus, Mammillaria dioica, Mammillaria mazatlanensis, Opuntia bergeriana, Opuntia fulgida, Opuntia puberula, Opuntia rileyi, Opuntia spraguei, Opuntia wilcoxii, Pachycereus pecten-aboriginum, Stenocereus alamosensis*, and *Stenocereus thurberi*. Additionally, the site supports habitat for the endemic cactus *Echinocereus sciurus* var. *floresii*.

SITE DETAILS

Overall, the site supports about 109 species, of 76 genera, belonging to 45 families (Balart et al. 1992, Gutiérrez-Barreras 1999). The site originated during the Pleistocene as a product of the action of the old delta of the Río Fuerte on the rocks of the Sierra Navachiste (Phleger and Ayala-Castañares 1969). During that time the area was tectonically active, which gave rise to the current Sierra Navachiste. From inland, the Fuerte River poured its waters into the sea just to the north of Topolobampo, giving rise to a delta at its mouth, forming a bar that limits the Bay of Colorado and the Santa María Lagoon, currently called "Isla Santa María" (Olivares-Beltrán 1969, Phleger and Ayala-Castañares 1969). Topolobampo Bay is characterized by having several coves and points, flowing through a 700-m-wide channel to the Bay of Ohuira, and subsequently out to the Gulf of

FIGURE 4.20 Composite aerial image of Lagunas de Santa María-Topolobampo-Ohuira. Designated February 02, 2009; Area: 22,500 ha; Coordinates: 25°36′33″N 109°06′23″W; Ramsar Site number: 2,025.

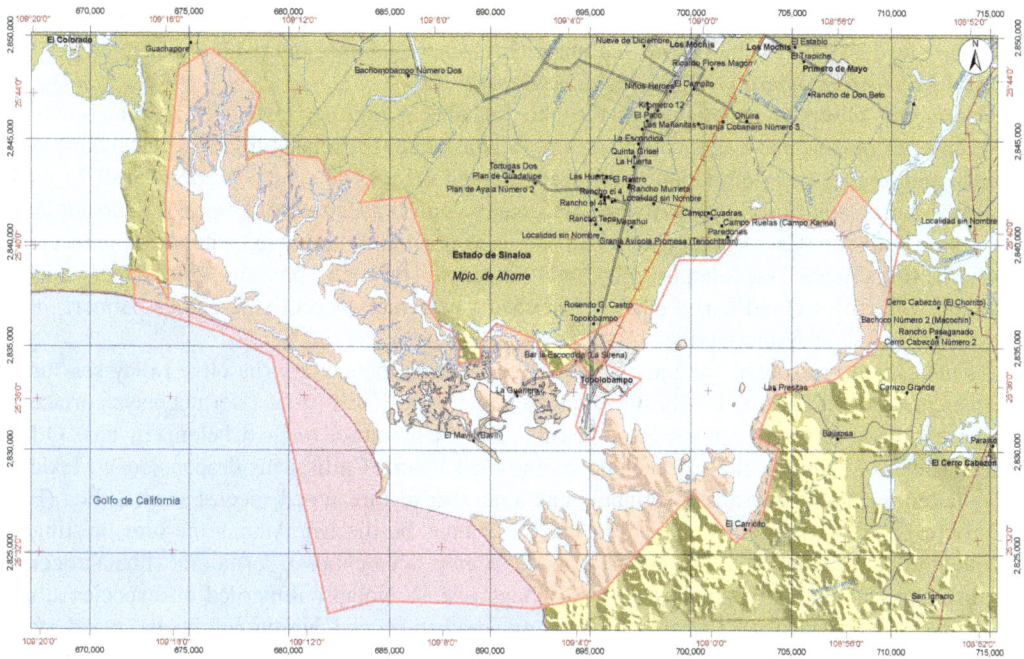

FIGURE 4.21 Map of Lagunas de Santa María-Topolobampo-Ohuira. Designated February 02, 2009; Area: 22,500 ha; Coordinates: 25°36′33″N 109°06′23″W; Ramsar Site number: 2,025.

California (Olivares-Beltrán 1969). The Bay of Santa María is the smallest of the three lagoon bodies at the site, with an area of approximately 40 km² (Phleger and Ayala-Castañares 1969, Escobedo-Urías 1997). To the west, the bay is separated from the Gulf of California by a barrier island (Isla Santa María) that is 21.7 km by 1.6 km wide. Ohuira Bay is the largest coastal embayment in the area, with an area of approximately 125 km².

The Río Fuerte is one of the most important hydrological resources of the North Pacific slope, and it plays an important role in the site's ecology. Its origin is located in the Sierra Tarahumara, from which it flows through the Municipality of Ahome, from northwest to southwest, from San Miguel Zapotitlán, then passing through Higuera de Zaragoza, on to the Gulf of California. Flowing downstream, these surface waters make their way through a multi-use dam system, "Luis Donaldo Colosio," thereby augmenting the once-natural hydraulic regime of the river.

Along the Bahía Santa María estuary, coastlines are extensive mangroves. The submerged flora of the areas, a bit further off the shore, are represented by *Thalassia* sp. and various species of macroalgae. Aguilar and López (1985) recorded the presence of a population of *Halodule wrightii* in Topolobampo Bay, 490 km from the first and only record in Punta Chueca, Sonora, occurring sympatrically with *Zostera marina* and *Ruppia maritima*. The site also supports several small terrestrial vertebrates such as the raccoon, kangaroo rat, coyotes, black iguana, and rattlesnakes, as well as marine vertebrates such as the California sea lion (Zalophus californianus), common bottlenose dolphin (Tursiops truncatus), and numerous species of fish. The site is a very important area for nesting and resting for birds, such as cormorants, hawks, pelicans, and herons, as well as other species subject to protection under NOM-059-SEMARNAT-2001 and/or -2010, such as Heermann's gull (Larus heermanni), the reddish egret (Egretta rufescens), wood stork (Mycteria americana), yellow-footed gull (Larus livens), the lesser tern (Sternula antillarum), and the elegant tern (Thalasseus elegans). The site's lagoons and the adjacent marine areas are important for feeding and development of four species of sea turtle: the olive ridley sea turtle (Lepidochelys olivacea), the leatherback (Dermochelys coriacea), the green sea turtle (Chelonia mydas), and the hawksbill sea turtle (Eretmochelys imbricata), all endangered per the Official Mexican Standard (NOM-059-SEMARNAT-2001).

Commonly observed vegetation at the site includes mangrove trees (Rhizophora mangle, Avicennia germinans, and Laguncularia racemosa); mesquite (Prosopis juliflora); coastal dune vegetation; and to a lesser extent halophytic vegetation and xerophytic scrub (Rzedowski 1978), such as agave (Agave angustifolia), brazilwood (Haematoxylum brasiletto), torote prieto (Bursera laxiflora), palo colorado (Caesalpinia platyloba), palo brea (Cercidium praecox), pitahaya (Stenocereus thurberi), sina (Stenocereus alamosensis), wild pineapple (Bromelia pinguin), prickly pear cactus (Opuntia wilcoxii), twisted barrel cactus (Ferocactus herrerae), bledo (Amaranthus palmeri), and Mexican yellowshow (Amoreuxia palmatifida).

As mentioned previously, the most noted species within the site are the olive ridley sea turtle (Lepidochelys olivacea), the critically endangered leatherback sea turtle (Dermochelys coriacea), the hawksbill sea turtle (Eretmochelys imbricata), and the green sea turtle (Chelonia mydas). Other types of reptiles found at the site include zebra-tailed lizard (Callisaurus draconoides), Mexican spiny-tailed iguana (Ctenosaura pectinata), and a number of threatened species, such as boa (Boa constrictor) and Mexican west coast rattlesnake (Crotalus basiliscus). Among the birds nesting at the site, the most abundant are brown pelican (Pelecanus occidentalis), cormorant (Phalacrocorax auritus), and magnificent frigatebird (Fregata magnificens). Notably imperiled bird species at the site include Heermann's gull (Larus heermanni), reddish egret (Egretta rufescens), wood stork (Mycteria americana), yellow-footed gull (Larus livens), lesser tern (Sternula antillarum), and the elegant tern (Thalasseus elegans). Mammals at the site include raccoon (Procyon lotor) and common bottlenose dolphin (Tursiops truncatus). The site is one of the most important fishing areas in the State of Sinaloa, specifically for the capture of, principally, brown shrimp (Penaeus californiensis), as well as the Pacific sierra (Scomberomorus sierra), grouper (Paralabrax maculatofasciatus), snapper (Lutjanus argentiventris), and crappie (Eucinostomus dowii).

The region has an important history, at the heart of the geographic region where, in the 19th century Albert K. Owen (1848–1916), a utopian reformer born in Chester, Pennsylvania, founded a 'co-operative community' in Topolobampo. Eventually Owen's dreams faded and he eventually abandoned the ideals of "Integral Co-operation." The site is presently threatened by the impacts of pollution produced by industrial, urban, and agricultural sources, as well as saline intrusion (CNA 2000). Scientific research in the Topolobampo area has been carried out mainly by the Topolobampo Oceanographic Station, since 1984, focusing primarily on pollution, biology, and meteorology of the area. Universidad de Occidente Unidad Los Mochis, Universidad Nacional Autónoma de México (UNAM), and other academic organizations are engaged in research at the site. The Regional Center for Fisheries Research in Mazatlán, Sinaloa also conducts research on water quality and monitors shrimp populations at the site. A number of tourist activities occur at the site as well, including boat rides to the estuary, dolphin observing, and sport fishing. The greatest influx of visitors to the site is during Holy Week and during the summer, mainly by visitors from the surrounding towns and cities, like Los Mochis.

SISTEMA LAGUNAR SAN IGNACIO-NAVACHISTE-MACAPULE, SINALOA: OVERVIEW

Sistema Lagunar San Ignacio-Navachiste-Macapule (Figures 4.22 and 4.23) is situated in the Gulf of California and is habitat for 21 endangered species, and considered of great importance for maintaining biological diversity in the region. A total of 87 species of terrestrial and halophytic plants have been observed at the site. The mangrove species at the site are *Laguncularia racemosa*, *Avicennia germinans*, and *Rhizophora mangle*. Key fauna at the site include common bottlenose dolphin (Tursiops truncatus), California sea lion (Zalophus californianus), and three marine turtles (Chelonia mydas,

FIGURE 4.22 Composite aerial image of Sistema Lagunar San Ignacio-Navachiste-Macapule. Designated February 02, 2008; Area: 79,873 ha; Coordinates: 25°26'N 108°49'W; Ramsar Site number: 1,826.

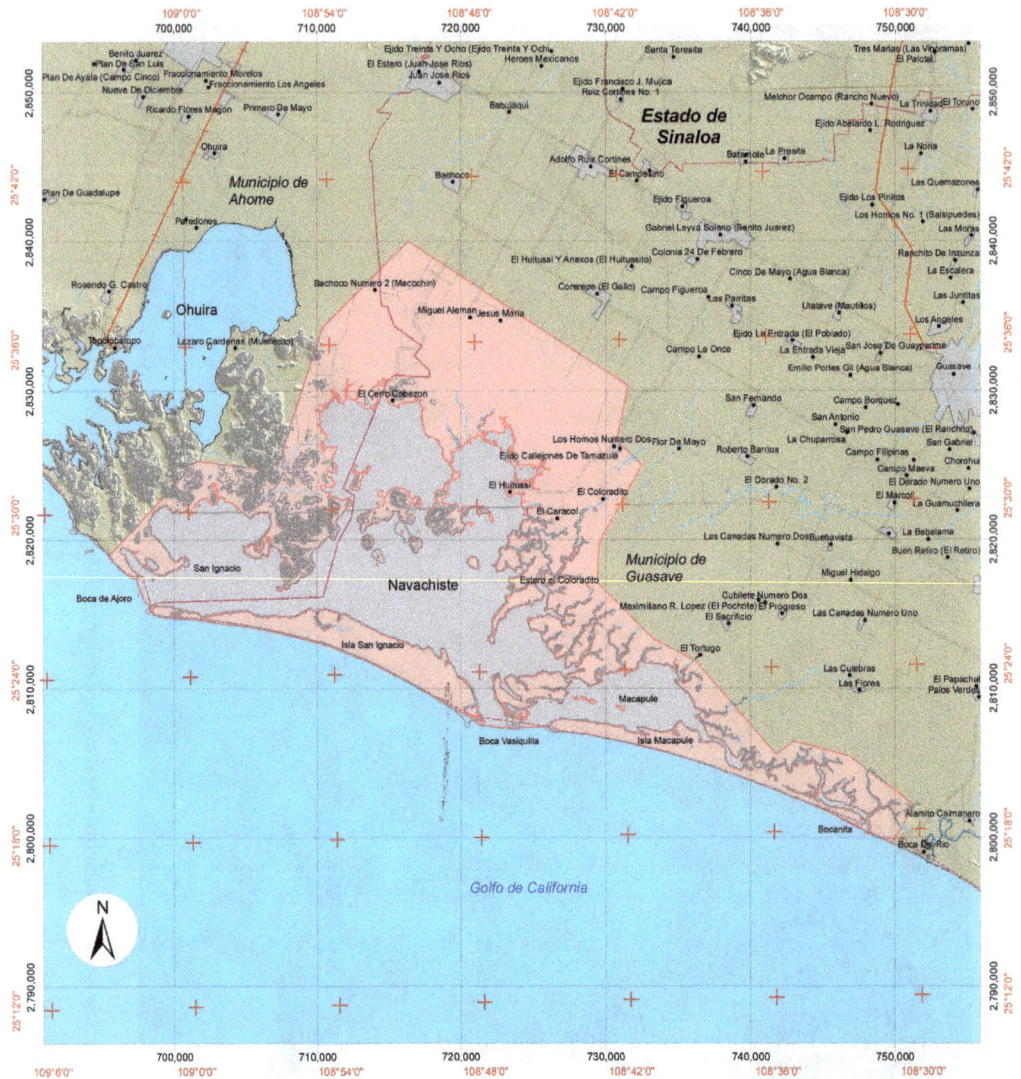

FIGURE 4.23   Map of Sistema Lagunar San Ignacio-Navachiste-Macapule. Designated February 02, 2008; Area: 79,873 ha; Coordinates: 25°26′N 108°49′W; Ramsar Site number: 1,826.

Eretmochelys imbricata, and Lepidochelys olivacea). The site is also considered an Important Bird Area (National Audubon Society), supporting nesting species such as *Phalacrocorax olivaceus*, *Fregata magnificens*, and other bird species such as *Ardea herodias herodias*, *Cathartes aura*, *Pandion haliaetus*, and *Caracara cheriway*. Traditional fishing, and shrimping (e.g., for Litopenaeus styliros- tris, L. vannamei, Farfantepenaeus californiensis, and F. brevirostris) are the predominant human activities at the site. In general, the site supports 21 endangered species, and an important additional diversity of flora and fauna, including 99 species of molluscs, 43 species of birds, 14 species of reptiles, 22 species of crustaceans, 9 species of mammals, and approximately 140 fish species. Among these species, those with commercial value are blue shrimp (Litopenaeus stylirostris), whiteleg shrimp (L. vannamei), yellow leg shrimp (Farfantepenaeus californiensis), and crystal shrimp (F. brevirostris). Certain charismatic species at the site exist, including common bottlenose dolphin (Tursiops trunca- tus), California sea lion (Zalophus californianus), and the three species of turtles mentioned previously (Chelonia mydas, Eretmochelys imbricata, and Lepidochelys olivacea). Flora at the site are also very diverse, with 87 species of terrestrial plants and halophytes, represented mainly by mangrove forest, halophyte plants, fleshy stemmed scrub, as well as 32 species of macroalgae.

SITE DETAILS

The site is a coastal lagoon ecosystem that includes three zones named San Ignacio Bay, Navachiste Bay, and Macapule Bay. Other areas of the lagoon complex are Babaraza, Algodones, El Cuchillo, El Coloradito, and El Tortugo or Tortuguero estuaries. The site has its main axis parallel to the coast, with many islands. The lagoon ecosystem has four entrances: La Boca de Ajoro; Basicilla's mouth that unites the Bay of Navachiste with the sea; the mouths of Macapule that unites the north of the Bay of Macapule with the sea; and the bocanita, which connects the sea with the southern part of Macapule Bay, through the estuary called El Esterón. Average depth of the lagoon is approximately four meters with the exception of the main channel that reaches a depth of 11.5 m.

The lagoon ecosystem is substantially surrounded by coastal flood plains and thus the site serves as a receptor and regulator of incoming water, both from the upland and from the sea. The mangroves of the site also function as a sediment trap, serving to stabilize the coastline and preventing it from eroding, as well as acting as a buffer during storm events. Mangroves are detrital ecosystems, and thus are important as major sources of organic material exported to coastal areas to support fish, fish nurseries, and other organisms that benefit from this detrital organic material.

The flora of the site are represented mainly by 87 species of terrestrial and halophytic plants, as well as 32 species of algae. Scrub predominates in the 0 m to 20 m above sea level upland areas, represented mainly by *Bursera* sp., *Jatropha* sp., *Cercidium microphyllum, Ipomea arborescens, Cercidium sonorae, Ziziphus sonorensis,* and *Opuntia* spp. Dune vegetation at the site includes *Heliotropium curassavicum, Cenchrus echinatus, Abronia maritima, Palafoxia rosea, Distichlis spicata, Asclepias subulata, Jatropha cinerea,* and *Croton punctatus.* Macroalgae at the site is predominated by *Ulva lactuca, Enteromorpha* sp., *Gracilaria* sp., and *Caulerpa* sp. Sea grass (Halodule wrightii and Zostera marina) are also abundant in lagoon waters.

The site is important habitat for resting and/or feeding for migratory birds; 44 of these species, corresponding to 20 families, have been recorded at the site. The family with the greatest diversity at the site is Ardeidae with nine species. Crustacean species at the site include shrimp (Litopenaeus stylirostris, L. vannamei, Farfantepenaeus californiensis, and F. brevirostris), warrior swimming crab (Callinectes bellicosus), and the arched swimming crab (C. arcuatus). The 99 species of molluscs at the site are of 39 families; the most representative of these species, by abundance and distribution, within the lagoon are *Cerithium stercusmuscarum, Neritina* sp., *Nerita funiculata, Crassostrea corteziensis, Crucibulum spinosum, Saccostrea palmula,* and *Nassarius luteostomus.* In the lagoon ecosystem, there are resident groups and occasionally visiting common bottlenose dolphin (Tursiops truncatus), which are well represented throughout the Gulf of California, being one of the most common cetaceans in this area of the sea. Other occasional visitors to the site are California sea lions (Zalophus californianus) and the fin whale (Balaenoptera physalus).

The recorded catch of gastropods at the site is approximately 200 tons per year, on average. The importance of molluscs for the indigenous inhabitants of the region is recorded in two petroglyphs found near the site, which have two beautiful spirals, representing snail shells. In addition to serving as sustenance for fishers themselves, especially when shrimp and crab fishing is closed, gastropod harvesting is very important to people of the region because this activity serves as food for surrounding residents and can provide income. As mentioned above, fishing is the main activity of the communities that inhabit the areas surrounding San Ignacio Navachiste Macapule Lagoon, and this activity has traditionally provided an important contribution to the economy of the municipality of Guasave. There are seven communities dedicated to fishing activities: El Huitussi, El Cerro Cabezón, El Caracol, El Coloradito, El Tortugo, La Pitahaya, and La Boca del Río. Fishing

is carried out by over a 1,000 members that are organized into approximately 25 fishing cooperatives, as well as approximately 1,000 freelance fishers. Fishing serves as their family's livelihood,
which amounts to more than 10,000 people. The main source of food for these communities is thus
shrimp, crab, mullet, sardines, mojarra, bass, and corvina from the region and the site. For reference, the annual fish production for the area reported for the 1992–1996 timeframe was an average
of 3,342 tons per year. Residents also benefit almost exclusively from resource extraction because
in the surrounding areas there are no plants that process their products. These products are thus
only sold fresh, with the exception of shrimp, which is packaged and frozen, and crab, whose pulp
is sold as cooked, packaged, or canned. There are also 12 freezer plants in the region that generate
approximately 1,300 jobs.

Although there is very little comprehensive archeological research on the topic, there is an
archaeological zone at a locale called Las Ventanas, part of the Navachiste lagoon, where petroglyphs by humans have been made on several large rocks, with drawings of crosses and other
indecipherable figures. Among these markings are beautiful spirals, resembling snail shells, about
a half-meter in diameter, which may have been made by Cahíta fishers and gatherers who may
have inhabited the site. There are also caves with other paintings at the site. Aside from the numerous fishing communities in the region, which include Cerro Cabezón, El Huiussi El Tortugo, La
Pitahaya, Boca del Rìo, El Caracol, and El Comalito, there is also active industrial agriculture in
and around the site, where approximately 30 products are grown, predominantly corn, beans, wheat,
safflower, soybeans, chickpeas, alfalfa, and mango.

Adverse factors affecting the ecology of the site include changes in land use (including water use),
and the direct impacts of development projects, such as clearing of deciduous forests to prepare land for
agriculture, and the excavation of the ponds, reservoirs, and channels for shrimp farms. Overexploitation
of the main fishing resources, and the deterioration of habitat and water quality due to the use of more
than 100 associated products in the operation of the shrimp farms, as well as the introduction of the
exotic white shrimp, can cause tremendous impacts on the native ecosystem conditions of the site.
Introduction of ice plant (Mesembryanthemum sp.) and mallow (Malva parviflora) to the site has displaced native plant and associated animal species from many of the site's islands, and have now become
dominant. The loss of native vegetation, impacts on water quantity and quality, as well as the physical
presence of shrimp farms at the site are posited to be associated with bird mortality at the site.

A number of research organizations have focused on the site, including the Regional Center
for Fisheries Research in Mazatlán, studying shrimp farming activities; the Marine Sciences and
Limnology Station, studying nutrient loads in the lagoon ecosystem as well as land use change with
remote sensing; the National Polytechnic Institute; the Center for Interdisciplinary Research for
Integral Regional Development (CIIDIR 2006); and the Food and Development Research Center,
studying land use change. Allied work is occurring with the Interdisciplinary Research Center for
Integral Regional Development, which occasionally provides talks in schools located in fishing
communities, about the conservation of turtles, birds, and other species observed at the site. A number of poets and artists from other areas meet annually at the site and have created an event called
the "Inter-American Poetry Meeting," which in recent years has been expanded to also include
Creation Workshops, presentation of books, and sculpture *in situ*. These events take place during Holy Week each year, and other than these activities the tourism of the area is not intense, but
indeed it does occur in smaller groups of people, who come to the site to enjoy nature.

LAGUNA PLAYA COLORADA-SANTA MARÍA LA REFORMA, SINALOA: OVERVIEW

Laguna Playa Colorada-Santa María La Reforma (Figures 4.24 and 4.25) consists of three bays
with wide mouths that open to the sea, featuring 153 islands, 25 marshes, and 18,700 ha of mangrove. The endangered black-vented shearwater (Puffnus opisthomelas), the brant (Branta bernicla), and numerous waterfowl species inhabit the site. The intertidal flats of the site also host very

FIGURE 4.24 Composite aerial image of Laguna Playa Colorada-Santa María La Reforma. Designated February 02, 2004; Area: 53,140 ha; Coordinates: 25°02'N 108°09'W; Ramsar Site number: 1,340.

FIGURE 4.25   Map overlay of aerial image of Laguna Playa Colorada-Santa María La Reforma. Designated February 02, 2004; Area: 53,140 ha; Coordinates: 25°02′N 108°09′W; Ramsar Site number: 1,340.

large numbers of shorebirds, with past counts reaching over 300,000 individuals. Economically important fish species at the site include mullets (Mugil cephalus and M. curema), bullseye puffer (Sphoeroides annulatus), Peruvian mojarra (Diapterus peruvianus), Pacific sierra (Scomberomorus sierra), snappers (Lutjanus spp.), and snooks (Centropomus spp.). This is a significant area along the Pacific coast of Mexico for fishing, with approximately 2,000 small fishing boats navigating its waters and over 10,000 ha of shrimp farms. Conservation International and the Universidad Autónoma de Sinaloa are working with the local community on sustainable management practices to reduce the impacts from the fishing and farming industries at the site.

Laguna Playa Colorada-Santa María La Reforma consists of three bays: Playa Colorada, Bahía Calcetín, Bahía Santa María. The site provides habitat for more than 600 species: 303 bird species; 185 species of fresh, brackish, or marine fish; 11 amphibian species; 24 reptile species; and

62 mammal species. In total, 46 of these species are imperiled, and included in the national list of species with some category of risk.

SITE DETAILS

Bahía Santa María is included as one of the Important Areas for the Conservation of Birds, which the National Commission for the Knowledge and Use of Biodiversity (CONABIO) established in 1998. This designation is due to the importance of the site for wintering of *Branta bernicla* and for the hibernation of *Pelecanus erythrorhynchos, P. occidentalis, Anas crecca, A. acuta, A. clypeata, Aythya americana, A. affinis, Bucephala albeola, Mergus serrator, Anser albifrons; Pandion hali-aetus* and *Fregata magnificens*, which have all been observed at the site. The site also supports approximately 500,000 shorebirds. About 23% (185) of the fish species in the Gulf of California live permanently or temporarily at the site.

The lagoon has marshes all around it, and of particular note is Malacataya, which is a winter refuge for migratory waterfowl; its main areas of wooded intertidal wetlands are located on the margins of the Playa Colorada Bay, to the south of Santa María Bay, associated with Talchichilte and Altamura Islands. A significant amount of wetlands, of approximately 18,700 ha, exists across the islands of Saliaca, Garrapata, Las Tunitas, El Mero, El Otate, and Otatito, with additional significant area of wetlands in the southeast region of Bahía Santa María.

The main species of flora at the site are *Rhizophora mangle, Laguncularia racemosa, Avicennia germinans,* and *Conocarpus erectus,* the four species of mangroves at this site; grass species *Salicornia* sp., *Sesuvium portulacastrum,* and *Atriplex barclayana*; halophytic plants *Salicornia pacifica, Lycium brevipes, Batis maritima, Atriplex barclayana, Coccoloba uvifera,* and *Coccoloba goldmanii;* salt pine forest species *Tamarix juniperina;* spiny forest plant species *Acacia cochliacantha, Acacia farnesiana, Agave angustifolia, Caesalpinia cacalaco, Prosopis juliflora, Ziziphus sonorensis, Pachycereus pecten, Acanthocereus tetragonus, Rathbunia alamosensis, Mammillaria occidentalis, Ferocactus herrerae, Stenocereus thurberi,* and *Opuntia* sp.; and the dune (invasive) species, *Ipomoea pes-caprae.*

The main ichthyofaunal species of the site include warrior swimming crab (Callinectes bellicosus) and the arched swimming crab (C. arcuatus); California venus (Chione californiensis) and ark clam (Anadara sp.); and mullet (Mugil cephalus and M. curema), pufferfish (Sphoeroides annulatus), Peruvian mojarra (Diapterus peruvianus), Pacific sierra (Scomberomorus sierra), corvina (Cynoscion reticulatus), triggerfish (Pseudobalistes spp.), snappers (Lutjanus argentiventris, L. colorado, L. guttatus, and L. griseus), and snook (Centropomus spp.). Thirty-one bird species at the site are at risk: one endangered, eight threatened, and 22 subject to special protection; within the last two categories, there are four species that also have endemic distribution. The bird species most representative at the site are: *Ardea herodias, Anas clypeata, Pelecanus occidentalis, Anser albifrons, Buteo jamaicensis, Quiscalus mexicanus, Pelecanus erythrorhynchos, Phalacrocorax olivaceus, Mimus polyglottos, Platalea ajaja, Bubo virginianus, Amazilia violiceps, Sula nebouxii, Sula leucogaster, Larus heermanni,* and *Rallus limicola.* The most common mammals of the site are *Didelphis virginiana, Mephitis macroura, Sylvilagus audubonii, Dasypus novemcinctus, Lepus alleni, Marmosa canescens,* and *Urocyon cinereoargenteus.* The most common reptiles at the site are *Agkistrodon bilineatus, Rhinoclemmys pulcherrima, Trachemys scripta, Sceloporus clarkii, Sceloporus horridus, S. nelsoni, Urosaurus bicarinatus, Holbrookia maculata,* and *Crotalus basiliscus.* Amphibians at the site include *Smilisca baudinii, Scaphiopus couchii, Bufo marmoreus, Bufo punctatus, Gastrophryne olivacea, Leptodactylus melanotus, Pachymedusa dacnicolor, Rana forreri,* and *Rana magnaocularis.*

At Playa Colorada, the first human settlement occurred in the 19th century. At the beginning of the 20th century, the site was used for ship repair and for exporting chickpeas, hides, minerals, brazilwood, mezcal, butter, and corn, and for the import of flour, clothing, footwear, and tools. Subsequently, fishing and agriculture became large industries for the area, and most recently shrimp farming. In addition, there is a small salt mine in the area, which produces approximately 5,000 tons of salt per year. The five towns surrounding the lagoon (La Reforma, Costa Azul, Dautillos, Playa Colorada, and Yameto) are heavily dependent upon fishing and the agricultural industry, the latter of which produces primarily corn, wheat, chickpea, sorghum, beans, and tomatoes for export. The area also supports a number of rangeland areas for cattle, goats, and sheep.

Siltation is a major impact to the site, generated by clearing of deciduous forest areas upslope/ upstream, to prepare land for agriculture, and by the excavation of ponds, reservoirs, and channels for shrimp farms. Further deterioration of the site is occurring due to: the use of more than 100 chemical products in the operation of shrimp farms (e.g., pesticides, carbamates, phosphorous, chlorinated compounds, herbicides, and fungicides), as well as the discharge of municipal waters from the city of Guamúchil without any treatment, to the site; salinization of water caused by effluents from agriculture and shrimp farming; eutrophication from the use of nutrients in shrimp farming; and reduction of fresh water volume from the Mocorito River's Eustaquio Buena Dam. Other major threats to the ecology of the site include overfishing, degradation of mangroves, and the concomitant increased likelihood of pathogens and invasives incoming to the site due to increasingly intense shrimp farming. Although there are water quality standards, bird-hunting limits, and fishing limits for the site, conservation measures at the site are only as good as the level of enforcement, which is a major challenge.

A number of research and conservation actions are ongoing at the site, including by the Regional Center for Fisheries Research, which has a permanent program that investigates impacts related to fisheries and shrimping areas. The Institute of Marine Sciences and Limnology of the UNAM has also carried out studies on the geomorphology of the lagoon, and Conservation International with the Autonomous University of Sinaloa have collaborated for a number of years to conserve and manage coastal wetlands in Bahía Santa Maria. The Center for Research in Food and Development, Mazatlán Unit in Aquaculture and Environmental Management, evaluated changes in plant cover and land use of the region in Bahía Santa María Lagoon, using remote sensing. Other institutions contributing knowledge on the biodiversity of the site include the Faculty of Marine Sciences of the Autonomous University of Sinaloa. The Sinaloense Foundation for the Conservation of Biodiversity has also carried out studies on the fauna of the islands of Las Tijeras, Las Tunitas, Garrapata, Saliaca, Altamura, and Tachichilte. The Culiacán Botanical Garden has been involved in conducting studies about the diversity of plants on Talchichilte Island.

Conservation International and the Autonomous University of Sinaloa have proposed workshops and a low-impact eco-tourism center to foster preservation and protection of the cultural and natural resources of the region, while enabling a certain degree of activities such as beach tourism, hunting, visits to bird sanctuaries, kayaking, sailing, sport fishing, and underwater fishing. Fishers in the area have also adapted to conduct guided tourist trips, on occasion.

ENSENADA DE PABELLONES, SINALOA: OVERVIEW

Ensenada de Pabellones (Figures 4.26 and 4.27) is on the Gulf of California coast and includes a series of lagoon complexes, estuarine waters, swamps, and marshes. The site supports more than 292 species of migratory and resident bird species. Because of its location in the Pacific Migration Corridor, the site is classified as a Priority Wetland in Mexico, and an IUCN Wilderness; CONABIO classifies the site as an Area of Importance for the Conservation of Birds, a Priority Hydrological Region, a Priority Land Region, and a Priority Marine Region. The site is home to more than 400,000 birds during peak season. The site is of regional importance for the American avocet (Recurvirostra americana), supporting 10% of the total world population of this species. Mangrove species *Rhizophora mangle*, *Avicennia germinans*, *Laguncularia racemosa*, and the shrub *Guayacum coulteri* are amongst the noteworthy flora found at the site. The main land uses of the site are aquaculture and fishing, with the negative impacts of the latter activity evident at the site.

FIGURE 4.26 Composite aerial image of Ensenada de Pabellones. Designated February 02, 2008; Area: 40,639 ha; Coordinates: 24°26'N 107°34'W; Ramsar Site number: 1,760.

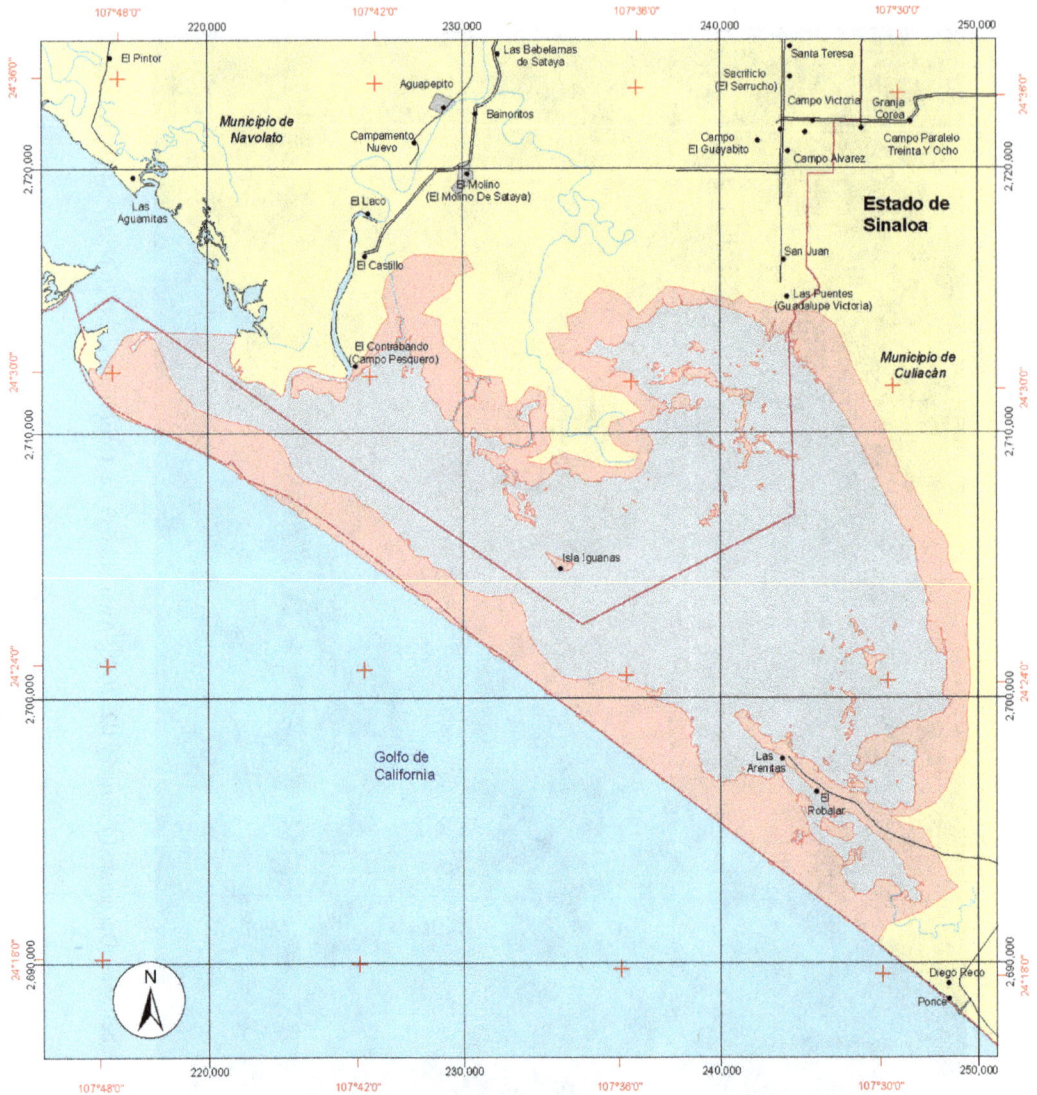

FIGURE 4.27   Map of Ensenada de Pabellones. Designated February 02, 2008; Area: 40,639 ha; Coordinates: 24°26′N 107°34′W; Ramsar Site number: 1,760.

SITE DETAILS

The site is located in the central coastal area of the State of Sinaloa (Northwest Mexico), northwest of the Municipality of Culiacán, and southwest of the Municipality of Navolato. The site can be accessed via different highways and neighboring roads. Communities within the close vicinity of the site include Las Arenitas, Municipality of Culiacán, with 1,800 inhabitants, Las Puentes, with 900 inhabitants, and El Castillo Municipality of Navolato, with 2,900 inhabitants.

Duck populations in the area have been counted in the hundreds of thousands, including 23 species, such as *Anas americana*, *A. crecca*, *A. platyrhynchos*, *A. discord*, *A. clypeata*, *A. cyanoptera*, *A. americana*, *A. collaris*, *Bucephala albeola*, and *Chen caerulescens*. The area is also critical for the survival of at least 23 species of birds, including long-billed curlew (Numenius americanus), marbled godwit (Limosa fedoa), surfbird (Aphriza virgata), western sandpiper (Calidris mauri), and the short-billed dowitcher (Limnodromus griseus).

The lagoon ecosystem is irregular, elongated, and parallel to the general orientation of the coast, except in its southeastern portion where it adopts a lobed form. The site is made up of two relatively shallow areas that correspond to the Altata lagoons to the northwest and Pabellones lagoons to the southeast, partially separated from each other by a pronounced narrowing (Ayala-Castañares et al. 1994).

The natural vegetation of Ensenada Pabellones is made up of spiny forest, mixed spiny forest with shrubby secondary vegetation, vegetation of coastal dunes, gallery forest, mangrove swamp, and halophytic vegetation. More than 300 species of birds are observed at the site, of which 235 are neotropical; 112 nest in the area, and 29 are found in the Official Mexican Standard NOM-059-SEMARNAT-2010, such as *Ixobrychus exilis*, *Ardea herodias*, *Egretta rufescens*, *Anser albifrons*, *Rallus longirostris*, and *Sternula antillarum*. Although relatively few formal reports for other species exist, red lynx (Mephitis macroura), large American opossum (Didelphis marsupialis), antelope jackrabbit (Lepus alleni), desert cottontail (Sylvilagus audubonii), and collared peccary (Tayassu tajacu), as well as olive ridley sea turtle (Lepidochelys olivacea), blackfin silverside (Atherinella crystallina), Pacific river goby (Awaous transandeanus), and halfbeak (Hypo rhamphus) have all been observed at the site.

There is evidence of the Capacha people in the coastal area of the site, near El Dorado, in the basin of the San Lorenzo River. The peoples of the Capacha culture are known in the Jalisco Sierra Madre Occidental, and the Colima Valley of Mexico, back to 800 BCE. Many traditional uses of the fauna of the site are known to currently exist, which specifically involve:

- Traditional use of wildlife, such as wild cat, opossum, and raccoon for food and medicinal purposes. During cooking, the fat of coyote is released and this is used for rheumatic pains, directly on the skin of the affected area;
- Armadillo, deer, collared peccary, rabbit, squirrel, hare, and badger species that are used for food;
- Mockingbird, cardinal, parakeets, parrots, and sparrows that are used for their feathers;
- Doves, quail, and collared peccary, which are hunted for food, and in the case of the cuichi (also known as the chachalaca bird) is domesticated to genetically improve breeds of fighting cocks;
- Vultures, cooked and the residues used, mainly as a broth, for people to drink to aid with stomach ailments;
- Scorpion, which is pulverized, mixed with other ointments, so as to be used when a person is bitten by a poisonous animal;
- Rattlesnake, which is dried in the sun and ground to be used instead of salt, and also used to combat respiratory diseases, leprosy, and cancer or simply as a vitamin supplement;
- The fat of the land turtle, which is used in people who suffer epileptic seizures;
- The meat of the river turtle, which is used as food, its shell used as decoration, and its blood used for medicine;
- Frog legs, which are used for food, when in season; and
- Scorpion, which is drowned in alcohol and the liquid that is generated is used to aid with poisonous animal bites.

Fishing and aquaculture within site are the major human activities at the site, as well as on the periphery, and in the broader vicinity of the site. In general, the indiscriminate utilization of resources, for example from poaching in fishing areas, is a major impact on the ecology of the site. The agricultural valley of Culiacán and Navolato that surround the site provide input of agrochemicals to the site, leading to eutrophication of the site. Conversion of natural areas to agricultural fields has resulted in fragmentation of the landscape and thus the degradation of the coastal ecosystems in the area, which are an integral part of the beauty and ecological functionality of the site. In addition, wastewater discharge from the cities of Culiacán and Navolato poses threats to the ecological integrity of the site, as does the ever-present threat of invasive species, such as cattail

(Typha angustifolia) and common water hyacinth (Eichhornia crassipes). Pronatura-Noroeste, a non-profit conservation organization, has worked over several years to coordinate research and education/outreach activities at the site (https://pronatura-noroeste.org/en/home-eng/). Tourist activities at the site are relatively minimal, and primarily focused within the Chiricahueto Lagoon, due to a hunting concession with the Pichigüila Gun Club.

SISTEMA LAGUNAR CEUTA, SINALOA: OVERVIEW

Sistema Lagunar Ceuta (Figures 4.28 and 4.29) is comprised of lagoon complexes and marshes, with mangroves *Rhizophora mangle*, *Laguncularia racemosa*, *Avicennia germinans*, and button mangrove (Conocarpus erectus). Designated as a Site of Regional Importance by the Western Hemisphere Shorebird Reserve Network in Mexico, and a National Sanctuary, the site supports habitat conditions for tens of thousands of birds, including *Calidris mauri*, *Phalaropus tricolor*, *Recurvirostra americana*, *Charadrius alexandrinus*, and *Sterna maxima*. Olive ridley sea turtles (Lepidochelys olivacea) also utilize the site for nesting. The primary human activities at the site are fishing, aquaculture, and agriculture.

SITE DETAILS

The site's Ceuta Lagoon is located in the south-central part of the coast of the State of Sinaloa, in the Municipality of Elota, which borders the Municipality of Culiacán to the north and the Municipality of Mazatlán to the south. Local communities include La Cruz, Potrerillo del Norote, and Ceuta, with an approximate total population of 43,000 people, all within the Municipality of Elota. The main surface water flowing to the site arrives via the San Lorenzo River, Elota River, Tecuichemona Stream, Arroyo del Tapón, Arroyo las Higueras, and the Arroyo del Norte (CNA 2006). Agricultural influences and urbanization in the watershed affect the input of agrochemicals and organic matter to the site, resulting in diminished water quality. The watershed is also influenced by the José López Portillo Dam, which has a storage capacity of 2,250 million cubic meters of water, and is used mainly for supporting agriculture (CNA 2006).

Numerous species at risk (per NOM-059-SEMARNAT-2010) exist at the site, including button mangrove (Conocarpus erectus), Mexican west coast rattlesnake (Crotalus basiliscus), western zebra-tailed lizard (Callisaurus draconoides), great blue heron (Ardea herodias), Bell's vireo (Vireo bellii), and yellow-rumped warbler (Dendroica coronata). The primary floral communities of the site comprise thorny forest, xeric scrubland, and aquatic and submersed vegetation. Primary floral species include *Rhizophora mangle*, *Laguncularia racemosa*, *Avicennia germinans*, *Conocarpus erectus*, *Abronia maritima*, *Ipomoea pes-caprae*, *Guayacum coulteri*, and *Maytenus phyllanthoides*. Main fauna at the site include a significant number of numerous species of birds, mammals, amphibians, and reptiles, including Caspian tern (Sterna caspia), red-spotted toad (Bufo punctatus), roseate spoonbill (Platalea ajaja), collared peccary (Tayassu tajacu), coyote (Canis latrans), black iguana (Ctenosaura pectinata), and boa (Boa constrictor).

The site has important social importance because of its contribution to self-consumptive fishing, which supports many families of the area. Agriculture and livestock are also major social values that the site provides, with a major crop of the area being tomatoes, which are farmed, harvested, and sold in the Sinaloa region. The establishment of aquaculture farms in the area of the site affects the natural ebb and flow of tides, thereby causing serious damage to the ecological functions of coastal ecosystems of the site. Research at the site has included an analysis of hypersaline soils and sediments in the Ceuta Lagoon; geochemistry of trace elements present in waters and sediments of the site; composition, structure, and stability of the phytoplankton community in lagoons of the site; influence of tidal flow on the migration of post-larval blue shrimp (Farfantepenaeus stylirostris) in lagoons of the site; zooplankton and the influence of hydrologic flows and alterations upon them; and conservation of the olive ridley sea turtle, *Lepidochelys olivacea*, in Ceuta Bay.

FIGURE 4.28 Composite aerial image of Sistema Lagunar Ceuta. Designated February 02, 2008; Area: 1,497 ha; Coordinates: 24°02′N 107°04′W; Ramsar Site number: 1,824.

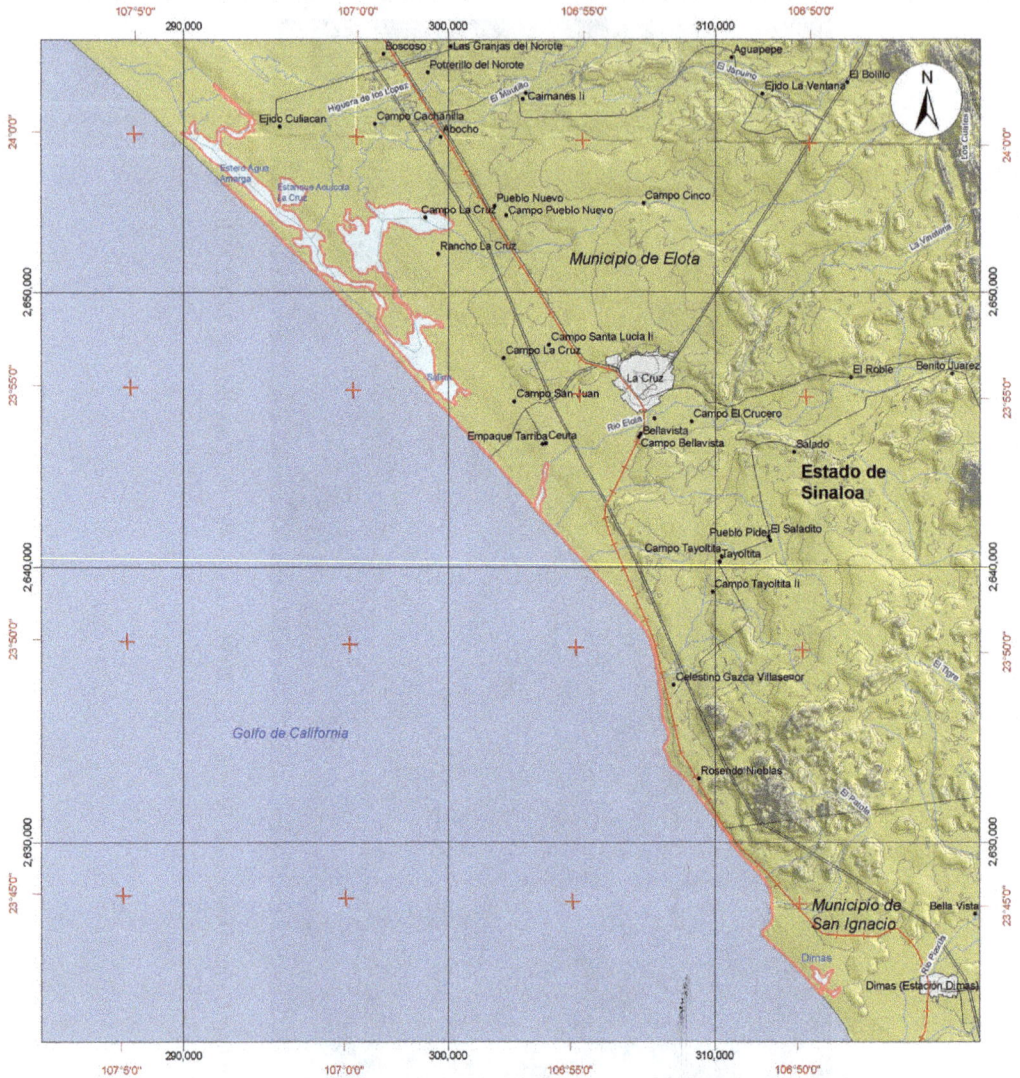

FIGURE 4.29   Map of Sistema Lagunar Ceuta. Designated February 02, 2008; Area: 1,497 ha; Coordinates: 24°02′N 107°04′W; Ramsar Site number: 1,824.

Elota Township has a sea turtle museum to aid in education and outreach to local community members, and other visitors to the site. The National Commission of Protected Natural Areas also carries out educational activities to promote conservation, specifically in communities of influence on the Ceuta Sanctuary, aimed at children at the primary school level, as well as visitors to the area. The Autonomous University of Sinaloa conducts workshops and awareness talks to visitors to the Playa Ceuta Sanctuary. Various events are held on the turtle conservation programs of the area, during the nesting season, and these events include the release of hatchlings, which typically involves the participation of children. There is a large influx of visitors to the site during sea turtle nesting season. Most of the visitors that come to the site live somewhere in the region, followed by national tourism, and to a lesser extent international tourists, who are all attracted to the site's beaches and the sea turtle museum.

PLAYA TORTUGUERA EL VERDE CAMACHO, SINALOA: OVERVIEW

Playa Tortuguera El Verde Camacho (Figures 4.30 and 4.31) is a 25 km long beach situated in the State of Sinaloa, Mexico. It represents critical habitat for the olive ridley sea turtle (Lepidochelys olivacea), a species that arrives in groups of 40–50 individuals per night to nest on the beach areas of the site. In addition to the olive ridley turtle, the beach is also feeding and migration habitat for other marine turtle species, such as the hawksbill sea turtle (Eretmochelys imbricata), green sea turtle (Chelonia mydas), and leatherback sea turtle (Dermochelys coriacea).

SITE DETAILS

The site falls under the jurisdiction of the Ministry of the Environment and Natural Resources (SEMARNAT), and abuts the jurisdiction of the municipality of Mazatlan (just to the south of the site). From west to east, the site ranges from the bathymetric isoline of 5 fathoms to the contour of the Mazatlán-Culiacán Highway. The site has been included in the list of turtle sanctuaries by the National Commission of Natural Protected Areas (CONANP), since 2001. Coastal lagoons of the site, which are protected behind the site's beach areas, are variably brackish and freshwater, providing refuge to resident and migratory birds and serving as nursery waters for economically important fish species, such as snapper (Lutjanus spp.). The dominant vegetation cover of the site includes mangrove forests, dominated by *Laguncularia racemosa* and deciduous tropical forests. The site also supports a diverse community of migratory bird species that arrive at the site for reproduction or to complete critical phases of their life cycle.

The site is an estuarine ecosystem with interspersed salt marshes, all of which are ultimately connected with the sea, providing biogeochemical inputs to the sea and adjacent coastal areas. The largest coastal wetland at the site is El Verde, which is fed by seasonal flow of the Quelite River. The wetlands of the site provide refuge to resident and migratory birds and serve as nursery waters for economically important fish species, such as snapper (Lutjanus spp.). The wetlands also support a diverse community of fauna, including deer. The site is home to a diverse community of migratory species that travel to this location to reproduce or to complete critical stages of their life cycle. The site is particularly renowned for its impressive array of bird species, boasting a total of 99 different species. This makes El Verde Camacho Beach one of the last refuges in Sinaloa that offers shelter and sustenance to a range of wildlife, including peccary, lynx, wild cat, and a healthy population of crocodiles. Additionally, the site is home to a variety of fish species, including snook, striped croaker, snapper, and mojarra. The economic importance of these fish species makes them a valuable resource for the local fishing industry. The site is also home to an impressive variety of flora and fauna. There are a total of 375 species of flora found at the site, and a diverse range of faunal species, which include turtles, fish, birds, and mammals.

The olive ridley turtle is a notable species at Tortuguera El Verde Camacho Beach. It arrives in groups of 40–50 individuals per night to nest on the beach. Olive ridley turtles are known for their synchronized nesting behavior, which results in a large number of turtles nesting in a single night. The turtles emerge from the sea and crawl up the beach to dig a nest in the sand, where they lay their eggs. The nests are subsequently, and carefully, covered up by the turtles, who then return to the sea. In addition to the olive ridley turtle, Tortuguera El Verde Camacho Beach is also a feeding and migration habitat for other marine turtle species. Hawksbill turtles are commonly found in the waters around the beach, and they use the beach as a corridor for migration. Green sea turtles are also found in the area and use the beach as feeding habitat. The leatherback sea turtle nests sporadically at the site.

Wetlands of the site, together with the coastal dunes, contribute to the stabilization and maintenance of the coastline, protection against floods and hurricanes, and safeguarding of human settlements near the site. Wetlands within the site thus also play an important role in supporting domestic

FIGURE 4.30 Composite aerial image of Playa Tortuguera El Verde Camacho. Designated February 02, 2004: Area: 6,454 ha; Coordinates: 23°24′N 106°32′W; Ramsar Site number: 1,349.

FIGURE 4.31 Map of Playa Tortuguera El Verde Camacho. Designated February 02, 2004; Area: 6,454 ha; Coordinates: 23°24′N 106°32′W; Ramsar Site number: 1,349.

consumption by towns and communities in the area, as well as for irrigation in the agricultural valley of the Quelite River.

Research at the site began in 1975, and has included the development of infrastructure for monitoring and fostering the population of olive ridley turtles. Additionally, researchers have conducted assessments of the wetland and terrestrial ecosystems at the site, for ecological and functional characterization, as well as for overall biodiversity assessment purposes. Environmental education programs at the site are available for school and community groups, with an interactive permanent exhibition room, and providing information to all visitors before they embark on walks along observation trails throughout the different habitat areas of the site. National and international volunteers are also invited to contribute to turtle recovery work, providing an opportunity for any visitors that desire conservation experiences, which also leverages volunteer's time to contribute to protecting local wildlife. While eco-tourism activities are continuously under development at the site, recreation is perennially popular at the site with local inhabitants, as well as national and international tourists, who all participate in fishing, walks on the beach, and water sports. The intensity of tourism at the site is customarily low, but there is potential for growth.

LAGUNA HUIZACHE-CAIMANERO, SINALOA: OVERVIEW

Located in the southeastern Gulf of California, Laguna Huizache-Caimanero (Figures 4.32 and 4.33) consists of a series of wetland types ranging from coastal and inland, to artificial. Due to its location along the Migratory Corridor of the Pacific, the site provides important habitat for birds such as the American white pelican (Pelecanus erythrorhynchos) and the roseate spoonbill (Platalea ajaja), as well as habitat for a wide range of fish, mammals, reptiles, amphibians, and invertebrates, including species at risk, such as the American crocodile (Crocodylus acutus), the Mexican beaded lizard (Heloderma horridum), the boa (Boa constrictor), and the olive ridley sea turtle (Lepidochelys olivacea). Upland edges of the site are mainly mangrove forest, including red (Rhizophora mangle), black (Avicennia germinans), and white (Laguncularia racemosa) mangroves. Fishing resources in the site's lagoon are the main protein sources for the surrounding communities; shrimp harvesting is the main economic activity of the area. Among the negative factors impacting the site are the sedimentation of the Presidio River and Baluarte River; deforestation; overfishing; and wastewater discharge into the lagoon.

SITE DETAILS

Huizache-Caimanero is a coastal lagoon, located between the Presidio and Baluarte Rivers, in the south of the State of Sinaloa, which is the least economically and socially developed coastal region of the area. The fishing resources of this body of water are the main source of animal protein for most of the neighboring towns. Shrimp harvesting is the main economic activity of the several fishing production cooperatives, and a significant number of illegal fishers from the surrounding towns. The lagoon is impacted by sediment deposition that originates from upland anthropogenic sources, and consists of two shallow basins with water levels that vary with rainfall and fluvial contributions. The site's main geomorphological characteristics are a long and narrow sandy barrier called Isla Palmito de la Virgin, with two mouths that are closed but that the fishers or the authorities dredge frequently to keep open. The lagoon provides habitat for 83 fish, along with shorebird populations. The site also provides habitat for penaeid (Penaeidae) shrimp and whiteleg shrimp (Litopenaeus vannamei).

The strategic location of the site's lagoon within the Pacific Migration Corridor makes it one of the most important wintering, resting, and foraging sites for birds along Pacific-coastal Mexico. Birds that come to winter at the site include white pelican (Pelecanus erythororchinchus), the roseate spoonbill (Platalea ajaja), and large numbers of wading birds in the muddy areas of the shallow lagoon, such as long-billed dowitcher (Limnodromus scolopaceus), the marbled godwit (Limosa fedoa), black-necked stilt (Himantopus mexicanus), and the American avocet (Recurvirostra americana). Other species of ecological importance at the site include various species of gulls and swallows, American coot (Fulica americana), and ducks such as the northern spatula (Spatula clypeata), mallard duck (Anas platyrhynchos), northern pintail (A. acuta), and the blue-winged teal (Spatula discors). The ichthyofaunistic communities of Huizache-Caimanero are approximately 8% freshwater fish; 8% estuarine fish; 31% marine fish visiting the estuary as adults and for feeding; 33% marine fish that use the estuary as feeding grounds; and 20% marine fish that are occasional visitors.

The basin of the Huizache-Caimanero lagoon is located between the lower basins of the Río Presido and the Baluarte River. Runoff into the Huizache Marsh and the Caimanero Lagoon have defined channels, and the lagoon regulates groundwater levels, upon which its plant communities depend. Together with the physicochemical and environmental fluxes created by these conditions, supportive habitat for protection and breeding of four shrimp species exists at the site, specifically for the 83 fish species, and other invertebrates of the site. The surface of the lagoon constitutes a catchment basin for rainfall, which may come in large boluses that can suddenly/intensely occur

FIGURE 4.32 Composite aerial image of Laguna Huizache-Caimanero. Designated February 02, 2007; Area: 48,283 ha; Coordinates: 22°50'N 105°55'W; Ramsar Site number: 1,689.

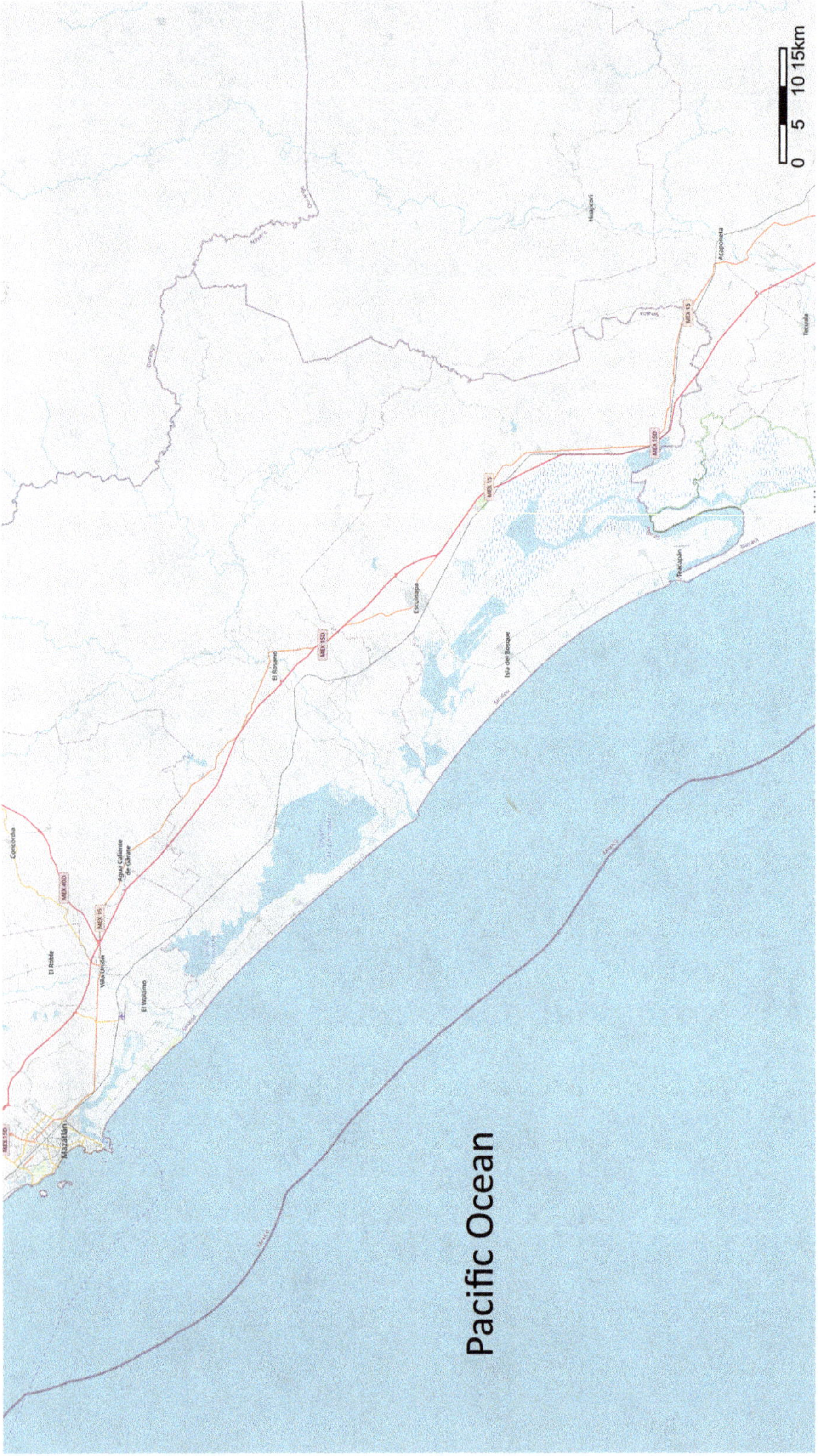

FIGURE 4.33   Map of Laguna Huizache-Caimanero. Designated February 02, 2007; Area: 48,283 ha; Coordinates: 22°50'N 105°55'W; Ramsar Site number: 1,689.

during storms, hurricanes, and cyclones, and thus the site attenuates both the flow of water and sediment within the site, and in the broader-landscape vicinity of the site. Such flows benefit the mangroves of the site, as well as the Presidio River and Baluarte River ecosystems. Mangroves also pay an important role in the hydrologic cycle, facilitating the recharging and discharging of groundwater, mediating the ebb and flow of water in the lagoon, reducing erosion, stabilizing shorelines, as well as trapping sediment and nutrients to maintain water quality.

The site is made up of the Huizache and Caimanero Sub-basins and several estuaries: El Oasis, Agua Dulce, Las Anonas, and Pozo del Caimán. Within the site is an island, Barra Palmito de la Virgen, approximately 40 km long and between 1.5 and 3.6 km wide. In the dry season, 80% of the lagoon's surface becomes a marsh. On the margins of the site are wooded intertidal wetlands, specifically on the margins of the mouth of the Presidio River, in the El Ostial and Agua Dulce estuaries. The lagoon itself receives flow from contributing rivers during the rainy season that provide nutrients and suspended material, which stimulates primary production, and further supports the phytoplankton communities of the site/region, represented by the diatom genera *Nitzschia, Cocconeis,* and *Cyclotella*; cyanophytes of the genera *Anabaena, Anabaenopsis,* and *Oscillatoria*; the chlorophytes (Chlamydomonas, Volvox, Pediastrum, Ankistrodesmus, and Descendesmus); and communities of neritic origin in the areas of marine influence, represented by the diatom genera *Rhizosolenia, Chaetoceros, Skeletonema,* and *Thalassionema.* During the dry season, algae develop in the shallow marginal areas of the basins, predominantly from the genera *Enteromorpha* and *Cladophora.* Blue algae are also common, as well as green algae *Chlamydomonas* sp., and diatoms such as genus *Nitzschia.* In the dry season, large muddy plains are developed at the site and are colonized by glasswort (Salicornia spp.). In the coastal plain areas of the site, halophytic grasses such as saltwort (Batis maritima) and shoregrass (Monanthochloe littoralis) are present. During the summer, widgeon grass (Ruppia maritima) is abundant, which covers a large area of the Caimanero Basin. In the Ostial and Agua Dulce estuaries, and in the riparian zones close to them, there are mangroves mainly made of three species in high densities: red mangrove (Rhizophora mangle), white or sweet mangrove (Laguncularia racemosa), and black or puyeque mangrove (Avicennia germinans). These three mangrove species extend for more than 50 m on both sides of the estuaries and channels. The mangrove species *Conocarpus erectus* (button mangrove) is less common but is present within the site, within areas of other terrestrial vegetation. All the above-mentioned species are important because they form the trophic base for the energy transfer and transformation for the different fauna in the lagoon. In addition, most of the vegetation provides protection to other organisms. The vegetation also has a close relationship with the fauna outside of the water, providing a number of microhabitats for the various faunal taxa that inhabit them. In addition to the natural vegetation at the site in the vicinity of the lagoon, there are large areas where vegetables such as corn, tomato, chili, mango, coconut palm, and beans are actively cultivated.

Main faunal species at the site include crustaceans: whiteleg shrimp (Litopenaeus vannamei), blue shrimp (L. stylirostris), yellowleg shrimp (Farfantepenaeus californiensis), crystal shrimp (F. brevirostris), warrior swimming crab (Callinectes bellicosus), and arched swimming crab (C. arcuatus); fish: mullet (Mugil cephalus and M. curema), anchovy (Anchoa panamiensis), catfish (Galeichthys caerulescens), corvina (Cynoscion reticulatus), sardine (Lile stolifera), botete (Sphoeroides annulatus), Peruvian mojarra (Diapterus peruvianus), Pacific sierra (Scomberomorus sierra), snapper (Lutjanus argentiventris), cochi (Pseudobalists spp.), and snapper (L. guttatus and L. griseus). Endemic species such as *Heloderma horridum, Ctenosaura pectinata,* and *Crocodylus acutus,* and endangered species such as *Felis concolor, Lynx rufus, Mimus polyglottos,* and *Carpodacus mexicanus* are noted at the site, as well as threatened species *Ctenosaura pectinata, Iguana iguana, Micruroides euryxanthus, Boa constrictor, Sylvilagus cunicularius, Sylvilagus graysoni,* and *Lepus callotis.*

The site was inhabited during pre-Hispanic times by an indigenous group known as Totorames, about eight centuries ago. In Chametla, at the southern mouth of the lagoon, objects such as stylized figures, pots, anthropomorphic figures, and mortuary urns have been studied. Agriculture, shrimping, and fishing have thus been the socio-economic mainstays in the lagoon region for a very long period of time, which continues today. The Chametla region caught the attention of the Spaniards because of the way in which the indigenous people fished and because of the enormous quantity of fish, oysters, and shrimp that were collected there. Indeed, these activities provide very important dietary elements for people of the towns in the area, including Agua Verde, Pedregosa, Matadero, El Cerro, Potrerillos, Guásimas, Vázquez Moreno, Guajote and Zopilote, Walamo, Ejido Nuevo, Amapa, Barrón, Francisco Villa, and Los Pozos. Shrimp extraction is the main economic activity of the fishing production cooperatives of the region, and a significant but indeterminate number of fishers of the aforementioned towns or other areas in the region, which due to the poverty conditions in which they live and the high price of shrimp, come to the site to illegally fish.

The position of the lagoon within the landscape makes it a receiving area for sediments transported by rivers, influenced by the dams along the Presidio and Baluarte Rivers, as well as the felling of trees and removal of other vegetation in the surrounding watershed. The associated siltation has reached a level where 85% of the lagoon is dry during the dry season. Other adverse factors at the site include the discharge of effluent from agriculture, aquaculture, and municipal wastewater; the use of bait (dog food) to fish for shrimp; and overfishing. Another adverse impact of shrimping activities on the ecology of the site are from the methods used to disperse birds, such as making loud noises, alerting calls, installation of wires and conspicuous objects, and explosions from 'acetylene cannons' and rockets.

Existing research activities are ongoing at the site by The Regional Center for Fisheries Research of the National Fisheries Institute, the UNAM Institute of Marine Sciences and Limnology, the Department of Marine Biology at the University of Liverpool, The Center for Research in Food and Development, The National Council of Science and Technology, The Autonomous University of Sinaloa, and the Instituto Tecnológico del Mar of the Ministry of Public Education. Relatively few tourist activities occur at the site, with primarily an air boat service that facilitates duck hunting (November to March).

REFERENCES

Álvarez-Borrego, S. and L.A. Galindo-Bect. 1974. Hidrología del Alto Golfo de California I. Condiciones durante Otoño. *Ciencias Marinas.* 1(1): 46–64.

Arreola Lizárraga., J.A. 1995. Diagnosis ecológica de bahía de Lobos, Sonora, México. IPN-CICIMAR, Tesis de Maestría. 120pp.

Arriaga, L., Espinoza, J.M., Aguilar, C., Martínez, E., Gómez, L. and E. Loa. 2000. *Regiones Terrestres Prioritarias de México.* RTP-15. Bahía de San Jorge. Comisión Nacional para el Conocimiento y uso de la Biodiversidad, México, DF.

Audeves, S., Pérez., A.M., Rozo., G, and F. Enríquez. 1997. *Estudio de los moluscos en Bahía Las Guásimas, Sonora.* Res. VI Congr. de la Asoc. de Investigadores del Mar de Cortés, A.C. 58pp.

Ayala-Castañares, M. and M. Gutiérrez Estrada, 1994. *Geología Marina del Sistema Lagunar Altata-Pabellones, Sinaloa, México.* Anales del Instituto de Ciencias del Mar y Limnología.

Bahre, C.J. 1983. Human impact: the Midriff Islands (pp. 290–306). In: T. Case and M. Cody (Eds.) *Island Biogeography in the Sea of Cortez.* University of California Press, Berkeley. 508pp.

Balart, E. F., Castro Aguirre., J.L., and R. Torres Orozco. 1992. Ictiofauna de las bahías de Ohuira, Topolobampo, y Santa María, Sinaloa, México. *Investigaciones marinas CICIMAR.* 7(2): 91–103.

Barlow, G.W. 1961. Gobies of the genus *Gillichthys*, with comments on the sensory canals as a taxonomic tool. *Copeia.* 1961(4): 423–437.

Basurto, X. 2002. *Community-Based Conservation of the Callo de Hacha Fishery by the Comcaác Indians, Sonora, Mexico.* MS Thesis, School of Renewable Natural Resources, University of Arizona, Tucson, Arizona.

Basurto, X. 2005. How Locally Designed Access and Use Controls Can Prevent the Tragedy of the Commons in a Mexican Small-Scale Fishing Community. *Society and Natural Resources.* 18: 643–659.

Bourillón-Moreno, L. 2002. *Exclusive fishing zone as a strategy for managing fishery resources by the Seri Indians, Gulf of California, Mexico. PhD Dissertation.* University of Arizona. 290pp.

Bowen, T. 1965. A survey of archaeological sites near Guaymas. *Sonora the Kiva.* 31(1): 14–36.

Bowen, T. 2000. *Unknown Island: Seri Indians, Europeans, and San Estebán Island in the Gulf of California.* University of New Mexico Press, Albuquerque, New Mexico. 548pp.

Braniff, B. 2009. La Historia Prehispánica de Sonora. *Arqueología Mexicana.* 27: 32–38.

Bruckner, A.W., Johnson, K.A., and J.D. Field. 2003. Conservation strategies for sea cucumbers: Can a CITES Appendix II listing promote sustainable international trade? *SPC Beche-de-mer Information Bulletin.* 18: 24–33.

Brusca, R. 1980. *Common intertidal invertebrates of the Gulf of California.* (2nd ed., revisada). Univ. of Arizona Press, Tucson.

Brusca, R.C., Cudney-Bueno, R., and M. Moreno-Báez. 2006. *Gulf of California Esteros and Estuaries Analysis, State of Knowledge and Conservation Priority Recommendations.* Final Report to the David and Lucile Packard Foundation by the Arizona-Sonora Desert Museum. 60pp.

Calderón, A.L.E. and J.F. Campoy. 1993. Bahía de Las Guásimas, Estero Los Algodones y Bahía de Lobos, Sonora (pp. 411–419). In: S. I. Salazar. and N. E. González (Eds.) *Biodiversidad Marina y Costera de México.* CONABIO y CIQRO, México, 865pp.

Campoy Favel., J.R. and L.E. Calderón-Aguilera. 1991. Observaciones ecológicas de las comunidades bentónicas de tres sistemas costeros de sonora, con énfasis en moluscos y crustáceos. III Congreso de la Asociación de Investigadores del Mar de Cortés. ITESM-Campus Guaymas.

Case, T. J. and M. L. Cody (Eds.). 1983. *Islands biogeography in the Sea of Cortez.* Univ. of California Press, Berkeley. 508pp.

Cervantes, M. 1994. *Guía regional para el conocimiento, manejo y utilización de los humedales del noroeste de México.* Instituto Tecnológico de Estudios Superiores de Monterrey - Campus Guaymas, Guaymas, Sonora.

CIAD. 2006. *Programa Especial Concurrente para el Desarrollo Rural Sustentable.* Distrito de Desarrollo Rural 139-Caborca, Sonora.

CIIDIR. 2006. *Monitoreo de los Recursos Naturales en las Islas de Navachiste.Informe.* Centro Interdisciplinario de Investigación para el Desarrollo Integral de la Región. I.P.N.

CITES (Convention on International Trade in Endangered Species of Wild Fauna and Flora). 2005. *Text of the Convention* (Appendices I, II, and III). United Nations Environment Program, Geneva, Switzerland. 48pp.

CNA (Comisión Nacional del Agua). 2000. *Expediente Técnico justificativo del Acuífero del Río Sinaloa para la publicación de la disponibilidad.* El Diario Oficial de la Federación.

CNA (Comisión Nacional del Agua). 2003. *Determinación de la disponibilidad de agua subterránea en el acuífero Valle del Yaqui, estado de Sonora.* Comisión Nacional del Agua. Gerencia de Aguas Subterráneas, México. 24pp.

CNA (Comisión Nacional del Agua). 2006. Subgerencia de Información Geográfica del Agua (SIGA).

CONABIO. 2007. Fichatécnica para la evaluación de los sitios prioritarios para la conservación de los ambientes costeros y oceánicos de México. Corredor pesquero Bahía Guásimas-Estero Lobos, clave de sitio 23. Mesa de trabajo Golfo de California. Grupo GAP análisis.

CONANP. 2007. *Programa de Conservación y Manejo de la Reserva de la Biosfera Alto Golfo de California y Delta del Río Colorado.* Comisión Nacional de Areas Naturales Protegidas, México, D.F. 319pp.

Contreras, E. F. 1993. *Ecosistemas Costeros Mexicanos. Departamento de Hidrología División de Ciencias Biológicas y de la Salud. Comisión Nacional para el Conocimiento y uso de la Biodiversidad.* Universidad Autónoma Metropolitana Unidad Iztapalapa, México. 415pp.

Cortez-Lara, A. and M.R. García-Acevedo. 2000. The lining of the All-American Canal: the forgotten voices. *Natural Resources Journal.* 40: 261–279.

DOF (Diario Oficial de la Federación). 2002. *Norma Oficial Mexicana NOM-059-SEMARNAT-2001. Proteccion ambiental-Especies nativas de Mexico de flora y fauna silvestres-Categorias de riesgo y especificaciones para su inclusion, exclusion o cambio-Lista de especies en riesgo.* Ministry of Environment, Natural Resources and Fisheries. 84pp.

DOF (Diario Oficial de la Federación). 2010. *Norma Oficial Mexicana NOM-059-SEMARNAT-2010, Protección ambiental-Especies nativas de México de flora y fauna silvestres-Categorías de riesgo y especificaciones para su inclusión, exclusión o cambio-Lista de especies en riesgo.* Official Gazette. Ministry of Environment, Natural Resources and Fisheries. 78pp.

Enríquez O., L.F. and L.E. Calderón A. 1990. *Análisis de la poliquetofauna de la bahía de Lobos, Sonora.* Res. del Congreso Nacional de Oceanografía. Octubre 1990, Mazatlán, Sinaloa.

Escobedo-Urías, D. 1997. *Hidrología, nutrientes e influencia de las aguas residuales en la Laguna de Santa María, Sinaloa.* Tesis de Maestría. CICIMAR-IPN. 87pp.

Ezcurra, E. 1984. The Vegetation of El Pinacate, Sonora. A quantitative study. PhD Dissertation. University College of North Wales, Bangor, Gwynedd, UK. 117pp.

Ezcurra, E., Equihua, M., and J. Lopez-Portillo. 1987. The Desert Vegetation of El-Pinacate, Sonora, Mexico. *Vegetatio*. 71(1): 49–60.

Felger, R.S. 2000. *Flora of the Gran Desierto and Rio Colorado of Northwestern Mexico*. The University of Arizona Press, Tucson, Arizona. 673pp.

Felger, R.S. and B. Broyles. 2007. *Dry Borders. Great Natural Reserves of the Sonoran Desert*. The University of Utah Press, Salt Lake City. 799pp.

Felger, R.S. and M.B. Moser. 1985. *People of the Desert and Sea. Ethnobotany of the Seri Indians*. The University of Arizona Press. 454pp.

Felger, R.S. and M.B. Moser. 1987. Sea turtles in Seri Indian culture. *Environment Southwest*. 519: 18–21.

Findley, L.T. 1976. Aspectos ecológicos de los esteros con manglares en Sonora y su relación con la explotación humana (pp. 95–105). In: B. Braniff and R.S. Felger (Eds.) *Sonora Antropología del Desierto*. Instituto Nacional de Antropología e Historia, D.F. Serie Científica No. 27, Mexico.

Fleishman, A.B. 2011a. Birds of the Bahia Kino Region, Sonora, Mexico: Coastal Areas from Puerto Lobos to San Carlos and the Midriff Islands. Version 2.0. Prescott College Center for Cultural and Ecological Studies, Bahía Kino, Sonora.

Fleishman, A.B. 2011b. *Nesting Waterbirds Along the Hermosillo Coast*. Prescott College Centro de Cultural and Ecological Studies, Bahía Kino, Sonora.

Fleishman, A.B. and N.S. Blinick. 2011. Northerly Extension of the Breeding Range of the Roseate Spoonbill in Sonora, México. *Western Birds*. 42: 243–246.

Gilmartin, M. and N. Revelante. 1978. *The Phytoplankton Characteristics of the Barrier Island Lagoons of the Gulf of California*. Center for Marina Studies, University of Maine, Orono, U.S.A. 29–47pp.

Glenn, E.P., Nagler, P.L., Brusca, R.C., and O. Hinojosa-Huerta. 2006. Coastal wetlands of the northern Gulf of California: Inventory and Conservation status. *Aquatic Conservation: Marine and Freshwater Ecosystems*. 16: 5–28.

Grijalva-Chon, J., Nunez-Quevedo, S., and R. Castro-Longoria. 1996. Ichthyofauna of La Cruz coastal lagoon, Sonora, Mexico. *Marine Sciences*. 22(2): 129–150.

Gutiérrez-Barreras, A. 1999. *Ictiofauna de fondos blandos de la Bahía de Topolobampo, Sinaloa, México*. Tesis de Maestría. IPN-CICIMAR. 108pp.

Hannah, D. 2008. *Community Vegetation Mapping of Estero Santa Cruz. Plant Ecology Class*. Prescott College Center for Cultural and Ecological Studies, Bahía Kino, Sonora.

Harrington, B. A. 1993. A coastal, aerial winter shorebird survey on Sonora and Sinaloa coast of México, January 1992. *Wader Study Group Bulletin*. 67: 44–49.

Hartmann, W. 2002. The mystery of the Hohokam Origins and Disappearance. *Noticias del CEDO*. 10(1): 36–41.

Hinojosa-Huerta, O., Iturribarría-Rojas, H., Calvo-Fonseca, A., Butrón-Méndez, J., and J.J. Butrón-Rodríguez. 2004. *Caracterización de la avifauna de los humedales de la Mesa de Andrade, Baja California, México*. Reporte de Pronatura Noroeste-Dirección de Conservación Sonora al Instituto Nacional de Ecología.

Honan, E. and P.J. Turk-Boyer. 2001. *Reports of Olive Ridley "Lepiodochelys olivacea" nesting in the northern Gulf of California*. Report presented at Congreso de la Asociación de Investigadores del Mar de Cortés, Ensenada, B.C. 6pp.

Huckell, B. 1996. The archaic prehistory of the North American Southwest. *Journal of World Prehistory*. 10: 305–373.

INEGI (Instituto Nacional de Estadística Geografía e Informática). 2010. Censo de población y vivienda 2010.

Juárez Romero, L. 2007. *Determinación de agentes causales de alta mortalidad en los cultivos del Ostión Japonés, Crassostrea gigas, de las costas de Sonora*. Instituto de Acuacultura del Estado de Sonora. Comité de Sanidad Acuícola del Estado de Sonora, Hermosillo, Sonora.

Kasper-Zubillaga, J.J. and A. Carranza-Edwards. 2005. Grain size discrimination between sands of desert and coastal dunes from northwestern Mexico. *Revista Mexicana de Ciencias Geológicas*. 22: 383–390.

Lancín, M. 1985. Universidad Nacional Autónoma de México. *Instituto de Geología. Revista*. 6(1): 52–72.

Lavin, M. F. and S. Organista. 1988. Surface Heat Flux in The Northern Gulf of California. *Journal of Geophysical Research*. 93(11): 14033–14038.

Luque, D. and A. Robles. 2006. *Naturalezas, saberes y territorios comcaac (seri)*. INE-SEMARNAT. México, D.F. 360pp.

Luque, D. and E. Gómez. 2007. La construcción de la región del Golfo de California desde lo ambiental y lo indígena. *Universidad Autónoma Indígena de México. Ra Ximhai*. 3(1): 83–116.

Luque, D. and S. Doode. 2007. Sacralidad, territorialidad y biodiversidad comcaac (seri). Los sitios sagrados indígenas como categorías de conservación ambiental. *Relaciones*. 28(112): 157–184.

Martínez-Yrízar, A. 2006. Los Esteros: un importante hábitat costero. *Nuestra Tierra*. 5: 3–8.

Mellink, E. and A. Orozco-Meyer. 2006. Abundance, Distribution, and Residence of Bottlenose Dolphins (*Tursiops truncatus*) in the Bahia San Jorge Area, Northern Gulf of California, Mexico. *Aquatic Mammals*. 32: 133–139.

Mellink, E. and E. Palacios. 1993. Note on breeding coastal waterbirds in northwestern Sonora. *Western Birds*. 24: 29–37.

Mitchell, D.R. and M.S. Foster. 2000. Hohokam shell middens along the Sea of Cortez, Puerto Penasco, Sonora, Mexico. *Journal of Field Archaeology*. 27: 27–41.

Morzaria-Luna, H., Iris-Maldonado, A., and P. Valdivia. 2007. Reporte Final Técnico de CEDO Intercultural para el Proyecto "Estudio Integral Para la Determinación de Agentes Causales de Alta Mortalidad en los Cultivos del Ostión Japonés, Crassostrea gigas, de las Costas de Sonora." Centro Intercultural de Estudios de Desiertos y Océanos, A.C., Puerto Peñasco, Sonora.

Morzaria-Luna, H., Polanco-Mizquez, E., López-Alvirde, S. and S. Reyes-Fiol. 2007. *Caracterización de la vegetación de los esteros de Bahía Adair, Estero Morúa, y La Salina y predios circundantes*. ANEXO. Centro Intercultural de Estudios de Desiertos y Océanos, A.C., Puerto Peñasco, Sonora. Reporte Final. Diagnóstico Social y Ambiental para la Aplicación de Herramientas Legales de Conservación de Tierras Privadas y Sociales en torno a tres esteros del Golfo Norte de California, Sonora. A. Castillo López, responsable técnico.

Mosiño, P. and E. García. 1974. The climate of Mexico (pp. 345–404). In: R. A. Bryson and F. K. Hare (Eds.) *World Survey of Climatology, Vol. 2*, Climates of North America. Elsevier. New York.

Nelson, E.W. 1921. *Lower California and its natural resources*. Memoirs of the National Academy of Sciences, Washington, D.C.

Okin, G.S., Mahowald, N., Chadwick, O.A., and P. Artaxo. 2005. Impact of desert dust on the biogeochemistry of phosphorus in terrestrial ecosystems. *Global Biogeochemical Cycles*, 18(2): 1–9.

Olivares-Beltrán, G. 1969. *Acceso a la Bahía de Topolobampo, Sinaloa, México*. Lagunas Costeras, Un Simposio. Mem. Simp. Intern. Lagunas Costeras. UNAM-UNESCO. 407–420pp.

Orozco-Meyer, A. 2001. Uso del hábitat por la tonina (Tursiops truncatus) y su relación con las mareas en la Bahía de San Jorge, Sonora. Tesis de Maestría. CICESE. 78pp.

Parker, R. 1964. *Zoogeography and Ecology of Macro-invertebrates: Gulf of California and the Continental Slope off Mexico*. JGF Publishers. 178pp.

Phleger, F.A. and A. Ayala-Castañares. 1969. *Marine Geology of Topolobampo Lagoons, Sinaloa, México*. Lagunas Costeras, Un Simposio. Mem. Simp. Intern. Lagunas Costeras. UNAM-UNESCO. 101–136pp.

Robledo-Mejía, M.L., Navarro-Reyes, C., and Y. Ramírez-Pastrana. 2012. *Environmental Education Program Final Report*. Prescott College Center for Cultural and Ecological Studies, Bahía Kino, Sonora.

Rzedowski, J. 1978. Vegetación de México. Limusa. México. 432pp.

Scott, D.A. and. M. Carbonell. 1986. *Inventario de Humedales de la Región Neotropical*. IWRB Slimbridge & UICN. Cambridge.

Secretaría de Pesca (SEPESCA). 1987. El Municipio y la Producción Pesquera, Encuentro de Presidentes Municipales. Secretaría de Pesca, Secretaría de Gobernación, México, D.F. 27pp.

Sheridan, T. E. 1999. Empire of Sand: The Seri Indians and the Struggle for Spanish Sonora, 1645-1803. The University of Arizona Press, Tucson, Arizona. (https://open.uapress.arizona.edu/read/empire-of-sand/section/fc9e9dc4-6381-48c0-839d-8e11db3caf31)

Tershy, B. 1998. Sexual dimorphism in the brown booby. PhD Dissertation. Cornell University. Ithaca, New York. 157pp.

Tershy, B. R., Breese, D., Angeles-P., A., Cervantes-A., M, Mandujano-H., M., Hernández-N, E., and A. Córdoba-A. 1992. Historia Natural y Manejo de la Isla San Pedro Mártir Golfo de California. Reporte a Conservation International, A.C. Programa Golfo De California. Guaymas, Sonora.

Thompson, D.A., Findley, L.T., and A.N. Kerstitch. 1979. Reef Fishes of the Sea of Cortez: The Rocky Shore Fishes of the Gulf of California. University of Arizona Press, Tucson, Arizona. 302pp.

Torre-Cosío, J. 2002. *Inventory, monitoring and impact assessment of marine biodiversity in the seriindian territory, Gulf of California, Mexico*. PhD Dissertation. University of Arizona, Tucson, AZ.

Torre-Cosío, J. and L. Bourillón-Moreno. 2000. Inventario y monitoreo del Canal del Infiernillo para el comanejo de los recursos marinos en el territorio Seri, Golfo de California. *Conservation International, México, A.C. Informe final SNIB-CONABIO proyecto No. L179. México D.F.* 76pp.

Valdés, C.C., Tordesillas, B. M., Alan, B. B., Chavarría, C. E., Oriza, B. A., and N.E, Bravo. 1994. *Evaluación y Requerimientos de Manejo de los Humedales Costeros del Sur de Sonora: Recursos Naturales, Actividades Humanas y Educación Ambiental. Unidad de Información Biogeográfica. Centro de Conservación y Aprovechamiento de los Recursos Naturales*. Instituto Tecnológico y de Estudios Superiores de Monterrey-Campus Guaymas, Guaymas, Mexico. 197pp.

Valdes-Casillas, C., Carrillo-Guerrero, Y., Zamora-Arroyo, F., Hinojosa-Huerta, O., Camacho-López, M., Delgado-García, S., and M. Moreno-Báez, M. 1999. *Mapping and management of coastal wetlands of Puerto Peñasco, Sonora: A Multinacional Project. Center for Conservation of Natural Resources (CECARENA), Instituto Tecnológico y de Estudios Superiores de Monterrey-Campus Guaymas (ITESM-CG); Pronatura.* Arizona State University, Sonora.

Villalobos-Hiriart, J.L., Nates-Rodriguez, J.C., Diaz-Barriaga, A.C., Valle-Martinez, M.D., Hernandez-Flores, P., Lira-Fernandez, E., and P. Schmidtsdorf-Valencia. 1989. *Listados Faunisticos de Mexico.* I Custaceos estomatopodos y decapodos intermareales de las Islas del Golfo de California, Mexico. Universidad Nacional Autonoma de Mexico Instituto de Biologia, UNAM, Mexico, D.F.

Villegas O., R.E., Gracia A., B.D., and M.P. Barcelo L. 1985. *Estudio de la contaminación por residuos de plaguicidas en las bahías de Yávaros y Lobos, Sonora. Tesis de Licenciatura.* Depto. de Ciencias Químico-Biológicas. UNISON. 89pp.

Wilcox, B.A. 1980. *Aspects of the biogeography and evolutionary ecology of some island vertebrates. PhD Dissertation.* University of California, San Diego, California.

Wittman, E. 2012. *Waterbird Monitoring Program, Estuarine Nesting Season Summary: Estero Santa Cruz, Estero Santa Rosa, and Estero Cardonal.* Prescott College Center for Cultural and Ecological Studies, Bahía Kino, Sonora.

Yepiz V., L.M. 1990. *Diversidad, distribución y abundancia de la ictiofauna en tres lagunas costeras de Sonora, México.* Tesis de Maestría. UABC.

Zavala González, A. 1990. *La población del lobo marino Zalophus californianus (Lesson 1828) en las Islas del Golfo de California, México. Tesis De Licenciatura. Facultad de Ciencias.* Universidad Nacional Autonoma de México, México, D.F. 253pp.

Zuria, I. and E. Mellink. 2005. Fish abundance and the 1995 nesting season of the Least Tern at Bahia de San Jorge, northern Gulf of California, Mexico. *Waterbirds.* 28: 172–180.

Zuria-Jordan, I. 1996. *Disponibilidad de alimento y reproducción de Sterna antillarum (Aves : Laridae) en dos colonias del Noroeste de México. Tesis de Maestría.* CICESE. 88pp.

5 Mexico
States of Nayarit and Jalisco

The 12 Ramsar wetlands along the western coast of north-central mainland Mexico, in the States of Nayarit and Jalisco, are numerous and diverse in their type and ecological functions, from the volcanic Isla Isabel to the critical sea turtle habitat of Laguna Chalacatepec. These 12 coastal wetlands (Figure 5.1) are:

- Marismas Nacionales
- Parque Nacional Isla Isabel
- La Tovara
- Islas Marietas
- Sistema Lagunar Estuarino Agua Dulce-El Ermitaño
- Estero el Chorro
- Estero Majahuas
- Laguna Xola-Paramán
- Laguna Chalacatepec
- Reserva de la Biosfera Chamela-Cuixmala
- Estero la Manzanilla
- Laguna Barra de Navidad

MARISMAS NACIONALES, SINALOA AND NAYARIT: OVERVIEW

Marismas Nacionales (Figures 5.2 and 5.3) is a large network of brackish coastal lagoons, mangroves, swamps, and saltmarshes fed by several rivers. The site includes estuaries, the most extensive mangroves of Mexican Pacific coastal areas (i.e., 20% of all the mangroves in the country), timber-grade forests, and pastures.

At least 60 species of nationally or internationally endangered vertebrates occur at the site, including 51 endemic species, 36 of which are endemic birds. The *Orbignya* palm forests on sand bars of the site constitute a threatened community type. Numerous creeks at the site have been transformed into large prawn farms, with continued pressure of this type of human use likely occurring into the future. Other human activities at the site include traditional fishing and cattle ranching, as well as limited stocking of pigs, fowl, and bees. Fruit and seeds from the site are harvested by industries. The site incorporates the regions known as Las Cabras, Teacapán, Agua Brava, National Marsh, and San Blas Marsh. The site is fed by seven rivers and streams: The Baluarte, Cañas, Acaponeta, San Pedro, Bejuco, Santiago, and San Blas. The site is located on the south coast of Sinaloa and the north coast of Nayarit; this region has 113,000 ha of mangroves and estuaries, tropical timber forests, and pastures. There are 14 species of native flora that are at risk (i.e., endemic, threatened, and/or endangered) at the site. The fauna species in the region total 99 endemics, with 73 threatened or endangered species.

SITE DETAILS

Mangroves are characteristic of the shores of the estuaries at the site, as well as at the mouth of rivers and other water bodies, dominated by *Laguncularia racemosa*, *Rhizophora mangle*, *Avicennia germinans*, and *Conocarpus erectus*. Other conspicuous species at the site are the ciruelillo tree

DOI: 10.1201/9781003046394-5

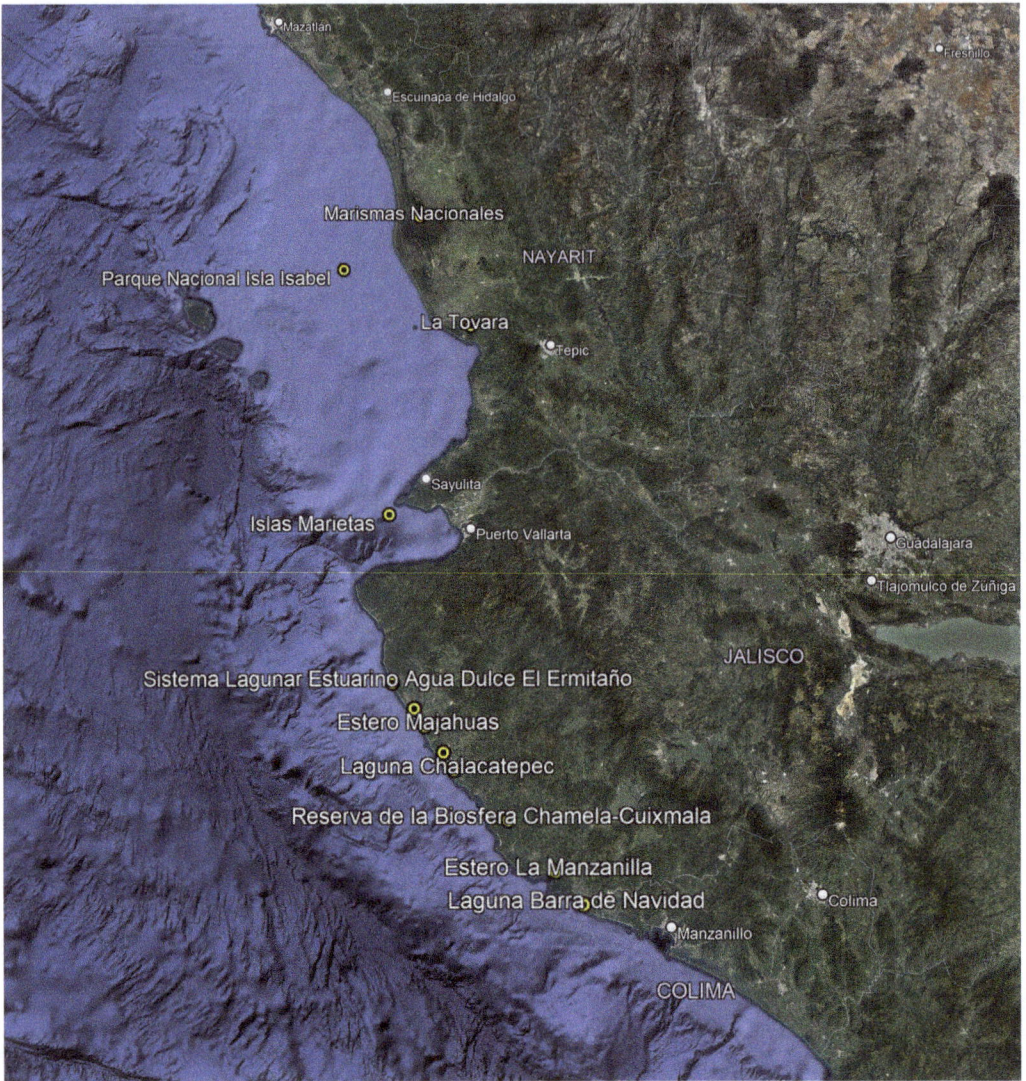

FIGURE 5.1　Referenced Ramsar wetlands generally shown along the western coast of north-central mainland Mexico, in the States of Nayarit and Jalisco.

(Phyllanthus elsiae), Guiana chestnut (Pachira aquatica), and the pond apple (Anona glabra). *Orbignya* palm forests of the site typically occur in disturbed sites, close to the coast, on sandy deep and well-drained soils. The dominant species in these areas of the site is *Orbignya guacuyule,* although other species such as *Ficus* spp. are sporadically present as well. The site's evergreen forests are distributed in patches throughout the lowlands, generally in areas with greater water availability and from 0 m to 1,000 m above sea level. The most conspicuous species in these areas of the site are jobo de lagarto (Sciadodendrom excelsum), breadnut (Brosimum alicastrum), coquito palm (Orbignya guacuyule,), and primavera tree (Tabebuia donnell-smithii). The fruits and seeds of *Orbignya* are exploited for the production of fats and soaps, and to the local people as food; logs from these forests are often used for construction of houses, but undoubtedly the greatest benefit is obtained from the leaves of this tree, which constitute a favorite material for house roofing, and are widely used for weaving bags, hats, and other handicrafts. Areas of *Orbignya guacuyule,* are often replaced by coconut plantations.

FIGURE 5.2 Composite aerial image of Marismas Nacionales. Designated January 22, 1995; Area: 200,000 ha; Coordinates: 22°08′N 105°32′W; Ramsar Site number: 732.

FIGURE 5.3 Map of Marismas Nacionales. Designated January 22, 1995; Area: 200,000 ha; Coordinates: 22°08′N 105°32′W; Ramsar Site number: 732.

The most common halophytic plant species at the site are *Salicornia* spp., *Batis* spp., *Sesuvium portulacastrum*, *Suaeda brevifolia*, *S. ramosissima*, and *Salicornia europaea*. The aquatic vegetation at the site is distinguished by three types of communities: tule, rooted to the bottom in shallow, flowing bodies of slow water, with the most frequent associations dominated by *Typha* spp., *Scirpus* spp., and *Cyperus* spp.; floating leaved vegetation, either rooted or unrooted, distributed in fresh or slightly brackish currents, dominated by *Eichhornia crassipes* and *Nymphaea* spp.; and algae, dominated by *Bostrychia radicans* and *Enteromorpha plumosa*.

Main faunal species in the coastal plain, outside the aquatic environment, are iguanas, bats, jaguar, armadillos, hares, rabbits, foxes, and deer (Anguiano 1992). Fauna at the site are classified as neotropical, are very diverse, and have a considerable number of endemic species. In Sinaloa and Nayarit, 408 (total) and 343 (vertebrate) species have been reported, respectively. Of these, at least 60 are endangered due to overexploitation and habitat destruction, of which 51 are endemic, including jaguar (Panthera onca), American crocodile (Crocodylus acutus), lilac-crowned parrot (Amazona finschi), green macaw (Ara militaris), and four species of sea turtle (Chelonia mydas, Dermochelys coriacea, Eretmochelys imbricata, and Lepidochelys olivacea). In the coastal region of Sinaloa-Nayarit, 98 species of mammals have been recorded (22% of the total within Mexico), which represent 8 orders, 21 families, 75 genera, and 165 species (National Atlas of Mexico 1990). Of the total number of species, 86 have been recorded in Sinaloa and 79 in Nayarit. At least 12 species are endemic to Mexico and 9 (10% of the total in Mexico) are in danger of extinction, as follows: river otter (Lutra canadiensis), the collared peccary (Tayassu tajacu), puma (Felis concolor), jaguar (Felis onca), ocelot (Leopardus pardalis), margay (Leopardus wiedii), and white-tailed deer (Odocoileus). Likewise, the jungles of the Pacific coast are the only winter habitat for 110 species of birds; there are a total of 252 species of birds at the site, of which 60% are residents, and the remaining bird species are migratory. The aquatic migratory bird species include those from groups such as plovers, waders, ducks, and pelicans. These species are generally winter visitors and their local distribution is restricted to bodies of water that provide shelter and food. Other noted species at this site are arboreal ducks (Dendrocygna autumnalis), stork (Mycteria americana), and osprey (Pandion haliaetus). The migratory species at the site are comprised of around 110 species of Passerines; densities of these migratory birds in the lowland jungle of the site are the highest recorded in the world. In addition, there are 36 endemic bird species at the site, among which are the lilac crowned parrot (Amazona finschi), the orange-fronted parakeet (Eupsittula canicularis), and the Mexican parrotlet (Forpus cyanopygius). The site regularly hosts 20,000 waterfowl at any one time. The species of reptiles and amphibians at the site are diverse and have key habitat types at the site that they depend upon, at least nine of which are endemic and 13 of which are in danger of extinction. Among these endangered species at the site are the Mexican beaded lizard (Heloderma horridum), the green iguana (Iguana iguana), the river crocodile, and the four species of sea turtles mentioned previously. The poisonous species in the region include the scorpion, the Mexican west coast rattlesnake (Crotalus basiliscus and C. atrox), the pit viper (Agkistrodon bilineatus), west Mexican coral snake (Micrurus distans), and the sea snake *Pelamys platurus*.

Aquaculture in the States of Sinaloa and Nayarit is a major human activity that affects the ecology of the site, particularly those impacts from shrimp farming. Although the shrimp farming itself can be an activity of economic importance to the people of the area, development without appropriate planning where the conservation of the environment is considered a priority, in particular mangroves and water quality, can be counterproductive to the goals of traditional fishers and fisheries, as well as for shrimp farming itself. Livestock cattle operations, and in the northern areas of the site, pig farming, poultry, and beekeeping are also major activities that have a negative impact on the ecology of the site. Nayarit has 900,000 head of cattle and Sinaloa has more than 2,000,000. Together these areas of cattle production represent more than 4% of the total number of head of cattle in the country of Mexico. The main tourist center of Nayarit is San Blas, and for Sinaloa the main tourist center is Mazatlán, which is outside of the site but receives hundreds of thousands of tourists per year, which affects the ecology of the site as well, due to proximity and intensity of the tourism industry. In the area surrounding the catchment basin of the site, there is the Aguamilpa Hydroelectric Project that is responsible for carrying out various conservation activities, focused on reforesting, and otherwise mitigating for, the areas negatively affected by the construction of the project. The extensive construction of aquaculture farms generated by national and international pressure to pursue this industrial track causes hydrologic pattern alterations, with consequences for the quantity and quality of freshwater runoff into the site, and to the sea. Also, shrimp farming itself has caused the elimination of resident and migratory birds from lands, indiscriminately.

Conservation measures adopted at the site include a project "linking communities, wetlands, and migratory birds," directed by Humedales Internacional and the federal delegation of SEMARNAT in Nayarit, which unites four sites in the North American Shorebird Reserve Hemispheric Network, which are sites that host migratory shorebirds on their journey from Mexico to Canada. These sites include Marismas Nacionales in Mexico, the Great Salt Lake in Utah, and Chaplin and Quill Lakes in Saskatchewan, Canada. This project focuses on the linkage of sites through education, communication, and conservation, with multiple local communities participating. Ongoing research activities at the site include the River Crocodile Conservation and Rescue Project, and the Reproductive Center of Crocodiles La Palma. Among tourist activities at the site are those of La Tovara Tours.

PARQUE NACIONAL ISLA ISABEL, NAYARIT: OVERVIEW

Parque Nacional Isla Isabel (Figures 5.4 and 5.5) is a volcanic island in the Pacific Ocean, 70 km off the port of San Blas. Deciduous tropical forest cover over lava soils, with grasses and bare rock are the dominant vegetational characteristics at the site. Cliffs, sandy beaches, coral reefs, and a crater lake with hypersaline water comprise this unique environment. Isla Isabel is one of the main nesting islands for seabirds in the Pacific, including large colonies of the magnificent frigatebird (Fregata magnificens), blue-footed boobies (Sula nebouxii), brown boobies (Sula leucogaster), and sooty terns (Sterna fuscata). Marine life at the site is rich, with over 24 shark and ray species, three species of sea turtles, the California sea lion (Zalophus californianus), humpback whale (Megaptera novaeangliae), and the killer whale (Orcinus orca). The island is not inhabited by humans, though there is a seabird reproductive research program run by the UNAM there, and tourists visit periodically to camp, fish, and bird watch. After having depleted a huge colony of sooty terns, which in 1978 had over 150,000 pairs, feral cats were finally eradicated from the island in 1995 and the terns are slowly recovering. The impact of hurricanes, fishing activities, and tourism are the primary concerns with regard to ecological restoration and conservation of the site.

SITE DETAILS

The site is located off of the coast of the State of Nayarit, 30 km from the Boca de Camichín Ranch in the Municipality of Santiago Ixcuintla, and 70 km from the port of San Blas in the Municipality of the same name. The port of San Blas is closest to Isabel Island and has a total population of 32,000 inhabitants. The site stands as a key refuge for seabirds, which, due to their life history characteristics (low reproductive rates, long life spans, few mechanisms against predators, specialized diets, and very specialized foraging techniques), are very vulnerable to disturbance. These birds depend almost exclusively on the islands to reproduce. A total of 92 species of birds have been noted for the site, highlighting nine species of seabirds, of which eight nest in large colonies, i.e., the blue-footed booby (Sula nebouxii), the brown booby (Sula leucogaster), the brown pelican (Pelecanus occidentalis), the magnificent frigatebird (Fregata magnificens), the red-billed tropicbird (Phaethon aethereus), Heermann's gull (Larus heermanni), the sooty tern (Sterna fuscata), the brown swallow or loggerhead swallow (Anous stolidus), and the red-footed booby (Sula sula).

The formation of the island is based on the overlapping of layers, made up of large basalt fragments, medium-sized basalt fragments, small basaltic fragments, and volcanic ash accumulated around the cone area. The layout of the volcanic materials of the island suggests that there were various volcanic activities in the past. The first phases of volcanism correspond to basaltic lava flows that formed the base of the island, extending several kilometers to the east, while to the north, west, and south the flows covered only a few 100 meters. In general, the island's rocks are lavic and pyroclastic, with the soils derived from these originating materials (Ruíz 1977). The formation of soil is also strongly related to precipitation, since there is a strong transport of materials into the inner-central depressions and the island's Crater Lake.

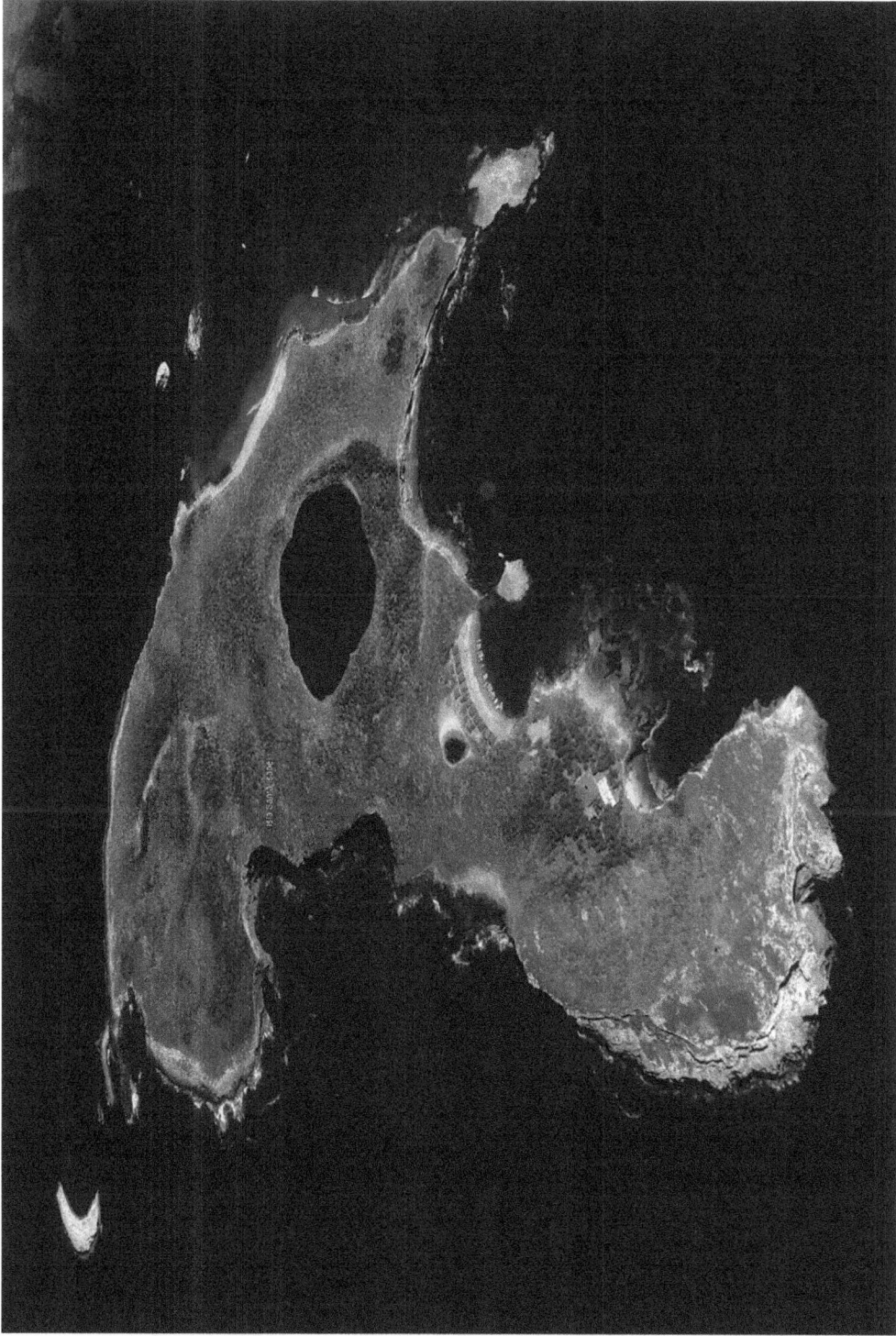

FIGURE 5.4 Composite aerial image of Parque Nacional Isla Isabel. Designated November 27, 2003; Area: 94 ha; Coordinates: 21°51′N 105°53′W; Ramsar Site number: 1,324.

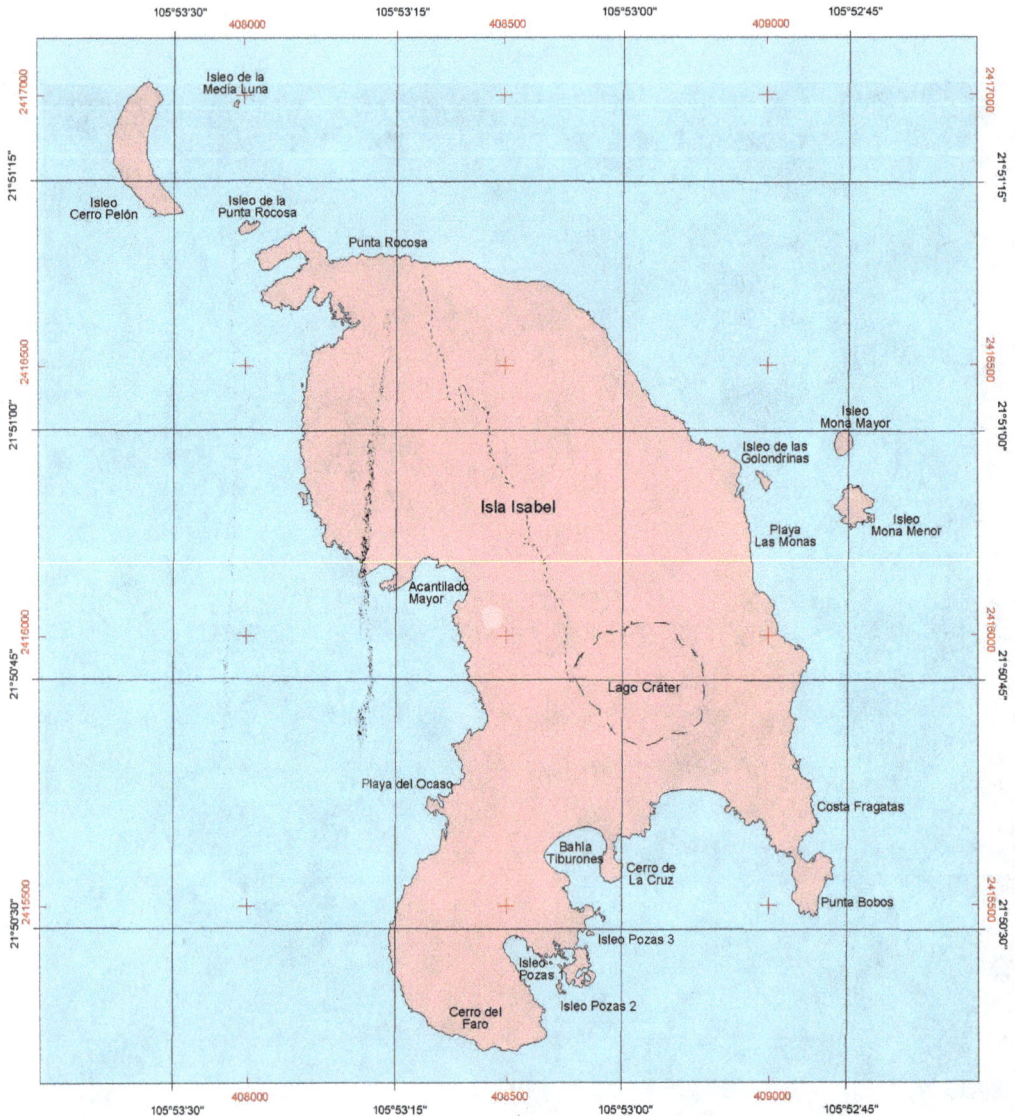

FIGURE 5.5   Map of Parque Nacional Isla Isabel. Designated November 27, 2003; Area: 94 ha; Coordinates: 21° 51′N 105°53′W; Ramsar Site number: 1,324.

73% of the entire surface of Isabel Island is covered with vegetation, distinguished by three main types of coverage, which occupy different proportions of the surface: deciduous tropical forest (62%); bare rock, islands, and rocky coastline (22%); grassland, prairie, and halophytic vegetation (9%); without vegetation beaches, inland water bodies (5%); and introduced vegetation (2%).

The spider flower tree (Creteva tapia), which is the most abundant plant species on the island, is a good colonizer. It has a high intrinsic rate of growth, achieved by it slow mortality rate and high reproductive rate. On Isabel Island, the breeding sites of the most abundant marine bird, *Fregata magnificens*, are linked to the distribution of *C. tapia* trees, where they establish their nests and rest. Also using *C. tapia* for nesting is the brown pelican and the red footed booby. Other plant species of ecological importance at the site are the native grass *Cyperus ligularis*, since in certain parts of the north and northwest of the island, sooty terns (Sterna fuscata) build their nests preferably in shelters, where these grasses, in the form of clumps, are folded or gathered to form a cavity, leaving a small entrance at ground level. No endemic floristic species have been reported for the island, nor floristic species registered under any category of risk.

Terrestrial fauna at the site include a total of 101 species, including 1 amphibian, 6 reptiles, 92 birds, 1 bat (Natalus stramineus), and 1 introduced species of rat (Rattus rattus). Domestic cats have been eradicated from the site. The following are the distribution patterns of the most abundant vertebrate species at the site: 92 species of birds, including the especially ecologically important *Fregata magnificens*; blue-footed booby (Sula nebouxii); brown booby (Sula leucogaster); red-footed booby (Sula sula); red-billed tropicbird (Phaethon aethereus), which is listed as a threatened species (NOM-059-SEMARNAT-2001 and/or -2010); Heermann's gull (Larus heermanni); sooty tern (Sterna fuscata); and brown swallow (Anous stolidus).

Six species of reptiles have been recorded at the site, none of them endemic. Marine species at the site include 22 representatives of the macroalgae, and 167 species of phytoplankton, of which 70 belong to the diatoms, 87 to dinoflagellates, 8 to cyanophytes, and 2 to silicoflagellates. A total of 34 species of macroalgae are reported at the site, corresponding to 27 genera, 21 families, 16 orders, and 4 divisions. From the records of benthic macroinvertebrates in the intertidal zones and shallow subtidal areas at the site, 11 species of crustaceans, 9 species of echinoderms, and 26 species of molluscs are noted at the site.

In the subtidal habitats of Isla Isabel, approximately 105 species belonging to five Phyla (Porifera, Cnidaria, Annelida, Mollusca, and Echinodermata) and one Subphylum (Crustacea) are supported. Decapod crustaceans represent the greatest richness at the site, with 24 species, followed by gastropod molluscs with 22 species; gorgonaceans-cnidarians with 20 species; and echinoderms with 19 species. Annelids-polychaetes and sponges (Porifera) are represented at the site by six species, and two species, respectively. Opisthobranch molluscs are represented by 25 different species at the site. Isabel Island's coral reef is composed of seven hermatypic coral species and at least two ahermatypic species: three species with branched growth, of genus *Pocillopora* (Pocillopora verrucosa, P. meandrina, and P. capitata), and several species with massive growth, of three genera: *Porites lobata, P. panamensis*; *Pavona gigantea*, and *Psammocora stellate* (an encrusting species). The ahermatypic corals present at the site are equatorial *Tubastraea coccinea* and *Astrangia*. Pérez (2001) reports the presence of 24 species of sharks and rays in waters near the island. There are also many sea turtles at the site, mainly olive ridley (Lepidochelys olivacea), green (Chelonia mydas), and hawksbill (Eretmochelys imbricata). Isla Isabel is also located among the vital routes of the humpback whale (Megaptera novaeangliae), which can be seen in winter as it journeys to the south. There are also sightings of California sea lion (Zalophus californianus), and numerous dolphins and killer whale (Orcinus orca) at the site.

For more than 80 years and prior to the creation of the site as a National Park (as of December 8, 1980), the island was used as a base of operations for fishers from the local communities, and as a refuge for maritime safety. The island is visited by groups of tourists who come to camp, fish, and participate in tours to observe and photograph birds. Isla Isabel falls under the jurisdiction of the Ministry of the Interior as a National Park, and is almost completely devoid of infrastructure and facilities. The number of fishing boats that can be found simultaneously at the site varies between 2 and 100, and fishing camps at the site are typically made of multiple palapas built of wood. Outside the limits of the site, the main fish species with commercial value captured are snapper, dogfish, and shark. On a smaller scale, other species that are used for self-consumption or to be used as bait are horse mackerel, skipjack, corvina, sardine, sole, mullet, mackerel, sea bass, and grouper. Isabel Island is constantly visited by national and foreign tourists who arrive via sailboats, yachts, or other small boats. The number of visitors to Isabel Island for recreational purposes is very limited, due to its remoteness from the coast and transportation costs; accordingly, and in general, there is not a large influx of visitors to the site, with an average of 1,000 visitors/year. Scientific research at the site consists of the UNAM seabird studies, as well as similar work by those at the Autonomous University of Nayarit and the University of Guadalajara.

Adverse factors affecting the ecology of the site include changes in land use and development projects in the region. Many of the activities carried out by fishers can affect the conditions of the environment at the site, causing problems related to the accumulation of domestic garbage and waste from fish, so constant vigilance is required in the associated camps to prevent proliferation of rats and flies. On the island, there are also several plant species that have been introduced by humans, including pineapple, banana, purslane, cane, lemon, and papaya, as well as exotic grasses. Due to the lack of permanent surveillance, the transit of tourists in bird nesting areas is another important cause of concern, as these disturbances can directly affect the success of bird colonies.

LA TOVARA, NAYARIT: OVERVIEW

La Tovara (Figures 5.6 and 5.7) is a transitional area, from continental uplands to oceanic waters, comprised of mangroves of four species: *Rhizophora mangle*, *Laguncularia racemosa*, *Avicennia germinans*, and *Conocarpus erectus*, all of which are under special protection per Mexican legislation. *Mastichodendron capiri* and *Chamaedorea pochutlensis* are threatened tree species found at the site, the latter (a palm) also endemic to Mexico. Bird species under special protection at the site include *Amazona finschi*, *Ara militaris*, *Ardea herodias*, *Cyanocorax beecheii*, *Forpus cyanopygius*, *Icterus spurius*, *Melanotis caerulescens*, *Tigrisoma mexicanum*, and *Vireo atricapillus*. The site is home to six cat species found under special protection. Agriculture, livestock, forestry, fishing, aquaculture, and tourism are the main human activities at the site.

The site is located in the State of Nayarit, municipality of San Blas, an area where continental and oceanic waters combine through the phenomenon of 'tidal mixing' that favors the combination of waters in the coastal zone, including sediment, and/or the removal of particles in those areas, as well as the entry and exit of organisms within estuaries and lagoons. This process promotes the exploitation of nutrients from coastal areas to the open sea, which has a direct influence on primary productivity and the fishing potential of contiguous marine areas offshore. Sources of income for communities in the area include cultivation of corn (Zea mays), banana (Musa paradisiaca), mango (Mangifera indica), plum (Spondias purpurea), water coconut (Cocos nucifera), and papaya (Carica papaya).

SITE DETAILS

The volume of water that can be stored within the ecosystem and surrounding areas allows for regulation and retention of groundwater tables, upon which the vegetational communities depend, such as mangroves, riverside forests, tules, and the semi-deciduous forests. Crops in the area also benefit from groundwater, retained soil moisture, and even the flooding of lowlands, which allows for sustained agricultural and livestock production. Mangroves and other aquatic plant communities also carry out important ecosystem services at the site, such as retention of sediments, elimination and sequestration of some pollutants, protection of nearby communities from strong winds and tidal surges, and preventing the alteration of the coastline, and the shores of bodies of water adjacent to the coast.

The site has a total of 704 species, of which 187 belong to plants and 517 to fauna. The plants of the site are grouped into 71 families; 57 trees, 31 shrubs, and 99 herbaceous species; faunal species are comprised of 199 bird species, 90 mammals, 22 reptiles, 9 amphibians, 160 insects, 31 fish, 3 molluscs and 3 crustaceans. The most representative trees at the site are *Aphananthe monoica*, *Ficus petiolaris*, *Guarea glabra*, *Brosimum alicastrum*, *Bursera simaruba*, *Castilla elastica*, *Ceiba pentandra*, *Enterolobium cyclocarpum*, *Ficus glabrata*, and *Calliandra magdalenae*. Other notable plant species at the site include *Byrsonima crassifolia*, *Cecropia peltata*, *Croton panamensis*, *Guazuma ulmifolia*, *Helicteres guazumaefolia*, *Pithecellobium dulce*, *Waltheria americana*, *Mimosa albida*, and *M. pigra*. The submersed and emergent plant communities at the site include *Spirodela polyrhiza*, *Pistia stratiotes*, *Lemna aequinoctialis*, *Typha domingensis*, *Sagittaria*

FIGURE 5.6 Composite aerial image of La Tovara. Designated February 02, 2008; Area: 5,733 ha; Coordinates: 21°35′N 105°15′W; Ramsar Site number: 1,776.

FIGURE 5.7   Map of La Tovara. Designated February 02, 2008; Area: 5,733 ha; Coordinates: 21°35′N 105°15′W; Ramsar Site number: 1,776.

lancifolia, Crinum sp., *Acrostichum danaeifolium, Hydrocotyle mexicana, Phragmites australis, Sorghastrum* sp., and *Paspalum paniculatum*. The four mangrove species that define this ecosystem are *Rhizophora mangle, Laguncularia racemosa, Avicennia germinans*, and *Conocarpus erectus*.

Per NOM-059-SEMARNAT-2010, 26 species of birds are imperiled at the site: 6 threatened, 17 under special protection, and 3 in danger of extinction, inclusive of the lilac-crowned parrot (Amazona finschi), green macaw (Ara militaris), great blue heron (Ardea herodias), purple-backed jay (Cyanocorax beecheii), Mexican parrot let (Forpus cyanopygius), chestnut lark (Icterus spurius), blue mockingbird (Melanotis caerulescens), bare-throated tiger heron (Tigrisoma mexicanum), and the black-capped vireo (Vireo atricapillus). Of the 90 species of mammals, 13 are in the NOM-059-SEMARNAT-2010: 8 threatened, 4 in danger of extinction, 1 in special protection, and 5 endemic. Among these are 6 species of cats present in the area, including: the jaguarundi (Herpailurus yagouaroundi), which is threatened; the jaguar (Panthera onca); the ocelot (Leopardus pardalis); the endangered margay (Leopardus wiedii); the puma (Puma concolor); and the bobcat (Linx rufus), which indicates a degree of conservation success in the area, despite a history of poaching and deforestation. Of 22 species of reptiles at the site, 11 are reported in NOM-059-SEMARNAT-2010, 9 under special protection and 2 threatened. Among these are the threatened boa (Boa constrictor) and the Mexican spiny-tailed iguana (Ctenosaura pectinata); the specially protected green iguana (Iguana iguana), American crocodile (Crocodylus acutus), Mexican west coast rattlesnake (Crotalus basiliscus), and coral snakes (Micrurus distans and Micrurus proximans). Four of the nine amphibian species at the site are under special protection per NOM-059-SEMARNAT-2010: *Gastrophryne usta, Rana forreri, Eleutherodactylus pallidus*, and *Eleutherodactylus modestus*, of which the last two are endemic.

Of the 199 bird species recorded at the site, 70 are aquatic and 129 are terrestrial; 29 aquatic birds are migratory and only 4 of the terrestrial birds are migratory. There are more than 25 species of fish at the site, belonging to 20 families: Carangidae (toritos, horse mackerel), Sciaenidae (drums), Engraulidae (anchovy), Centropomidae (sea bass), Lutjanidae (snails), and the Cupleidae (sardines). Among the dominant species are *Mugil curema, Centropomus pectinatus, Lutjanus colorado*, and *Galeichthys* sp. Among the commercially important species are *Centropomus* sp. (bass), *Lutjanus* sp., *Mugil* sp. (mullet), *Penaeus* sp. (shrimp), and the oysters (Crassostrea iridescens and C. corteziensis).

In 1587, Viceroy Álvaro Manrique de Zúñiga ordered his forces to continue exploration of the Mexico-Guadalajara route until reaching the sea. Accounts hold that Friar Pedro Gutiérrez also brought indigenous peoples from mountainous regions in 1607 to further populate areas at the site, where they continue to reside today (Cardenas 1968). During the month of January 1768, steps were taken to install the port of San Blas, where Captain Croix appointed Don Manuel Ribero as Commander, so that together with the population of San Blas the port was established. Due to the war of independence in 1810, the port fell into disrepair and disarray, yet artisanal fishing has been carried out in the bodies of water at the site ever since, by a fisher's cooperative (primarily men) of San Blas with 120 members, plus a group of 25 women who fish in El Conchal. Much of the land in the area is actively used for agriculture, livestock, forestry, fishing, aquaculture, and tourism.

Modification of the environment has impacted the site through the destruction of habitat, deforestation, and the desiccation and burning of the mangroves. Change in the hydrologic patterns of the site have caused salt intrusion into adjacent areas, both towards the plain (cultivation areas) and towards the mangroves. Additional urban and agricultural wastewater pollution, as well as from garbage and agrochemicals, has impacted the ecology of the site and vicinity. Overuse and uncontrolled extraction of fish, crustaceans, and other vertebrates through poaching is also an ongoing risk to the ecology of the site. An awareness campaign called Conservation Strategies for Priority Areas, coordinated with the Nayarita Institute for the Sustainable Development, Ministry of Public Education-Nayarit, Conservation International Mexico, and the Public Education Services of the State of Nayarit has raised awareness among the population about the importance of conserving and managing the site sustainably through talks, conferences, workshops, and print media.

ISLAS MARIETAS, NAYARIT: OVERVIEW

Islas Marietas (Figures 5.8 and 5.9) are located off the coast of Nayarit, in Bahía de Banderas, an archipelago that consists of two small islands and additional islets, all of volcanic origin. The wind, sun, rain, and waves have transformed the substrate of the site, creating various areas that produce an enormous amount of local biodiversity. Islas Marietas are home to a large variety of resident and migratory seabirds, including brown boobies (Sula leucogaster), seagulls, and pelicans. The archipelago is also important for the breeding of marine species like the humpback whale (Megaptera novaeangliae) and the olive ridley sea turtle (Lepidochelys olivacea). The main threat to the site is the "Escalera Náutica del Golfo de California" project, which involves the construction of marinas and other associated infrastructure (hotels, airports, etc.), as well as the increase of the number of boats and visitors sailing around or coming to the islands. The archipelago has been declared a Special Biosphere Reserve and a National Park.

FIGURE 5.8 Composite aerial image of Islas Marietas. Designated February 02, 2004; Area: 1,357 ha; Coordinates: 20°42′N 105°34′W; Ramsar Site number: 1,345.

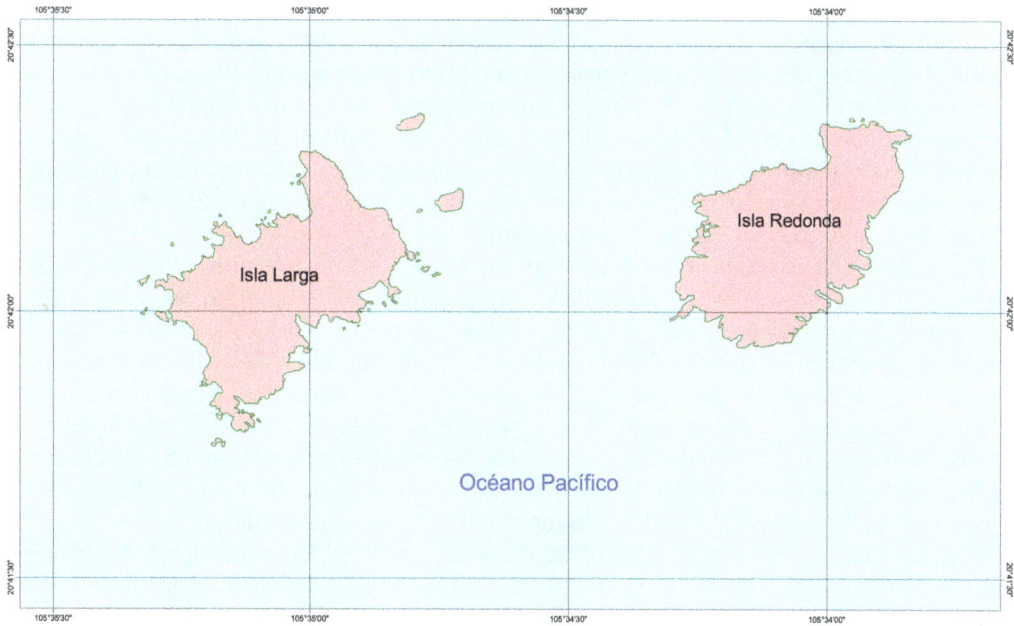

FIGURE 5.9 Map of Islas Marietas. Designated February 02, 2004; Area: 1,357 ha; Coordinates: 20°42′N 105°34′W; Ramsar Site number: 1,345.

SITE DETAILS

The Marietas Islands are surrounded by permanent shallow marine waters, coral reefs, and subtidal marine beds; its coasts are formed by cliffs, sandy beaches, rocky beaches, and several islets. The Marietas Islands (Long Island and Round Island) stand out for their ornithological and ichthyofaunistic richness, as well as being essential for the reproduction of populations of protected species, among which are the humpback whale, the olive ridley turtle, and several species of birds. The site supports the largest nesting colonies in Mexico of brown booby (Sula leucogaster), bridled tern (Sterna anaethetus), brown swallow (Anous stolidus), and the laughing gull (Larus atricilla) [Ribbon 1999]. In the marine zone, the presence of corals and a great variety of associated reef fauna are notable, as evidenced by the 115 species of reef fish present at the site.

On both islands the families Gramineae (Poaceae) and Cyperaceae predominate, with plants of short stature with rhizomatous, and semi-climbing habits. The shrub layer at the site is minimal and the tree layer is almost absent. Grasslands of the site are located mainly in flat areas, where slightly deeper soils are present. There are also open pastures in several areas, where the terrain becomes more irregular and with frequent outcropped areas of rocks. In the northern middle part of Long Island, the species *Tripsacum dactyloides* predominates, with the following other mixes found on the island in areas of shores and cliffs: *Cyperus ligularis* and *C. sanguineo-ater* associated with *Pennisetum setosum*, *Eragrostis prolifera*, *Hackelochloa granularis*, *Aristida ternipes*, and *Cyperus dentoniae*. In open grassland stands, *Chamaesyce thymifolia*, *Fimbristylis dichotoma*, *Phyllantus standleyi*, *Ophioglossum engelmannii*, and *Piriqueta cistoides* reside. In some places *Lygodium venustum* form large dense patches. Rare, isolated communities of shrubs *Opuntia wilcoxii*, *Waltheria americana*, *Physalis minuta*, *Commicarpus scandens*, and *Elytraria imbricata* can be observed. In the caves of the southern part of Long Island, *Phlebodium decumanum* and *Stenocereus standleyi* are found.

Main fauna species at the site are the blue-footed booby (Sula nebouxii), magnificent frigatebird (Fregata magnificens), brown pelican (Pelecanus occidentalis), Brandt's cormorant (Phalacrocorax penicillatus), Heermann's gull (Larus heermanni), and the royal tern (Sterna maxima) [Gavino and Uribe 1980, Rebón 1989]. The Islands represent the geographical limits and areas of expansion of distribution for reproduction of species that live in nearctic zones, such as *Phalacrocorax penicillatus*, *Larus heermanni*, and *Sterna maxima*, and for species that nest in neotropical areas such as *Sterna anaethetus*. Per NOM-059-SEMARNAT-2001 and/or -2010, of the 92 recorded birds at the site, four species are threatened and five are species subject to special protection. The presence of the humpback whale (Megaptera novaeangliae) in the Bay of Banderas represents a very significant role for the site, which supports their breeding during the winter months.

The main human activities linked to the Marietas Islands are coastal fishing (30%) and tourism (70%), upon which most of the local population of the communities are influenced, either directly or indirectly. Due to the increase in visitors arriving in Puerto Vallarta, tourist activities at this destination are continually increasing. This has resulted in a diversification of activities of the riverside fishers, who now participate more in tourist services, both for the development of tourism-based fishing activities, and for the transport of tourists who are interested in diving and snorkeling in the surroundings of the Marietas Islands, as well as for the observation of humpback whales during the months of November to March. The Marietas Islands are under federal jurisdiction with the Ministry of the Interior, and completely lacking in infrastructure and facilities.

Adverse factors affecting the ecology of the site include potential for the development of marinas and associated infrastructure (hotels, airports, etc.), as well as the increase in the number of vessels sailing in the Gulf of California and, consequently, an increase in the number of visitors, which can bring indirect impacts to the islands such as the introduction of exotic species, pollution, fire risks, impact on corals, and disturbances to nesting bird colonies. Specifically, in the State of Nayarit, the localities of Nuevo Vallarta, Jaltemba, and San Blas have been chosen as sites for the establishment of nautical stopovers by vacationers. Similarly, with the growth of tourism comes community development, which has potential for increasing urban, agricultural, aquaculture, and mining or other industrial impacts at the site.

The Islands of the Gulf of California are recognized worldwide for their beauty, biological richness, and productivity of the waters that surround them, which is why the Mexican Government has executed policies for its conservation, including the designation of Natural Areas (August 2, 1978) that establishes the area as a reserve area and refuge for migratory birds and wildlife on the islands within the Gulf of California. As of June, 2000, the site is also considered an official Protection Area for Flora and Fauna within the Islands of the Gulf of California. The Official Mexican Standard NOM-131-SEMARNAT-1998 (DOF 2000) establishes guidelines and specifications for the development of whale watching activities, as related to their protection and conservation of habitat. This protective measure is complemented by a notice in which it establishes specifications and prohibitions, highlighting the fact that the marine areas surrounding Marietas Islands, specifically to a radius of 1.5 km, and the strip located from Playa Litigú to the Ameca River, are a restricted zone. In this restricted zone, access and development of whale watching activities will only be allowed by scientists under the permits issued by SEMARNAT, and any other type of whale watching is prohibited. Additionally, Marietas Islands are a Protected Natural Area with the category of National Park, effective November 27, 2002. On June 5, 2003, the Advisory Council of the Isla Isabel National Park and the Marietas Islands was established to further enhance involvement of community members in sustaining Isabel Island and the Marietas Islands, and to advise on management strategies for the future of the sites.

Research at the site includes several projects, such as 'Monitoring of the Coral Communities of the Marietas Islands' (University Center of the Coast, University of Guadalajara; 'Unification of the catalog of photo identification of the humpback whale (Megaptera novaeangliae) in Banderas Bay' (Ecotours de Mexico); and 'Cetaceans of Bahía Banderas and adjacent waters, Coast of Nayarit, and Jalisco' (Technological Institute of the Sea). Education and outreach activities at the site have included annual brochures about the site's flora and fauna and other general information about the Marietas Islands, and the importance of conserving them, along with recommendations for sustainable visitation. As was mentioned above, tourism at the site is thriving, with activities including snorkeling, diving, boat touring, sport/commercial fishing, whale watching, and dolphin watching.

SISTEMA LAGUNAR ESTUARINO AGUA DULCE-EL ERMITAÑO, JALISCO: OVERVIEW

Sistema Lagunar Estuarino Agua Dulce-El Ermitaño (Figures 5.10 and 5.11) is a lagoon ecosystem and is considered the main coastal water body for the State of Jalisco, comprised of two water bodies: Agua Dulce Lagoon and El Ermitaño Estuary. The estuaries are interconnected with a channel, constructed in the 1960s with floodgates that regulate the entrance from the estuary to the lagoon. A total of 95 species of waterfowl have been reported at the site, of which 69 species are migratory and 26 that are residents of the site and its surroundings. The site is also home to species found under special protection per Mexican conservation statute, which includes *Heloderma horridum*, *Iguana iguana*, *Crocodylus acutus* (endangered), *Branta bernicla*, *Anas platyrhynchos*, and *Nomonyx dominicus* (threatened). The main threats that negatively affect the ecology of the site include the diversion of water for agricultural purposes, the use of agrochemicals in the surrounding crops, sewage water coming from rivers, and deforestation. The main human activities at the site and vicinity are fishing and tourism. The site is also located next to an important marine turtle sanctuary (El Playón de Mismaloya).

SITE DETAILS

The site is adjacent to the beach called El Playón de Mismaloya, along a coast that is characterized by small bays with sandy beaches at its southern end, and by rocky outcroppings and mountains that define the site's associated alluvial valleys. The central and central-north portion of the Jalisco coast is characterized by large beaches, occasionally fragmented by rocky hills, and bathed by rivers that discharge a large amount of sediment into the sea and beaches.

The site is dominated by red mangrove (Rhizophora mangle), white mangrove (Laguncularia racemosa), black mangrove (Avicennia germinans), button mangrove (Conocarpus erectus), all of which are included in NOM-059-SEMARNAT-2001 and/or -2010 as species subject to special protection. The flora on the margins of the site are represented by 19 families and 43 species. Mangrove areas are exploited for their diversity and productivity, by the local inhabitants mainly for firewood, construction wood, posts, medicines, and crafts. Plant species at the site are also used for obtaining or manufacturing food (Amaranthaceae, Compositae, Cucurbitaceae, Gramineae, and Leguminoceae), timber use (Combretaceae, Leguminoceae, Verbenaceae, Sterculiaceae, and Rubiaceae), and for medicinal uses (Aizoaceae, Boraginaceae, Compositae, Cucurbitaceae, Cyperaceae, Graminaceae, Leguminoceae, Malvaceae, Portulacaceae, and Rubiaceae).

Ninety-five aquatic bird species have been identified at the site, which represents 81% of the species identified along the entire coast of Jalisco. Among the species at the site are those included in NOM-059-SEMARNAT-2001 and/or -2010, such as: *Ardea herodias*, *Egretta rufescens*, *Mycteria americana*, *Larus heermanni*, *Sterna elegans*, *Sternula antillarum* (special protection); *Branta bernicla*, *Anas platyrhynchos*, and *Nomonyx dominicus* (threatened) [Hernández-Vásquez 2005]. The 95 species recorded at the site are also included in the Red List (IUCN 2007); five of the species noted are in the near threatened category (Puffinus griseus, Charadrius melodus, Numenius americanus, Sterna elegans, and Larus heermanni), and 90 are categorized as least concern. Among reptiles at the site, there are a number of endangered species, such as *Heloderma horridum*, *Iguana iguana*, and *Crocodylus acutus*, all protected by NOM-059-SEMARNAT-2001 and/or -2010. In the contiguous area of the site, to the southwest and west, is a marine turtle sanctuary (El Playón de Mismaloya), in which the following sea turtles nest: leatherback sea turtle (Dermochelys coriacea), olive ridley sea turtle (Lepidochelys olivacea), and green (Chelonia mydas), and all three are

FIGURE 5.10 Composite aerial image of Sistema Lagunar Estuarino Agua Dulce-El Ermitaño. Designated February 02, 2008; Area: 1,281 ha; Coordinates: 20°0′N 105°30′W, Ramsar Site number: 1,825.

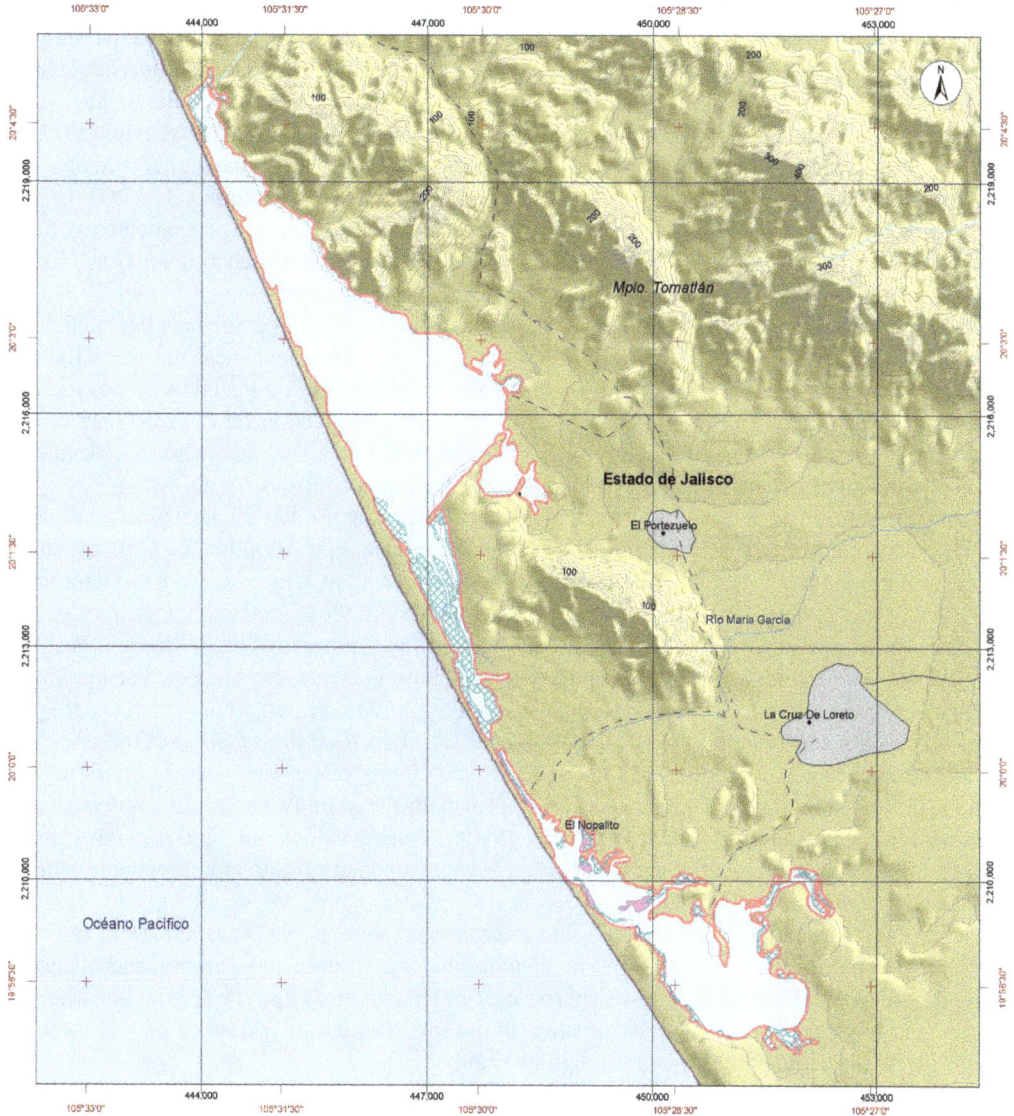

FIGURE 5.11 Map of Sistema Lagunar Estuarino Agua Dulce-El Ermitaño. Designated February 02, 2008; Area: 1,281 ha; Coordinates: 20°0′N 105°30′W, Ramsar Site number: 1,825.

classified as endangered species in danger of extinction in NOM-059-SEMARNAT-2001 and/or -2010, the Red List of the IUCN, as well as in the CITES.

The site also has 26 resident bird species, 13 of which breed in mangroves and sandy flats (Ardea alba, Butorides virescens, Bubulcus ibis, Cochlearius cochlearius, Egretta caerulea, Egretta thula, Egretta tricolor, Nyctanassa violacea, Nycticorax nycticorax, Phalacrocorax brasilianus, Himantopus mexicanus, Haematopus palliatus, and Rynchops niger) [Hernández-Vázquez 2005]. The lagoon also provides critical habitat for 79 species of fish, belonging to two classes, 15 orders, and 34 families. Five of these species reproduce within the lagoon, and 26 enter in early stages and grow within of the ecosystem, while seven species play an important role in stabilizing the ecosystem (i.e., are an important component of the food and/or nutrient web/cycle). The lagoon also provides refuge, feeding, and breeding habitat for shrimp belonging to the families Palaemonidae, Alpheidae, Hippolytidae, Proccesidae, and Penaeidae (Hendrickx 1988). Due to the above, the site is considered to be especially contributory to high ichthyological diversity along the western coast of Mexico (Aguilar-Palomino 2006). Some species of fish found at the site are crappie (Gerreidae), snappers (Lutjanidae), sea bass(Centropomidae), catfish or chihuiles (Ariidae), hammerhead sharks (Sphyrna lewini and Sphyrna zygaena), catfishes (Arius guatemalensis, Arius platypogon, Arius planiceps, Ariopsis seemanni, and Sciadeops troschelii), milkfish (Chanos chanos), mullets (Mugil cephalus, Mugil curema, and Mugil hospes), and Pacific fat sleeper and guavina (Dormitator latifrons and Guavina micropus) [Nelson 1994, Fisher et al. 1995, Aguilar-Palomino et al. 2006]. Other fauna at the site include opossum (Didelphis marsupials), armadillo (Dasypus novemcinctus), South American coati (Nasua nasua), and raccoon (Procyon lotor); and reptiles such as iguanas and lizards (Iguana iguana, Ctenosaura pectinata, Anolis nebulosus, and Sceloporus spp.).

The first major settlements on the Jalisco coast were established in what are now the municipalities of Tomatlán and Cihuatlán. The most important of human settlements near the site is La Cruz de Loreto, with approximately 1,900 inhabitants. This community is dominated by cattle ranching, agriculture, and fishing. The Cooperative Society of Fishery Production La Cruz de Loreto, with more than 50 members, has maintained itself, for fishing within the site; this fisher's association is a model in the management of fishing cooperatives in the State of Jalisco.

Adverse effects on the ecology of the site include the diversion of water for agricultural purposes, use of agrochemicals in the surrounding cultivation areas, water extraction for domestic use that is discharged into the rivers that supply the estuary, and deforestation. There is also a constant threat of upcoming tourist mega-developments on the margins of both the lagoon and the estuary.

Conservation measures at the site include the work of the Cooperativa de Producción Pesquera La Cruz de Loreto, with strict closed seasons for catching of some fish species, but mainly for shrimp. The University of Guadalajara has developed scientific research whose products are published in journals, and in support of bachelor's, master's, and doctoral theses. Some of the associated projects address key aspects of hydrology and climatic factors of the site, specifically in relation to the hydrological balance of the site, as associated with the salinity variations observed in the site's waters (Ocegueda et al. 1978, Ocegueda 1980). Olmedo-Valdovinos (2007) carried out an analysis of water and sediment of the site, finding acceptable conditions to culture whiteleg shrimp *Litopenaeus vannamei*.

The site is visited mainly by national and international tourists, who participate in the main recreational activities that take place there, including fishing, swimming, walking, and horseback riding on the beach. Boat tours with the purpose of observing birds and the scenic landscapes of the site help to support additional conservation at the site, through the exposure of visitors to the complexity of the ecology of the site, through its beauty.

ESTERO EL CHORRO, JALISCO: OVERVIEW

Estero El Chorro (Figures 5.12 and 5.13) is an estuarine ecosystem located along Mexico's western-central littoral zone. The mouth of the site is open to the sea for only 6 months of the year, and most of the lagoon is surrounded by spiny forest vegetation, as well as some mangroves of the species *Laguncularia racemosa* and *Conocarpus erectus*. The site provides ideal habitat for a variety of fish, molluscs, crustaceans, reptiles, as well as resident and migratory bird species. The site is important resting and feeding habitat for *Himantopus mexicanus*, *Calidris mauri*, *Catoptrophorus semipalmatus*, *Ardea alba*, and *Butorides virescens*. Human activity at the site includes artisanal fishing and guided tours. Mangrove deforestation and agricultural expansion are the main ecological threats to the site.

SITE DETAILS

The opening of the mouth of the estuary is manipulated by the fishers of the area; when the mouth of the lagoon is closed, the water levels gradually decrease due to the effects of evaporation and underground seepage. On the periphery of the lagoon, vegetation of deciduous forest and mixed forest

FIGURE 5.12 Composite aerial image of Estero El Chorro. Designated February 02, 2008; Area: 267 ha; Coordinates: 19°54′N 105°24′W; Ramsar Site number: 1,791.

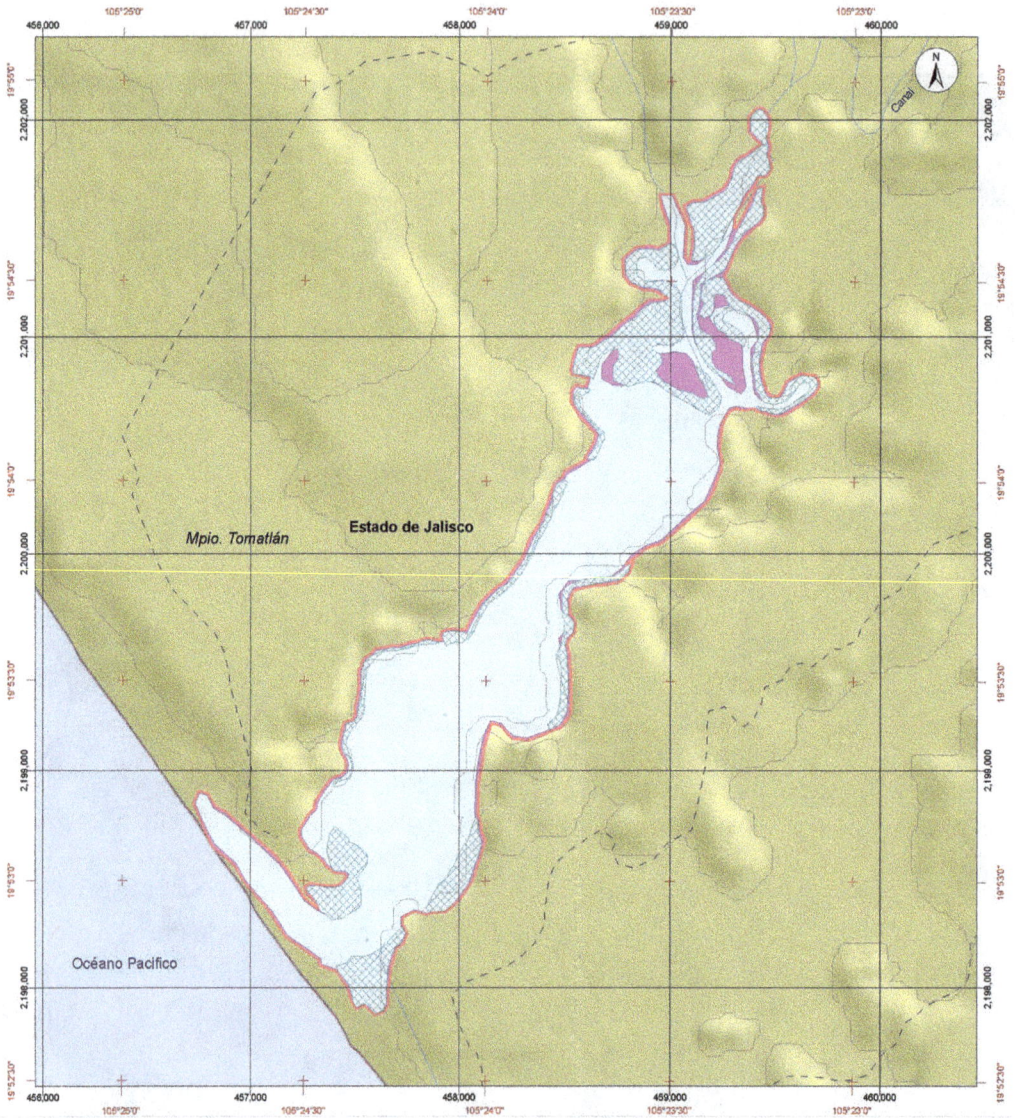

FIGURE 5.13   Map of Estero El Chorro. Designated February 02, 2008; Area: 267 ha; Coordinates: 19°54'N 105°24'W; Ramsar Site number: 1,791.

predominates, and in the northeastern part of the site there are mangrove patches with a predominance of *Laguncularia racemosa* and *Conocarpus erectus*, included in NOM-059-SEMARNAT-2001 and/ or -2010, as species subject to special protection. These types of vegetation favor the establishment of a great variety of organisms, such as fish, molluscs, crustaceans, reptiles, and birds, among which many are migratory species. The human use of resources in the area is based on fishing traditions, although some mangrove areas have been cut down in the last years for other purposes. The main portion of water at the site is protected by sand dunes, up to 15 m high in some areas.

The flora on the margins of Estero El Chorro are represented by 16 families and 37 species. The two mangrove species mentioned above stand out for their conservation importance, and yet the mangrove is used actively by local inhabitants for firewood, construction wood, medicines, as well as for crafts and ornamental uses. Also, some families of plants are used to manufacture food (Amaranthaceae, Compositae, Cucurbitaceae, Gramineae, Leguminoceae, among others).

According to a series of studies on waterfowl at the site, 66 species have been identified in the estuary, which represent about 56% of the species identified along the entire coast of Jalisco. Within this group of species, there are five that are included in NOM-059-SEMARNAT-2001 and/or -2010: *Ardea herodias, Mycteria americana, Larus heermanni, Sterna elegans*, and *Sternula antillarum*, which are subject to protection (Hernández-Vásquez 2005), and included in the Red List (IUCN 2007). In the lagoon, only two of the 20 resident waterfowl species have been observed during the reproductive season (Butorides virescens and Nyctanassa violacea). For shorebirds, areas with soft substrates provide feeding and resting sites during low tides (e.g., for Himantopus mexicanus, Calidris mauri, Catoptrophorus semipalmatus, and Numenius americanus); mangroves provide substrates for herons and seabirds to rest (e.g., for Ardea alba, Butorides virescens, Bubulcus ibis, and Pelecanus occidentalis); and other groups of birds like the ducks and seabirds use the site's bodies of water to feed and rest (Hernández-Vázquez 2005). Based on a series of studies on waterfowl of the coast of Jalisco, it has been observed that the El Chorro Estuary, as well as other relatively small wetlands that characterize this coast, are critical habitat along the migratory routes of these birds because they are used as stop over sites. Even these small wetlands (particularly Agua Dulce, El Ermitaño, and Paramán) can contain a substantial number of waterfowl species, albeit slightly less than those reported in larger wetlands located in the northern parts of coastal Mexico (Carmona and Danemann 1998).

The ichthyological fauna associated with El Chorro Estuary and its mangroves are made up of 56 species, belonging to two classes, 12 orders, and 26 families. Five of these species breed within the estuary, 22 enter early stages and grow within the ecosystem, while six species play an important role in stabilizing the ecosystem (i.e., are an important component of the food and/or nutrient web/ cycle). The lagoon also provides refugia, feeding, and breeding areas for shrimp belonging to the families Palaemonidae, Alpheidae, Hippolytidae, Proccesidae, and Penaeidae (Hendrickx 1988). The 56 recorded fish species feed in El Chorro Estuary, include catfish or chihuiles (Arius platypogon, Ariopsis seemanni, and Sciades troschelii), milkfish (Chanos chanos), mullets (Mugil cephalus, Mugil curema, and Mugil hospes), and Pacific fat sleeper and guavina (Dormitator latifrons and Guavina micropus) [Nelson 1994, Fisher et al. 1995, Aguilar-Palomino et al. 2006]. The terrestrial fauna of the margins of the estuary are represented mainly by opossum (Didelphis marsupials), armadillo (Dasypus novemcinctus), coati (Nasua), and raccoon (Procyon lotor). Reptiles such as iguanas and lizards (Iguana iguana, Ctenosaura pectinata, Anolis nebulosus, and Sceloporus spp.), as well as the Mexican beaded lizard (Heloderma horridum) have been noted at the site.

In the estuary area, local communities are dedicated mainly to agriculture, livestock, and fishing. Marine turtles have played an important role in local customs and traditions over many years, and represent a source of very important economic resources for the inhabitants of these underdeveloped communities near the El Chorro Estuary, and the nesting beach of Playón de Mismaloya. The estuary is regularly visited by inhabitants of nearby towns and the municipal seat of Tomatlán for entertainment and recreational purposes. At Easter it is traditional that temporary rustic camps are established at the site, where many families from various localities of the municipality come to vacation.

The main impacts to the ecology of the site are from the felling of mangroves in some parts of the estuary, and the deforestation of some areas of deciduous forests at the site. The relentless advance of agriculture in the area has led to the reduction of mangroves and jungle in recent years. There is always a risk of latent tourism and associated mega-developments in this area, and any concomitant lack of environmental planning and mitigation of negative environmental impacts. The beach adjacent to the site is protected, and thus the University of Guadalajara, through the Department of Studies for Sustainable Development of Coastal Zones has developed a turtle protection program that bolsters that aspect of conservation at the site. Associated with this work, the University of Guadalajara has carried out research on birds and fish at the site. The "La Gloria" turtle camp, operated by the University of Guadalajara, promotes the conservation of sea turtles and the natural resources of the area, and receives more than 500 external visitors per year, as well as approximately 500 local visitors that are from primary and secondary schools, as well as organized groups.

ESTERO MAJAHUAS, JALISCO: OVERVIEW

Estero Majahuas (Figures 5.14 and 5.15) is composed mainly of the mangrove species *Rhizophora mangle* and *Laguncularia racemosa*, both under special protection. The site supports around 60 species of aquatic birds, including *Fulica americana*, *Porphyrio martinica*, *Gallinula chloropus*, *Anas* spp., *Dendrocygna autumnalis*, *Chroroceryle americana*, *Ceryle alcyon*, *Larus heermanni*, *Sterna elegans*, *Mycteria americana*, *Egretta rufescens*, and *Ardea Herodias*. The site also supports American crocodile (Crocodylus acutus) and the olive ridley sea turtle (Lepidochelys olivacea), which uses the beaches during nesting periods. Traditional river fishing and eco-tourism are the main human activities undertaken at the site. Among the threats to this ecosystem are mangrove deforestation, which is often the result of agriculture and livestock area expansions.

Estero Majahuas is a body of water that is parallel to the coast, an estuarine environment with an intermittent opening to the sea (June to October) when the mouth-bar (connection with the sea) breaks and there is water exchange between the estuary and the sea. The estuary is primarily used for coastal artisanal fishing, and eco-tourism in certain seasons of the year; the neighboring beach is a reserve for the protection of sea turtles. The main contribution of continental water to the estuary is from the Tomatlán River.

SITE DETAILS

The flora on the margins of Estero Majahuas are represented by 17 families and 38 species. The mangrove species mentioned immediately above are used by the local inhabitants mainly for firewood, wood of construction, medicinal purposes, crafts, and ornamental uses. A number of plant families at the site are also used for food (Amaranthaceae, Compositae, Cucurbitaceae, Gramineae, Leguminoceae, among others); timber (Combretaceae, Leguminoceae, Verbenaceae, Sterculiaceae, and Rubiaceae); and medicines (Aizoaceae, Boraginaceae, Compositae, Cucurbitaceae, Cyperaceae, Graminaceae, Leguminaceae, Malvaceae, and Portulacaceae). Along the coast of Jalisco, 1,100 species of plants have been recorded, including 124 families. In the areas surrounding the site, the vegetation is primarily of deciduous forest type, with coconut palms.

The site supports 77 species of waterfowl, representing about 66% of the species identified along the coast of Jalisco. Among these species, seven are included in NOM-059-SEMARNAT-2001 and/or -2010, as follows: *Tigrisoma mexicanum*, *Ardea herodias*, *Mycteria americana*, *Larus heermanni*, *Sterna elegans*, and *Sternula antillarum* (subject to special protection); and *Charadrius melodus* (endangered) [Hernández-Vásquez 2005, Hernández-Vázquez et al. 2002]. The 77 species recorded in the Majahuas Estuary are also included in the Red List (IUCN 2007), with five of the species in the near threatened category (Puffinus griseus, Charadrius melodus, Numenius americanus, Sterna elegans, and Larus heermanni). The different habitats within the lagoon provide suitable conditions for different groups of birds, which meet their needs for food and rest; sandy areas surrounding the mouth of the estuary are used by several species of shorebirds to feed and rest (e.g., Himantopus mexicanus, Charadrius semipalmatus, Charadrius alexandrinus, and Catoptrophorus semipalmatus); and the extensive mangroves at the site provide suitable substrates for herons and seabirds, which can rest and locate their nests there (e.g., Egretta caerulea, Egretta thula, Egretta tricolor, and Nyctanassa violacea). Other groups of birds, such as Anatidae and seabirds, use the site's bodies of water to feed and rest (Hernández-Vázquez 2005, Hernández-Vázquez et al. 2002).

The ichthyological fauna associated with the estuary and its mangroves are made up of 58 species, belonging to two classes, 12 orders, and 26 families, considered to be of medium-high ichthyological diversity, particularly for the coast of western Mexico (Aguilar-Palomino 2006). Notable fish at the site are mullet (Mugilidae), crappie (Gerreidae), snapper (Lutjanidae), sea bass (Centropomidae), catfish or chihuiles (Ariidae), and sleeper gobies (Eleotridae), representing 80% of the estuarine fish of the coastal Mexican Pacific (Amézcua-Linares 1996). The lagoon also provides refugia, feeding, and breeding habitat for shrimp of the families Palaemonidae, Alpheidae,

FIGURE 5.14 Composite aerial image of Estero Majahuas. Designated February 02, 2008; Area: 786 ha; Coordinates: 19°50'N 105°21'W; Ramsar Site number: 1,791.

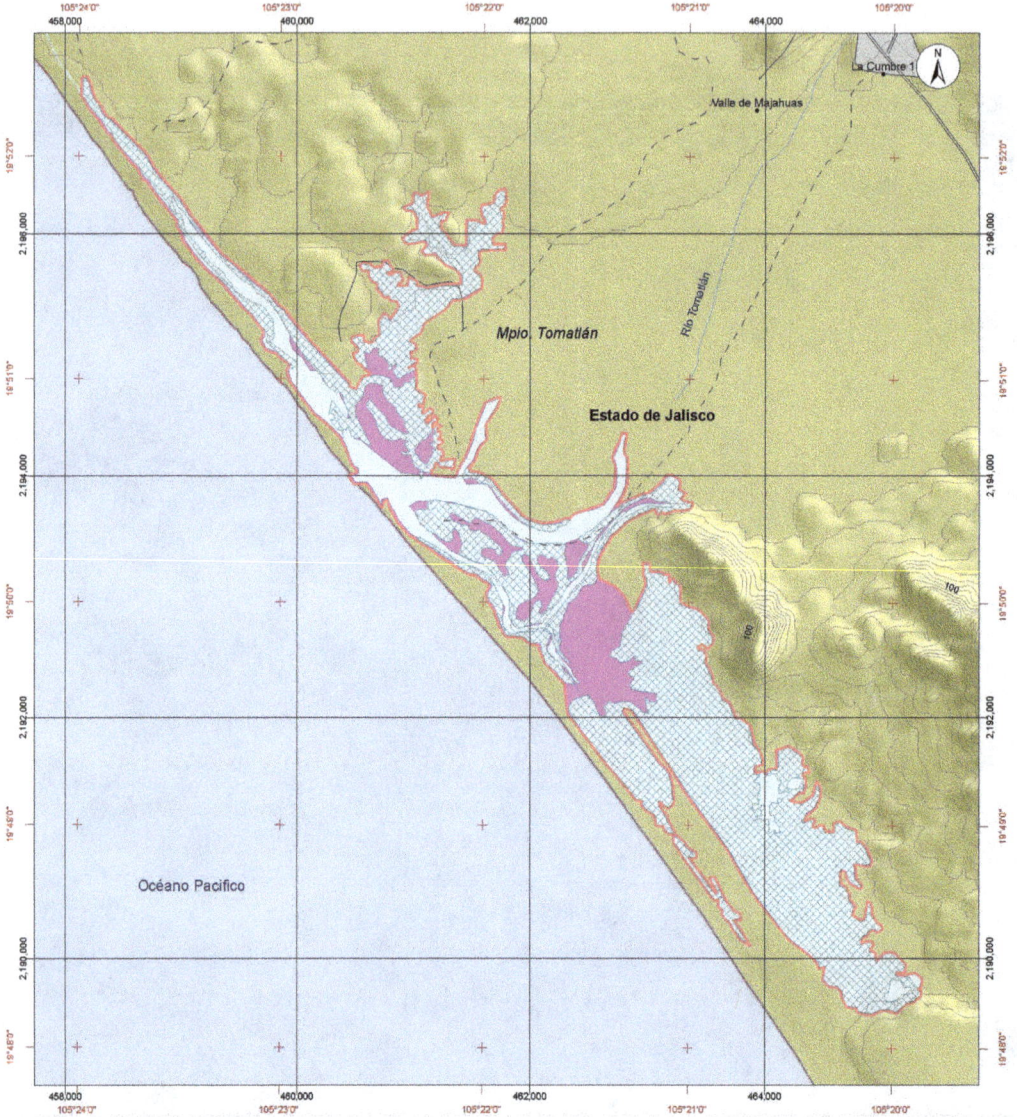

FIGURE 5.15   Map of Estero Majahuas. Designated February 02, 2008; Area: 786 ha; Coordinates: 19°50′N 105°21′W; Ramsar Site number: 1,791.

Hippolytidae, Proccesidae, and Penaeidae (Hendrickx 1988). Notable fish species at the site include *Gerres cinereus* and *Eucinostomus argenteus*, and crustaceans *Callinectes arcuatus* (arched swimming crab) and *Farfantepenaeus californiensis* (yellowleg shrimp). The terrestrial fauna of the margins of the estuary are represented primarily by small mammals such as opossum (Didelphis marsupials), armadillo (Dasypus novemcinctus), raccoon (Procyon lotor); and reptiles such as iguanas and lizards (Iguana iguana, Ctenosaura pectinata, Anolis nebulosus, and Sceloporus spp.).

The municipality of Tomatlán has grown relatively slowly, as the younger population migrates to Puerto Vallarta, Guadalajara, and the United States to find better socioeconomic opportunities. In the estuary area, local communities are dedicated mainly to the primary sectors of agriculture, livestock, and fishing. However, sea turtles have played an important role in local customs and traditions. The estuary itself is regularly visited by inhabitants of the nearby towns and the municipal seat of Tomatlán, primarily for entertainment and recreational purposes. During Holy Week, most families come to the site to vacation, hosting other family members from other parts of the country. In general, the greatest use of the site is for artisanal fishing and guided ecotours. Recreational activities at the site include camping, rowing canoes, swimming, and fishing. There are rustic facilities (palapas) made with materials from the region on the margin of the site, adjacent to the beach areas.

Mangrove clearing in some parts of the lagoon, as well as the deforestation of some areas of deciduous forest in the vicinity, represent a ceaseless advance of the agricultural frontier, which has an impact on the percent cover of mangroves, and thus the ecology of the site. Similarly, the ever-present possibility of mega-developments in this region poses a continual risk of improper planning and environment impacts on the ecology of the site. The beach adjacent to the estuary is protected by CONANP as a sea turtle sanctuary (El Playón de Mismaloya), which has potential for continued conservation in the face of development pressures.

Existing research at the site includes work by those at University of Guadalajara, particularly with regard to bird and fish communities. A number of these research projects have resulted in journal articles, bachelor's and master's theses, as well as doctoral dissertations. Additionally, there is an associated 'turtle camp' dedicated to the protection of these species, and the Government of the State of Jalisco operates a Technological Development Center for Marine Species, whose objective is to further their conservation. The turtle camp, operated by the University of Guadalajara, also generally promotes conservation of the natural resources of the area through awareness among local communities and the general public, who have the opportunity to release turtles to the sea when they leave their nests. There are also local community members that have organized panga rides through the estuary during the holiday seasons, mainly to observe birds and crocodiles. Lastly, during Holy Week, fishing tournaments and jet ski races are organized to provide recreation and tourism opportunities for visitors.

LAGUNA XOLA-PARAMÁN, JALISCO: OVERVIEW

Laguna Xola-Paramán (Figures 5.16 and 5.17) is a marine turtle sanctuary and coastal wetland/lagoon, surrounded by forest vegetation, characterized by small bays of sandy beaches and rocky areas that are associated with small alluvial valleys of the area. The site supports the reproductive success of migratory and resident waterfowl, including *Ardea herodias*, *Egretta rufescens*, *Mycteria americana*, *Larus heermanni*, *Sterna elegans*, and *Sternula antillarum*, all of which are under special protection. The site also supports notable flora including *Bursera* spp., *Eysenhardtia polystachya*, *Acacia pennatula*, and *Forestiera* spp. The marine turtles *Dermochelys coriacea*, *Lepidochelys olivacea*, and *Chelonia mydas* spawn on the beach next to the site's lagoon, which is under protection by the National Council of Protected Areas of Mexico, as a marine turtle sanctuary. Salt extraction and fishing are the main human activities carried out at the site. Primary impacts on the ecology of the site are mangrove deforestation, agricultural practices, and water pollution from agrochemicals.

SITE DETAILS

The site is located on the central-western Mexican coast of the Pacific Ocean, in the central portion of the coast of the State of Jalisco, within the municipality of Tomatlán. The site is adjacent to the beach called "Chalacatepec" that is southwest of the town of Campo Acosta (approximate population, 3,100), and approximately 120 km from the city of Puerto Vallarta, Jalisco. The site is surrounded by jungle vegetation, small patches of mangrove, and some natural pastures. Most of the lagoon is shallow, reaching maximum depths of 2 m. During the dry season the water level drops drastically, to the degree of near total desiccation, causing the salinity at the site to exceed 100 ppm.

The species of mangrove present at the site are white (Laguncularia racemosa), black (Avicennia germinans), and button (Conocarpus erectus). The three mangrove species present at the site are included within the NOM-059-ECOL-2001, as species subject to special protection. 69 species of seabirds have been identified in this lagoon, which represent 59% of the species identified along the entire coast of Jalisco, and among which seven species are included in NOM-059-ECOL-2001: *Ardea*

FIGURE 5.16 Composite aerial image of Laguna Xola-Paramán. Designated February 02, 2008; Area: 775 ha; Coordinates: 19°44′N 105°16′W; Ramsar Site number: 1,768.

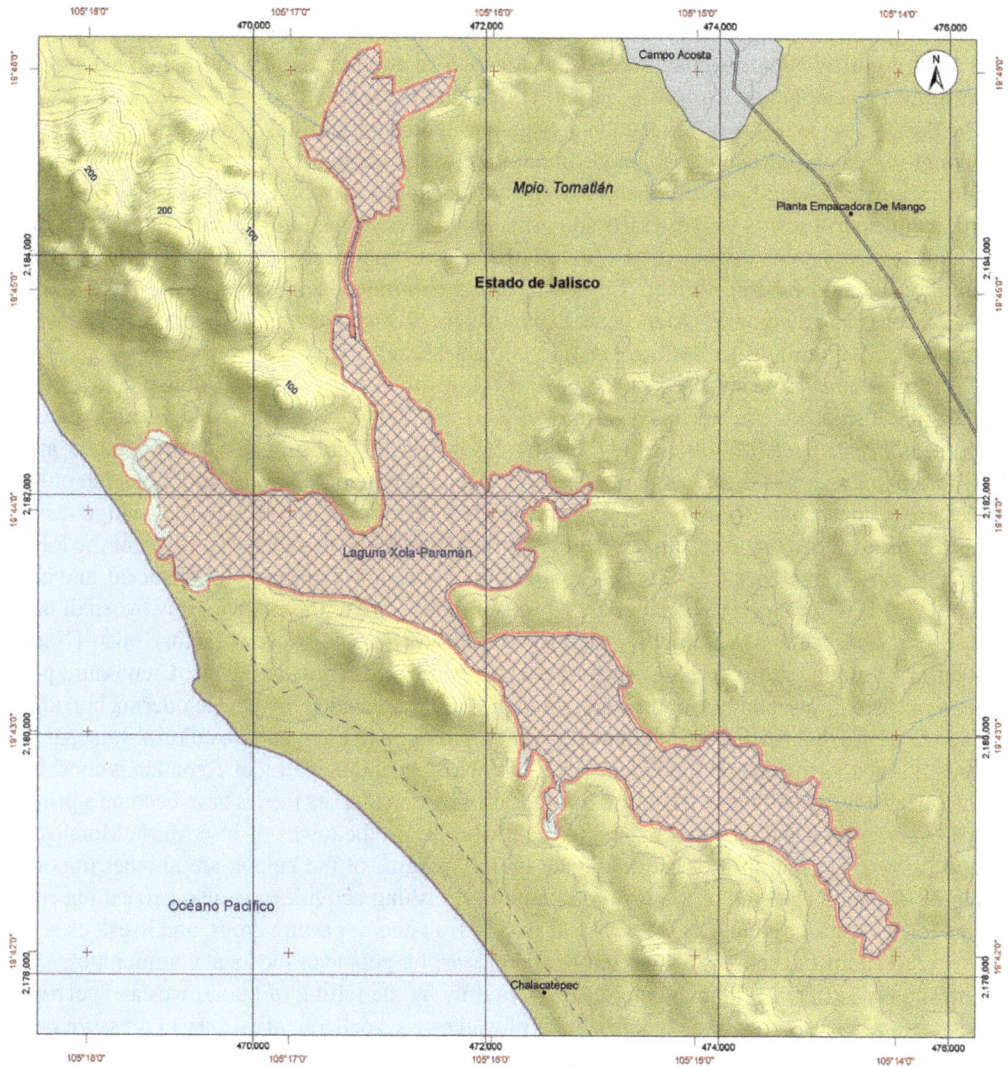

FIGURE 5.17 Map of Laguna Xola-Paramán. Designated February 02, 2008; Area: 775 ha; Coordinates: 19°44′N 105°16′W; Ramsar Site number: 1,768.

herodias, *Egretta rufescens*, *Mycteria americana*, *Larus heermanni*, *Sterna elegans*, *Sternula antillarum* (special protection), and *Cairina moschata* (threatened) [Hernández-Vázquez 2005]. These 69 species are also included in the Red List (IUCN 2007), three of which are near threatened (*Numenius americanus*, *Sterna elegans*, and *Larus heermanni*) and 66 categorized as least concern. Of special note are the 15 seabird species at the site that breed in the sandy areas (*Sternula antillarum*, *Himantopus mexicanus*, and *Charadrius wilsonia*) [Hernández-Vázquez 2000, Rodrigo-Esparza 2002]. The different habitats of the lagoon provide adequate conditions for groups of birds to satisfy their feeding and resting needs; for shorebirds, areas with soft substrates provide feeding and resting sites (e.g., for *Recurvirostra americana*, *Himantopus mexicanus*, and *Charadrius semipalmatus*); for seabirds, ducks, and herons open water provides adequate resting and feeding sites (Hernández-Vázquez 2000).

Numerous fish species are recorded at the site, which feed in the lagoon: three of these species reproduce within the site, i.e., catfish (*Arius platypogon*, *A. planiceps*, and *Ariopsis seemanni*); five fish species play an important role in stabilizing the ecosystem (i.e., are an important component of the food and/or nutrient web/cycle),. e.g., the milkfish (*Chanos chanos*), mullets (*Mugil cephalus* and M. curema), and Pacific fat sleeper and guavina (Dormitator latifrons and Guavina micropus); and 18 fish species enter the site at early stages and grow within the ecosystem (Nelson 1994, Fisher et al. 1995, Aguilar-Palomino et al. 2006). The lagoon also provides habitat for shrimp belonging to the families Palaemonidae, Alpheidae, Hippolytidae, and Penaeidae.

The flora on the margins of the lagoon are represented by 18 families and 40 species, including the mangrove species already mentioned. In addition to the ecological benefits of mangroves, the mangrove areas are used by local inhabitants too, primarily for firewood, construction wood, posts, and medicinal purposes, as well as crafts and ornamental uses. Some families of plants are also used for obtaining or manufacturing food (Amaranthaceae, Compositae, Cucurbitaceae, Gramineae, Leguminoceae, and others), timber (Combretaceae, Leguminoceae, Verbenaceae, Sterculiaceae, and Rubiaceae), and medicines (Aizoaceae, Boraginaceae, Compositae, Cucurbitaceae, Cyperaceae, Graminaceae, Leguminoceae, Malvaceae, and Portulacaceae).

Site ichthyofauna are comprised of 45 species, belonging to two classes, 11 orders, and 24 families. Notable fish at the site include *Gerres cinereus* and *Eucinostomus argenteus*, and the crustacean *Callinectes arcuatus* (arched swimming crab). One of the groups heavily studied at the Xola-Paramán Lagoon is that of aquatic birds: 69 species are recorded at the site, of which only 15 species are residents and 54 are migratory birds (Hernández-Vázquez 2000). In the contiguous area of the lagoon is the sea turtle sanctuary "El Playón de Mismaloya" in which sea turtles of the following species nest: leatherback (Dermochelys coriacea), olive ridley (Lepidochelys olivacea), and green (Chelonia mydas). Terrestrial fauna on the margins of the site are represented mainly by small mammals, such as opossum (Didelphis marsupials), armadillo (Dasypus novemcinctus), coati (Nasua), raccoon (Procyon lotor); and reptiles such as iguanas and lizards (Iguana iguana, Ctenosaura pectinata, Anolis nebulosus, and Sceloporus spp.), and the Mexican beaded lizard (Heloderma horridum).

The sea turtles that nest on the beach next to the Xola-Paramán Lagoon have cultural value, as well as food and economic value for many families in the area. The municipality of Tomatlán is considered one of the most marginalized in the entire State of Jalisco, such that sea turtles have become a primary source of sustenance and income of some of those who live in the towns of José María Morelos and Campo Acosta. The salt mines that developed on the margins of the lagoon are another important source of income for people of the local communities. Fishing activities are also carried out in the lagoon, for self-consumption. Local tourism as well as fruit and/or pasture crops, and livestock activities, serve as additional sources of sustenance and income for people of the local communities.

The site has suffered ecological degradation, mainly by the felling of mangroves and surrounding vegetation, as well as by the contamination of water by agrochemicals. Tourism pressure on the dune and beach areas, as well as the construction of an airstrip in the area, are notable impacts upon the ecology of the site, and specifically for migratory waterfowl. The beach adjacent to the Xola-Paramán Lagoon is protected by sea turtle sanctuary guidelines, however this has not prevented the establishment of tourism mega-developments bordering the lagoon. The site is considered a national asset and concessions for its use are administered by the National Water Commission, a decentralized body of the Secretariat of Environment and Natural Resources (SEMARNAT).

The University of Guadalajara has carried out research on birds and fish at the site, resulting in a number of published journal articles, bachelor's and master's theses, and doctoral dissertations/work. Additionally, existing communication, education, and public awareness activities have resulted from these studies. Local community members visit the turtle habitat on Chalacatepec Beach to fish, swim, and watch birds. And, lastly, there are ongoing developments at the site to expand eco-tourism projects.

LAGUNA CHALACATEPEC, JALISCO: OVERVIEW

Laguna Chalacatepec (Figures 5.18 and 5.19) is a coastal lagoon that is representative of the transitional regions where elements of both the Nearctic and Neotropical biogeographic regions come together. The site is within the San Nicolás-Cuitzmala River Basin, originating in the vicinity of

FIGURE 5.18 Composite aerial image of Laguna Chalacatepec. Designated February 02, 2008; Area: 1,093 ha; Coordinates: 19°40′N 105°13′W; Ramsar Site number: 1,818.

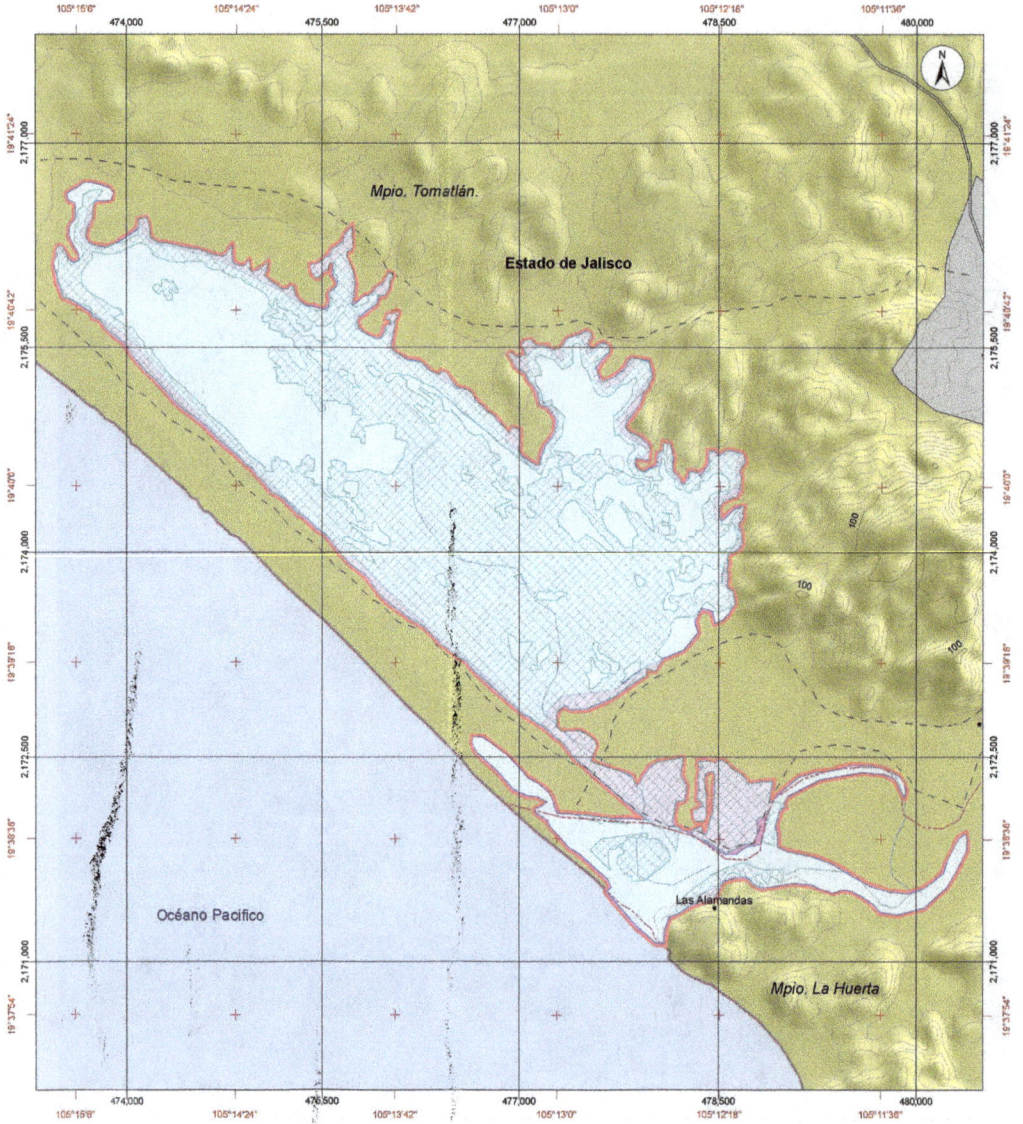

FIGURE 5.19 Map of Laguna Chalacatepec. Designated February 02, 2008; Area: 1,093 ha; Coordinates: 19°40′N 105°13′W; Ramsar Site number: 1,818.

Cerro Camalote, located 15 km northwest of the town of Villa Purificación, at an elevation of 1500 m above sea level. The basin covers approximately 1,141 km² and flows into the Pacific Ocean 5 km southeast of Punta Farallón. Mangroves are the predominant vegetation at the site, including *Laguncularia racemosa*, *Rhizophora mangle*, *Conocarpus erectus*, and *Avicennia germinans*. Populations of the American crocodile (Crocodylus acutus) have been found at the site, as well as the presence of the southern river otter (Lutra annectens), a threatened species according to the Mexican normative. This lagoon is essential for the reproduction of resident waterfowl, such as *Nyctanassa violacea*, *Ardea alba*, *Egretta thula*, and *Phalacrocorax brasilianus*, and is used by migratory waterfowl (Anas acuta, Fulica americana, and Dendrocygna autumnalis) as a staging point for feeding and resting along their long-distance journeys. The main human activities undertaken at the site are fishing, agriculture, and extensive livestock ranching in the surrounding areas.

The site lies adjacent to the marine turtle sanctuary El Playón de Mismaloya, dedicated mainly to the conservation of marine turtles, with activity at the site evidently becoming more and more popular as a tourist attraction, which also creates awareness regarding the importance of the protection of natural resources at the site and in the region.

SITE DETAILS

The Chalacatepec Lagoon is parallel to the coastline, with its brackish waters at a maximum depth of 2 m. Ecologically, this is an estuarine environment (Day et al. 1979), and geologically the site is classified as a coastal lagoon (Phleger 1969), being a flooded depression on the interior margin of the continental shelf, surrounded by terrigenous surfaces on its inner shore and protected from the sea by an external sandy bar, with an average width of 400 m. Until approximately the late-1980s, the lagoon was continuously connected to the San Nicolás River, receiving both marine and freshwater influences through that connection; however human interventions since then have minimized the entry of fresh and marine water into the ecosystem, leading to the current state of the hydrology in the lagoon.

Although geological, biological, and environmental characteristics present in the Chalacatepec Lagoon are also found in other wetlands along the northern portion of the west coast of Mexico, in the central and south-central portions of the country (i.e., States of Jalisco, Colima, Michoacán, and Guerrero) there are few coastal wetlands whose mangroves cover more than 50% of the total surface area of the wet areas of their wetland site. At this site, mangroves cover approximately 70% of the moist areas of the wetland. For this reason, the National Commission for the Use and Knowledge of Biodiversity (CONABIO) identifies the Chalacatepec Lagoon as one of the Priority Mangroves for its conservation and restoration efforts (CONABIO 2008).

Seventy-two aquatic bird species have been identified in this lagoon, which represents about 65% of the species identified along the entire coast of Jalisco. Five of these species are included in NOM-059-SEMARNAT-2001: *Ardea herodias, Mycteria americana, Larus heermanni, Sterna elegans*, and *Sternula antillarum* (special protection) [Hernández-Vásquez 2005]; the 72 species noted above are also included in the Red List (IUCN 2007), with three of these species as near threatened (*Numenius americanus, Sterna elegans*, and *Larus heermanni*), and 69 species in the category of least concern. The sea turtles that nest at the site are: leatherback (*Dermochelys coriacea*), olive ridley (*Lepidochelys olivacea*), and green (*Chelonia mydas*), classified as endangered in NOM-059-SEMARNAT-2001, in the IUCN Red List, and in the CITES. The Chalacatepec Lagoon, as well as the Chamela Cuixmala Ramsar Site and La Manzanilla Estuary, both in the State of Jalisco, support one of the most abundant populations of American crocodile (*Crocodylus acutus*) on the coast of Jalisco, listed within NOM-059-SEMARNAT-2001 (special protection). Another key species at the site is southern river otter (*Lutra annectens*), a threatened species within the NOM-059-SEMARNAT-2001.

Flora on the margins of the lagoon are represented by 18 families and 42 species. Some families of plants at the site are used for obtaining or manufacturing food (Amaranthaceae, Compositae, Cucurbitaceae, Gramineae, Leguminoceae, and others); timber (Combretaceae, Leguminoceae, Verbenaceae, Sterculiaceae, and Rubiaceae); and medicines (Aizoaceae, Boraginaceae, Compositae, Cucurbitaceae, Cyperaceae, Graminaceae, Leguminoceae, Malvaceae, and Portulacaceae). The Laguna Chalacatepec area shares some types of vegetation found in the Chamela-Cuixmala Reserve, i.e., deciduous forest, mangroves, and palms. Predominant vegetation at the site are button mangrove (*Conocarpus erectus*), white mangrove (*Laguncularia racemosa*), and red mangrove (*Rhizophora mangle*) across approximately 70% of the wetland area (Hernández-Vázquez 2005). The mangrove species at the site are threatened, per NOM-059-SEMARNAT-2001. The areas surrounding the lagoon are predominantly deciduous forest vegetation, with coconut palms.

The ichthyofauna of the site are comprised of 54 species, belonging to two classes, 14 orders, and 26 families. Noted at the site are fisheries species, such as decapod crustaceans, fish species

Gerres cinereus and *Eucinostomus argenteus*, and the crustaceans *Callinectes arcuatus* (arched swimming crab) and *Farfantepenaeus californiensis* (yellowleg shrimp). Aquatic birds at the site include 72 species, of which 23 species are visitors and 49 are resident birds within the lagoon and its surroundings (Hernández-Vázquez 2000, 2005). Due to its apparent abundance as well as being a top predator, the American crocodile (Crocodylus acutus) is a key resident species of the lagoon. The terrestrial fauna on the margins of the estuary are represented mainly by small mammals such as opossum (Didelphis marsupials), armadillo (Dasypus novemcinctus), South American coati (Nasua nasua), and raccoon (Procyon lotor); reptiles such as iguanas and lizards (Iguana iguana, Ctenosaura pectinata, Anolis nebulosus, and Sceloporus spp.) have also been observed at the site.

The lagoon is located within an area of high biological diversity and abundance where there are active plans for its use for development of new tourist centers with low ecological impact, and high-quality services; for example, the natural beauty of the site attracts growing number of tourists from the city of Puerto Vallarta. Part of the social and cultural value of the Chalacatepec Lagoon is that it is naturally linked to these turtle protections and conservation activities carried out on Chalacatepec Beach, adjacent to the site. These activities are being carried out to strengthen eco-tourism, attracting an increasing number of visitors and at the same time raising awareness about the care of the natural resources in the region. There is also a shipwreck located on the beach adjacent to the site, which is part of the 'legends of place' for the site. Another cultural and social value of the site is linked to the traditional/historical visitation of local populations during Holy Week and Easter Week, and to a lesser extent during summer holidays and New Year.

The Chalacatepec Lagoon is under federal jurisdiction, with responsibilities accordingly associated with the National Water Commission (CONAGUA). Fishing resources are exploited by local fishing production cooperatives, and their administration is the responsibility of the National Commission of Aquaculture and Fisheries (CONAPESCA). Concessions for the use of the federal maritime-terrestrial zone (20 m from the margins of the lagoon) are administered by the Secretariat of Environment and Natural Resources (SEMARNAT). Land ownership on the terrestrial surrounding areas is divided into small properties, mainly from adjoining and private tourist developments, as well as communal agricultural (i.e., "ejido") lands (Ejido José María Morelos), administered by the neighboring communities.

The lagoon is also a fishing site for the self-consumption of residents of neighboring communities, and commercial fishing by fishing cooperatives. The site has suffered degradation of natural habitat and its natural resources, mainly by the felling of mangroves, the alteration of the hydro-dynamics of the ecosystem, contamination by agrochemicals, and the lack of proper management of fishing activities. The increases in human population in the vicinity of the site and the development pressures of the region increase the risk if ecological impacts at site. Increases in the demand for land, and its conversion to agriculture, along with other built areas increases the risks of soil erosion, and associated direct or indirect degradation of water quality at the site. The University of Guadalajara has carried out some studies at the site, particularly of birds and fish, which have included analyses of general ecosystem dynamics. The products of these investigations have been published in journals, bachelor's and master's theses, and doctoral dissertations.

RESERVA DE LA BIOSFERA CHAMELA-CUIXMALA, JALISCO: OVERVIEW

Reserva de la Biosfera Chamela-Cuixmala (Figures 5.20 and 5.21) is comprised of a mountainous landscape, rocky coasts, and a deltaic plain on the Pacific coast of Mexico, notable for its deciduous tropical forest communities. The estuary of the Cuixmala River and the lagoons of El Corte and La Manzanillera are home to approximately 600 American crocodile, nesting habitat for several marine turtle species, and the southernmost colony of least tern (Sternula antillarum). The site's forests host pumas, ocelots, and jaguars and is the only known site for the Magdalena tree rat (Xenomys nelsoni). The site is essentially uninhabited and the vegetation is well preserved. Fishing, hunting, and scientific research by the UNAM are the main human activities at the site, thus making the site one of the best-known tropical areas in the area, in terms of its ecology.

FIGURE 5.20 Composite aerial image of Reserva de la Biosfera Chamela-Cuixmala. Designated February 02, 2004; Area: 13,142 ha; Coordinates: 19°29′N 104°59′W; Ramsar Site number: 1,334.

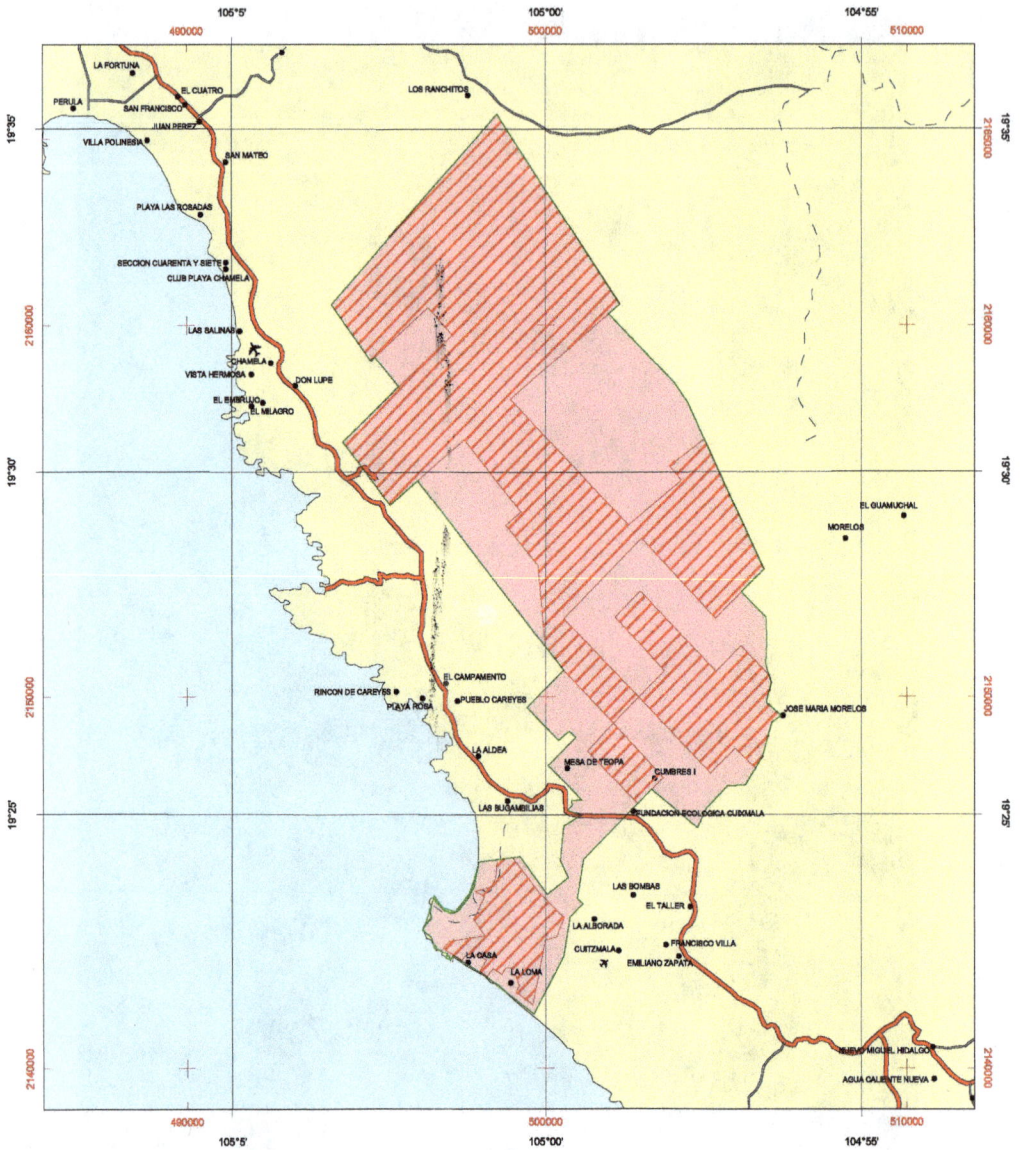

FIGURE 5.21   Map of Reserva de la Biosfera Chamela-Cuixmala. Designated February 02, 2004; Area: 13,142 ha; Coordinates: 19°29′N 104°59′W; Ramsar Site number: 1,334.

SITE DETAILS

In Mexico, the site is also known as the Chamela-Cuixmala Biosphere Reserve (RBCC), one of the protected areas that, to maintain deciduous tropical forests as a national priority. The site has several other types of vegetation, namely semi-evergreen forest, mangroves, scrub, riparian vegetation, reed beds, and aquatic vegetation within the site's wetlands. The site contains a delicate wetland, formed by the estuarine lagoons of Corte and La Manzanillera and by the Cuixmala River Estuary, within which aquatic birds depend, as well as numerous species of flora and fauna that reside in the deciduous forests, among which several are threatened, endangered, rare, and/or endemic.

The site's wetland area is located within the Cuitzmala River Basin. The Cuitzmala River originates in the vicinity of Cerro Camalote, located 15 km northwest of the town of Villa Purificación, at an elevation of 1500 m above sea level; during its journey of 85 km, the river retains a predominantly south-southwest direction. The basin covers an area of approximately 1,141 km² and flows into the Pacific Ocean at a point located 5 km southeast of Punta Farallon. There are no large permanent natural bodies of water along the river.

The site provides habitat for 10 types of vegetation, including 1,100 species of vascular plants, and more than 450 species of vertebrates. Dominant or very conspicuous species at the site include the leguminous tree *Caesalpinia eriostachys*, cuachalalate (Amphipterygium adstringens), cascalote (Caesalpinia sp.), prickly pear cactus (Opuntia excelsa), Spanish plum (Spondias purpurea), salmwood (Cordia alliodora), bocote (Cordia elaeagnoides), and yellow paper (Jatropha chamelensis). Characteristic fauna of this type of vegetation association are pygmy skunk (Spilogale pygmaea), squirrel (Sciurus colliaei), bat (Artibeus intermedius), white-tailed deer (Odocoileus virginianus), common potoo (Nyctibius griseus), Cooper's hawk (Accipiter cooperii), East Brazilian pygmy owl (Glaucidium minutessimum), cat-eyed snake (Psedoleptodeira uribei), tropical tree lizard (Urosaurus bicarinatus), and Gaige's thirst snail-eater (Dipsas gaigeae). Key forest species are jobo de lagarto (Sciadodendrom excelsum), breadnut (Brosimum alicastrum), cohune palm (Orbignya cohune), primavera tree (Tabebuia donnell-smithii), and fig (Ficus sp.). Typical forest fauna are the tree rat (Nyctomys sumichrasti), mice (Peromyscus perfulvus and Osgoodomys banderanus), gray vireo (Vireo vicinior), ruby-throated hummingbird (Archilochus colubris), green macaw (Ara militaris), anoles (Anolis schmidti), indigo snake (Drymarchon corais), Mexican spotted wood turtle (Rhynoclemmys rubida), grayish mouse opossum (Marmosa canescens), armadillo (Dasypus novemcinctus), and rat (Oryzomys melanotis). Once forests at the site were cleared, grasslands were established that typically contain the following plant species: *Cenchrus ciliaris*, *Chloris inflata*, *Eragrostis ciliaris*, *Panicum maximum*, *Rhynchelytrum repens*, *Sorghum bicolor*, *Diectomis fastigiata*, and *Digitaria ciliaris*. Fauna in these areas are typically white tailed rabbit (Sylvilagus cunicularius), rat (Sigmodon mascotensis), southern pygmy mouse (Baiomys musculus), lesser roadrunner (Geococcyx velox), Mexican dove (Columbina inca), rusty-breasted seedeater (Sporophila minuta), Mexican patch-nosed snake (Salvadora mexicana), and the Mexican west coast rattlesnake (Crotalus basiliscus). Mangroves at the site are dominated by *Laguncularia racemosa* and *Rhizophora mangle*. As is typical of these trees, they form dense forests at the site, and their height reaches approximately 10 m. Other conspicuous species in these areas are the ciruelillo tree (Phyllanthus elsiae) and pond apple (Anona glabra). In areas surrounding the mangroves one finds the manchineel tree (Hippomane mancinella), a tree that measures 15 to 17 m. Riparian vegetation is confined to the banks of the Cuixmala River and the Chamela Stream. Some typical trees of this vegetation type are the Chilean pencil willow (Salix chilensis), several species of genus *Ficus*, and *Asthianus viminalis*. Aquatic vegetation (mainly Typha latifolia and Scirpus sp.), common water hyacinth (Eichhornia crassipes), and dot leaf waterlily (Nymphaea ampla) are found in some of the more open water areas of the site.

The 72 species of mammals in the region are of tropical origin, with a high degree of endemism. At least 18 species are endemic to Mexico, including a marsupial, an insectivore, a carnivore, 9 rodents, a lagomorph, and 3 bats; 60% of endemic mammals of Mexico are found in this region. For the Magdalena tree rat (Xenomys nelsoni), this region constitutes the sole locality currently known where it is present. Other mammal species present at the site have very restricted local distributions; for example, the greater bulldog bat (Noctilio leporinus) and southern river otter (Lutra annectens) are only found in the lagoons and mangroves of the Cuixmala area. At least 22 species at the site are threatened, endangered, or otherwise imperiled. Among these species are the puma (Felis concolor), the jaguar (Panthera onca), the ocelot (Leopardus pardalis), and the margay (Leopardus wiedii). The Magdalena tree rat (Xenomys nelsoni), the nectar-eating bat (Musonycteris harrisoni), the collared peccary (Tayassu tajacu), and the white bat (Diclidurus albus) are also present at the site. Of the 271 species of birds in the region, 60% are residents and 40% are migratory visitors to the site.

Of special interest are the marine and freshwater aquatic habitats at the site because their diversity (Berlanga et al. 1987, Hutto 1989, Arizmendi et al. 1991, García and Ceballos 1995) supports the 36 endemic bird species in the region, among which are the lilac crowned parrot (Amazona finschi), the yellow-headed amazon (Amazona oratrix), and the Mexican parrotlet (Forpus cyanopygius). Historical overexploitation of such bird species has resulted in 28 species being listed as threatened or endangered. The site maintains water throughout the year in such a way that it is common to observe waterfowl, such as *Fulica americana, Porphyrio martinica,* and *Gallinula chloropus*; ducks (Anas spp., Dendrocygna autumnalis, and Cochlearius cochlearius); herons and kingfishers (Chroroceryle americana and Ceryle alcyon); flycatchers (Myozetetes similis and Pitangus sulphuratus); and swallows (Stelgidopterix serripennis, Hirundo rustica, and Tachycineta albilinea). Most reptiles and amphibians at the site have very narrow habitat requirements, so they are associated with particular types of habitats; about 80 of these species, combined, are resident at the site, of which at least 42 species are endemic to Mexico and 10 are in danger of extinction (García and Ceballos 1995). Among the endangered species at the site are the Mexican beaded lizard (Heloderma horridum), the green iguana (Iguana iguana), the American crocodile (Crocodylus acutus), and four species of sea turtles (Lepidochelys olivacea, Dermochelys coriacea, Eretmochelys imbricata, and Chelonia mydas). Poisonous species of the region include Mexican west coast rattlesnake (Crotalus basiliscus), the pit viper Agkistrodon bilineatus, the west Mexican coral snake (Micrurus distans), and the sea snake *Pelamys platurus.*

In the Chamela-Cuixmala region, pre-hispanic times saw the presence of the people of Nahuatl origin, a nomadic people of either warrior or farming lifestyle (De Ita 1983). Currently, in the vicinity of the Rivers Cuixmala, Chamela, and San Nicolás, there are numerous areas with archaeological remains. During the colonial period, the region was managed through haciendas dedicated to agriculture and the raising of livestock. The first massive settlements in the region occurred in Tomatlán and Cihuatlán. Chamela and Barra de Navidad were important ports from which goods were traded in those times. For example, Barra de Navidad was the port of departure of the expedition that discovered the Philippines. The Jalisco coast is a region that, until recently, remained practically isolated, due to the difficult access. With the construction of the Melaque-Puerto Vallarta Highway in 1970, that isolation ended and that process of development continues to this day. Nevertheless, social welfare levels along the entire coast are deficient, since the infrastructure of human settlements in the area are in poor condition (Sedesol 1991). However, as time has progressed, the economy of the region continues to be relatively diversified, expanding in agricultural and fishing production, to tourism.

The site and its vicinity is heavily impacted by human use, which contributes to Mexico's deforestation rates, which are among the highest in the world, devastating the country's jungles over the past 50 years (Rzedowski 1978, Toledo 1988). Much of this devastation is due to the fact that communities of the area have little capacity to generate productive and continuous jobs, causing high unemployment and underemployment and, in the best of cases, solely subsistence income is possible (Sedesol 1991). Resultant clearing of land in the alluvial plain of the Cuixmala River has occurred to expand the cultivation of various fruits and vegetables, such as papaya, banana, mango, watermelon, melon, pumpkin, corn, coconut, sorghum, onion, chili, and beans, with just a small portion of the population dedicated to mechanized agriculture. In addition, the hills are generally cleared to create pastures for extensive livestock. Among the remaining natural areas of the site, 120 plant species are used for food, medicinal purposes, or in the construction of houses, fences, and corrals. The fauna of the region is also intensively used as food, medicine, and ornamentation. The mouth of the Cuixmala River are used for fishing of various fresh water and brackish species, such as shrimp. Notwithstanding the intensive human uses of the area, the site's ecology is protected through international agreements assumed by Mexico with the United States and Canada. Sea turtle populations at the site are also protected under the Inter-American Convention for the Protection and Conservation of Turtles Laws (1996), of which Mexico has been a signatory since December 29, 1998.

The Chamela Biology Station at the Institute of Biology at the National Autonomous University of Mexico has facilities at the site, including laboratories and a library. In addition, the Cuixmala Ecological Foundation has facilities at the site that can accommodate visiting researchers. Education an outreach activities at the site include regular environmental education events, which implement projects focused on the sustainable use of resources, and social development with strict adherence to environmental regulations, which tiers and overlaps with tourist and recreational activities that are compatible with the conservation objectives of the reserve/site.

ESTERO LA MANZANILLA, JALISCO: OVERVIEW

Estero La Manzanilla (Figures 5.22 and 5.23) is located in Tenacatita Bay, along Mexico's Pacific coast, surrounded by approximately 200 ha of mangroves that are in relatively good condition, which include the presence of *Rhizophora mangle*, *Laguncularia racemosa*, and *Conocarpus erectus*. A variety of floral and faunal species abound at the site, with large populations of the American crocodile (Crocodylus acutus) present. The estuary is essential for the reproduction of several species of aquatic animals and supports the largest reproductive colony of the boat-billed heron (Cochlearius cochlearius) in the area. Fifty-five different species of aquatic birds have been identified at the site, and 42 species of fish, from ten orders and 21 families, which utilize the site for feeding. Since the 1970s, human activities such as urban growth and deforestation have negatively impacted the estuary, including the construction of a paved coastal road, which impacts the mangroves by limiting the flow of water to the estuary.

SITE DETAILS

The Estero La Manzanilla is an estuarine ecosystem, semi-parallel to the coastline with narrow and elongated morphology, nestled in the Bay of Tenacatita, which includes three beaches (Boca de Iguanas, Tenacatita, and La Manzanilla). This area is one of the five largest bays on the Pacific coast of Mexico. Along the beach areas, there are coconut palms (Cocos nucifera) and bayhops (Ipomoea pes-caprae). Of the species identified at the site, which represent 47% of the species identified along the entire coast of Jalisco, eight are included in NOM-059-SEMARNAT-2001 and/or -2010, as follows: *Tigrisoma mexicanum, Ardea herodias, Mycteria americana, Larus heermanni, Sterna elegans*, and *Sternula antillarum* (special protection); as well as the least storm-petrel (Oceano dromamicrosoma) and the red-billed tropicbird (Phaethon aethereus) [threatened] {Hernández-Vázquez 2000, 2005}. Fifty-five species of aquatic birds recorded in the Laguna Barra de Christmas are included in the Red List (IUCN 2007) as near threatened (Sterna elegans and Larus heermanni), and the rest as least concern. The Estero La Manzanilla, as well as the Chamela Cuixmala Ramsar Site and the Laguna de Chalacatepec, both in the State of Jalisco, support the largest populations of American crocodile (Crocodylus acutus) on the coast of the State, a species within the NOM-059-SEMARNAT-2001 and/or -2010 under special protection. One species to highlight at this estuary is the boat-billed heron (Cochlearius cochlearius), which reproduces at the site and has the largest population recorded to date on the coast of the State of Jalisco. The diverse habitats of the site provide adequate conditions for groups of birds to feed and rest; swamps provide suitable substrates for herons and seabirds to rest and locate their nests (e.g., Cochlearius cochlearius, Butorides virescens, Anhinga anhinga, and Phalacrocorax brasilianus).

The ichthyological fauna associated with the lagoon and its mangroves are comprised of 42 species, belonging to two classes, 10 orders, and 21 families. Notable fish species at the site are mullet (Mugilidae), crappie (Gerreidae), snappers (Lutjanidae), sea bass (Centropomidae), catfish or chihuiles (Ariidae), and sleeper gobies (Eleotridae). Forty-two species of fish utilize the lagoon as feeding habitat, and belong to 10 orders and 21 families. Four of these species breed in the lagoon: nurse shark (Ginglymostoma cirratum), and the catfishes *Arius platypogon, Arius planiceps, and Ariopsis seemanni*; five species play an important role in stabilizing the ecosystem (i.e., are an

FIGURE 5.22 Composite aerial image of Estero La Manzanilla. Designated February 02, 2008; Area: 264 ha; Coordinates: 19°18′N 104°47′W; Ramsar Site number: 1,789.

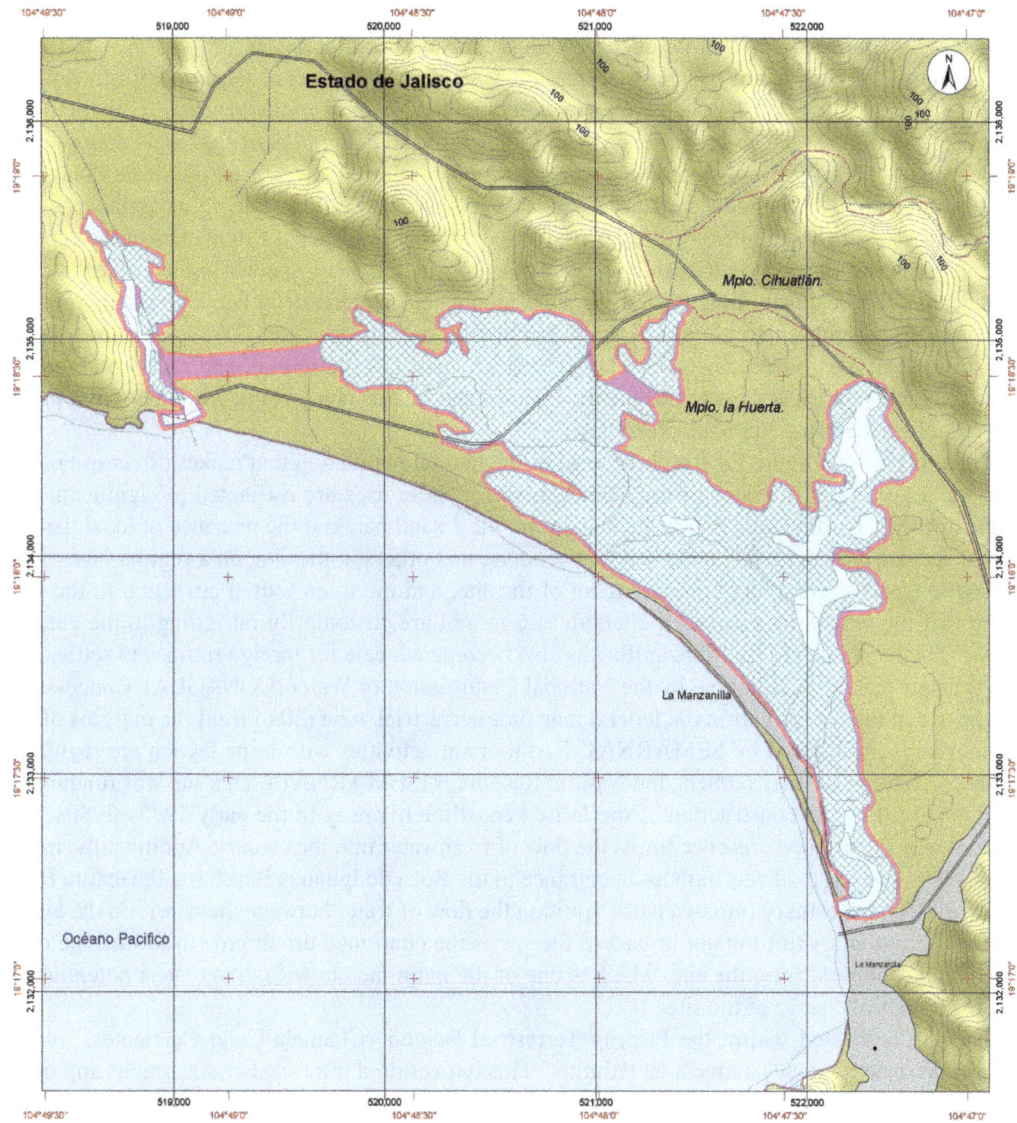

FIGURE 5.23 Map of Estero La Manzanilla. Designated February 02, 2008; Area: 264 ha; Coordinates: 19°18′N 104°47′W; Ramsar Site number: 1,789.

important component of the food and/or nutrient web/cycle): milkfish (Chanos chanos), the mullets *Mugil cephalus and Mugil curema*, as well as the Pacific fat sleeper and guavina (Dormitator latifrons and Guavina micropus); 18 other species (such as Centropomus robalito, Arius platypogon, Mugil curema, Eucinostomus currani, and Chanos chanos) enter the lagoon at early life stages, and then mature within the ecosystem (Nelson 1994, Fisher et al. 1995, Aguilar-Palominoet al. 2006). The lagoon also provides refuge and feeding/rearing habitat for shrimp belonging to the families Palaemonidae, Alpheidae, Hippolytidae, and Penaeidae.

The flora on the margins of the lagoon are represented by 17 families and 40 species. In addition to the ecological habitat value of the mangroves, they are also used by local community members for firewood, construction wood, poster board, as well as for craft and ornamental uses. Some families of plants are also used to obtain or manufacture food (Amaranthaceae, Compositae, Cucurbitaceae, Gramineae, Leguminoceae, among others); timber (Combretaceae, Leguminoceae, Verbenaceae,

Sterculiaceae, and Rubiaceae); and medicinal uses (Boraginaceae, Compositae, Cucurbitaceae, Cyperaceae, Graminaceae, Leguminoceae, Malvaceae, and Portulacaceae).

Also noted at the site are aquatic species such as decapod crustaceans (shrimp with freshwater and brackish affinities); fish species *Gerres cinereus* and *Eucinostomus argenteus*; Tilapia; and crustaceans *Callinectes arcuatus* (arched swimming crab) and *Farfantepenaeus californiensis* (yellowleg shrimp). Forty-seven waterfowl species have been reported at the site, of which 18 species are visitors and 29 are resident birds (Hernández-Vázquez 2000, 2005). Due to its abundance and biomass, as well as being one of the top predator species that regulates the dynamics of the ecosystem, the American crocodile (Crocodylus acutus) is considered a key species in the estuary, which is indicative of a relatively functional ecosystem food web. The terrestrial fauna on the margins of the estuary are represented mainly by small mammals, such as opossum (Didelphis marsupialis), armadillo (Dasypus novemcinctus), South American coati (Nasua nasua), and raccoon (Procyon lotor). Notable reptiles at the site include the iguanas and lizards (Iguana iguana, Ctenosaura pectinata, Anolis nebulosus, and Sceloporus spp.)

There is no exact figure for the percentage of the human population that makes direct use of the ecological goods and services of the lagoon, however these uses are estimated as significant due to the immediate urban area that is located on the sites' sand bar, and the presence of local fishers, tourist service providers, restaurants, street vendors, and others at the site, on a regular basis. The crocodile has become a recognized symbol of the site, and the main tourist attraction to the site, given that the crocodiles can reach a length of 6 m and are customarily intriguing to the general public. The local town of La Manzanilla has also become a locale for foreign retirees to settle. The site is under federal jurisdiction, by the National Commission of Water (CONAGUA). Concessions for the use of the waters within the federal maritime-terrestrial zone (20 m from the margins of the lagoon) are administered by SEMARNAT. Eco-tourism activities within the lagoon are regulated by the Secretariat of Environment and Natural Resources (SEMARNAT). This site was fundamentally impacted by the construction of the Jalisco coastline highway in the early 1970s–1980s, and the highway's continued presence limits the flow of freshwater into the estuary. Additionally, in the early 1980s a paved road was built as an entrance to the Boca de Iguanas Beach (in Tenacatita Bay), which divided the estuary into two parts, limiting the flow of water between them, within the larger lagoon ecosystem. A third major impact to the site is the continued urban growth toward the east, south, and southwest from the site, which is one of the main factors with the greatest potential for future ecological change at the site.

The site is located within the Priority Terrestrial Region "Chamela-Cabo Corrientes," in the Priority Marine Region "Chamela-El Palmito." This has resulted in several organizations and institutions becoming interested in the conservation of the Manzanilla Estuary mangroves, and associated flora and fauna described above. To this end, research has been carried out by a number of institutions and universities, such as the UNAM and the University of Guadalajara, which have conducted studies on the population dynamics of the American crocodile (Crocodylus acutus), and other studies of migratory waterfowl. In addition, researchers and students from the University of Nevada (USA) and the University of South Florida (USA) have carried out hydrological studies at the site in recent years, as well as on the floral and faunal communities in the estuary. Additionally, a civil social organization called "Tierralegre," which develops a garbage-recycling program in urban areas of the site, along with educational work on environmental issues, has facilitated conservation activities at the site, involving local community members and others.

LAGUNA BARRA DE NAVIDAD, JALISCO: OVERVIEW

Laguna Barra de Navidad (Figures 5.24 and 5.25) is an estuarine ecosystem along Mexico's western central littoral zone, with a permanent connection with the ocean. The vegetation found at the margins of estuary is comprised of the mangrove species *Rhizophora mangle*, *Laguncularia racemosa*, *Avicennia germinans*, and *Conocarpus erectus*. Sixty aquatic bird species have been identified at

FIGURE 5.24 Composite aerial image of Laguna Barra de Navidad. Designated February 02, 2008; Area: 794 ha; Coordinates: 19°11′N 104°40′W; Ramsar Site number: 1,817.

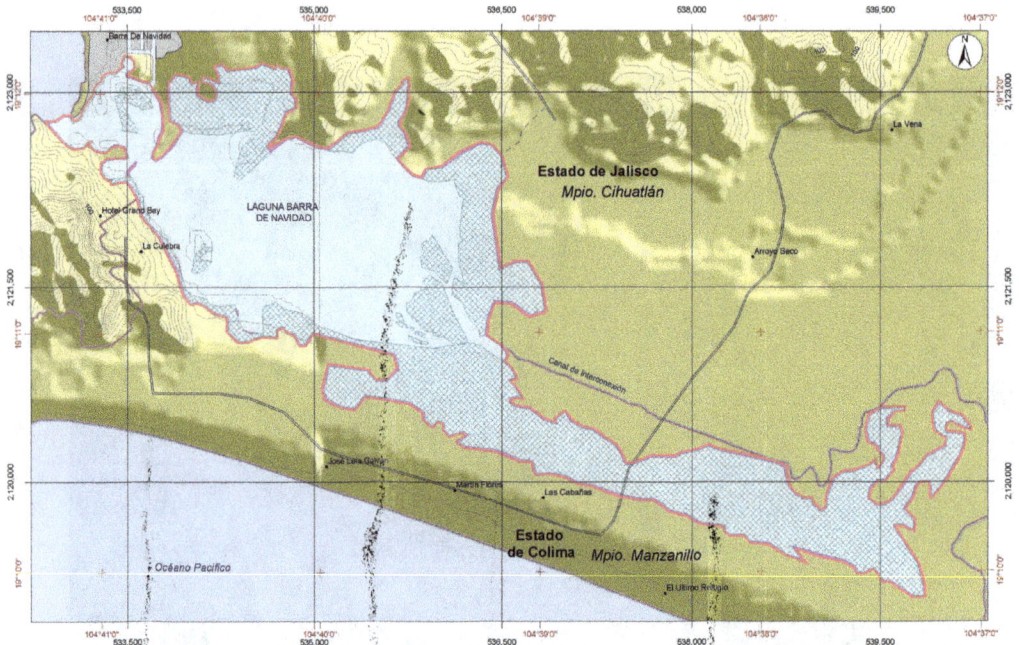

FIGURE 5.25 Map of Laguna Barra de Navidad. Designated February 02, 2008; Area: 794 ha; Coordinates: 19°11′N 104°40′W; Ramsar Site number: 1,817.

the site, representing 50% of the species identified for the coast of Jalisco, including *Ardea herodias*, *Egretta rufescens*, *Mycteria americana*, *Larus heermanni*, *Sterna elegans*, *Buteogallus anthracinus*, and *Nomonyx dominicus*, with all except the last on this list under legal protection. The snow goose (*Chen caerulescens*) has been observed in the site's lagoon area, which is a rare siting along the coast of Jalisco. The lagoon is also essential for the reproduction of migratory and resident waterfowl, and is also of great importance for fish species in different stages of their life cycles.

SITE DETAILS

The site is characterized at its southern portion by small bays with sandy beaches, with rocky and mountainous upland areas that lead to alluvial valleys, bathed by rivers and temporary streams. There are only two wetlands of this region that are considered to be, geologically, 'coastal lagoons' that have their mouths permanently open to the sea, one of which is the Barra de Navidad Lagoon and the other the Cuyutlán Lagoon, which is in the neighboring State of Colima. This site has its headwaters in the vicinity of Cerro Camalote, located 15 km northwest of the town of Villa Purificación, at an elevation of 1500 m above sea level. During the entire journey of upland water flow toward the site, along streams and channels for approximately 85 km, it maintains a predominantly south-southwest direction, ultimately reaching the site. The basin itself covers an area of approximately 1,141 km² and flows ultimately into the Pacific Ocean at a point located 5 km to the southeast of Punta Farallon.

The lagoon has an approximate 'water mirror' surface area of 3.7 km² and a maximum length and width of 3.5 km² and 1.5 km², respectively (DEDSZC 2007); it is permanently connected to the sea by a channel that is 80 m wide and with depths of up to 7 m, with a central portion that is periodically mechanically dredged to allow the passage of boats to marinas that serve tourism needs. The main supply of fresh water comes from the Marabasco River, and the Arroyo Seco River primarily during the rainy season. The lagoon has salinities no greater than 35%, and its tidal pattern is semidiurnal (CONABIO 2007).

The floral species at the site, within the lagoon, are represented by 19 families and 43 species. Mangroves at the site are used by the local inhabitants mainly for firewood, construction wood, posts, and crafts. Some families of plants at the site are also used for obtaining or manufacturing food (Amaranthaceae, Compositae, Cucurbitaceae, Gramineae, Leguminoceae, and others); timber (Combretaceae, Leguminoceae, Verbenaceae, Sterculiaceae, and Rubiaceae), and medicines (Aizoaceae, Boraginaceae, Compositae, Cucurbitaceae, Cyperaceae, Graminaceae, Leguminoceae, Malvaceae, Portulacaceae, and Rubiaceae). The Barra de Navidad Lagoon shares some types of vegetation found in the Chamela-Cuixmala Reserve (deciduous forest, mangroves, and palms).

The aquatic fauna associated with the lagoon and its mangroves are made up of 87 species of fish (belonging to two classes, 16 orders, and 44 families) [Aguilar-Palomino 2006]; 13 species of echinoderms; 74 species of decapod crustaceans (including seven Caridae shrimp); 110 species of molluscs; and 34 species of annelids (Rodríguez 1993). At least 23 species of fishing-interest exist at the site, belonging to three taxonomic groups, including scale fish, crustaceans, and bivalves, more specifically: the fish species *Gerres cinereus* and *Eucinostomus currani*, sardines (Anchoa mundeola and Anchovia macrolepidota), snook (Centropomus spp.), red snapper (Lutjanus spp.), mullet (Mugil spp.), catfish (Arius spp.), and botetes (Sphoeroides spp.); crustaceans such as the lesser blue crab (Callinectes crissum) and yellowleg shrimp (Farfantepenaeus californiensis); molluscs (Pinna rugosa, Atrina maura, and Megapitaria squalida) [Hernández-Cruz 2005, Aguilar-Palomino 2006, CONABIO 2007].

Other important groups of aquatic fauna in the lagoon are the seabirds (14 species), ducks (7 species), shorebirds (19 species), and herons (14 species) [Hernández-Vázquez 2005], in addition to the American crocodile (Crocodylus acutus), which is subject to special protection under NOM-059-SEMARNAT-2001 and/or -2010. The site also provides nesting habitat for leatherback sea turtle (Dermochelys coriacea), olive ridley sea turtle (Lepidochelys olivacea), and green sea turtle (Chelonia mydas), all three of which are endangered species in NOM-059-SEMARNAT-2001, the IUCN Red List, and in the CITES. Terrestrial fauna at the site are represented by small mammals such as opossum (Didelphis marsupials), armadillo (Dasypus novemcinctus), South American coati (Nasua nasua), and raccoon (Procyon lotor); and, beyond sea turtles, additional reptiles such as iguanas and lizards (Iguana iguana, Ctenosaura pectinata, Anolis nebulosus, and Sceloporus spp.) [Gonzalez-Guevara 2001].

Farmers and ranchers are traditional and historical users of the site, for cultivation of banana, mango, and coconut and for livestock paddocks and grazing. Fishers utilize the site for seafood extraction, and inhabitants of the area utilize the mangroves for firewood. There is no exact figure for the percentage of the population that makes direct use of the ecological goods and services of the lagoon, however it is likely quite high since the site is within an urbanized area with a substantial number of fishers, tourist service providers, restaurants, and vendors. The tourist infrastructure of Barra de Navidad comprises approximately 20% of the total lodging on the south coast of the State of Jalisco, providing employment to approximately 85% of the economically active population of the local community. The annual economic benefit associated with tourist uses of the site is substantial, which originates from tens of thousands of national and foreign visitors (Hernández-Cruz 2005).

The first expeditions of the Spanish to Jalisco took place in 1523, led by Gonzalo de Sandoval, and in 1,528 Hernán Cortés reported back to the King of Spain that the province of Cihuatlán was a land of women, and very rich in pearls and gold. Subsequently, the area was expanded, with a shipyard that contributed to the conquests of the Philippines in the early 1500s. In 1564, the shipyards were dismantled, and in the early 1800s the vicinity of the site was decreed by the President of the Republic as a port for foreign trade and passenger transport.

Within the site's marinas of Cabo Blanco and Puerto de La Navidad, numerous sailboats and recreational yachts visit each year. Formally organized groups that are directly dependent on water and resources of the lagoon include fishing cooperatives and family businesses. The uses of the lagoon are highly variable, and range from the activities of loading and unloading of fishery products, the provisioning of services, fishing (commercial, recreational, and sports), and transportation.

The groups operating within the lagoon do so with a fleet of hundreds of small vessels, operating inside and outside of the lagoon. The boats also transport passengers, tourists, and employees to the restaurants and hotels of Colimilla, offering other tourist services for those interested in walking tours in the lagoon and sport fishing (Hernández Cruz 2005). Accordingly, the main proximal uses of the site within the lagoon are artisanal fishing, subsistence fishing, aquatic and underwater activities (diving, kayaking, and swimming), and bird watching.

Since the early 20th century, the site has also been used for salt extraction, which has been the impetus for many of the built structures and infrastructure that now exist. The associated work has produced substantial modifications to the lagoon, including the felling of mangroves, filling of channels, construction of embankments, and a number of modifications of the hydrodynamics of the lagoon. Beginning in the 1950s, the commercialization of various fishery products led to negative impacts on sea turtles, as well as sharks (Galván-Piña et al. 2007). Beginning in the 1960s, additional negative impacts to the ecology of the site occurred from urbanization, such as the degradation of water quality, which impacted oyster beds. The lagoon underwent a third round of major modifications in the 1970s (Galván-Piña et al. 2007), including channel modifications and the construction of Federal Highway 200 (Manzanillo, Puerto Vallarta) and 80 (San Patricio (Melaque-Guadalajara). During all of these periods of development, extensive areas of mangroves were cleared in the area south and southwest of the lagoon (Galván-Piña et al. 2007).

The University of Guadalajara's Department of Studies for the Sustainable Development of Coastal Zones, and the Autonomous University of Guadalajara have developed research programs that are focused on monitoring and assessing the ecology of the lagoon, within the larger context of the southern coastal zone landscape changes. These studies include the recognition of the fact that the current tourism sector is the economic engine of the town of Barra de Navidad and of the entire area of Bahía de Navidad, with fishing, agriculture, and livestock as similarly important contributions to the economy of the area. Nautical tourism has also increased in recent years, with the development of marinas within the lagoon, with concomitant increases in this type of visitorship from the United States. Many of these visiting yachts and sailboats stay for weeks to months in the lagoon, between November and April.

REFERENCES

Aguilar Palomino, B., González Sansón, G. and F. Silva-Bátiz. 2006. *Inventario Ictiofaunístico de la Costa de Jalisco*. Ed. Centro Universitario de la Costa Sur, Universidad de Guadalajara. Jalisco, México. 150pp.

Amézcua-Linares, F. 1996. *Peces demersales de la plataforma continental del Pacífico central de México*. UNAM and CONABIO, Mexico. 113pp.

Arizmendi, C., Berlanga, H., Márquez, L., Navarijo, L., and F. Ornelas. 1991. Avifauna de la región de Chamela, Jalisco. *(Serie Cuadernos No.4)*. Instituto de Biología, UNAM, México.

Berlanga, H., Arizmendi, C., and F. Ornelas. 1987. Las aves. In: G. Ceballos (Ed.) *Reporte ecológico de la región de Cuixmala-El* Faro, Jalisco. Reporte inédito with the accent, Instituto de Biología, UNAM, México.

Carmona, R. and G. Danemann. 1998. Distribución espacio-temporal de aves en la salina de Guerrero Negro, Baja California Sur, México. Ciencias Marinas. 24: 389–408.

CONABIO. 2007. *Opinión técnica sobre el estado actual e impactos en la Laguna Barra de Navidad, Cihuatlán, Jalisco*. Dra. Patricia Koleff, Directora de Análisis y Prioridades/Comisión Nacional para el Conocimiento y Uso de la Biodiversidad. No. Oficio DTAP/255/2007. 25 de Septiembre de 2007. 6pp.

CONABIO. 2008. *Manglares de México. Comisión Nacional para el Conocimiento y Uso de la Biodiversidad*. México. 38pp.

Day, J. and A. Yañez-Arancibia. 1979. *Lagoon-estuarine environments as ecosystems*. Semin. Latinoam. Pric. Mét. Ecol. Lag. Costeras. UNAM- OEA. Cd. del Cármen 8pp.

De Ita, M.C. 1983. *Patrones de Producción agrícola en un ecosistema tropical estacional en la costa de Jalisco. Tesis de Licenciatura (Biología)*. Facultad de Ciencias, UNAM. 183pp.

DEDSZC. 2007. *Dictamen Técnico sobre Impactos Ambientales del Proyecto Turístico"Isla Primavera"*, 2007. Departamento de Estudios para el Desarrollo Sustentable de Zonas Costeras. Universidad de Guadalajara. Comisión Técnica: Dra. Carmen Franco Gordo, Dr. Salvador Hernández Vázquez, Dr. Jorge Arturo Rojo Vázquez, Dr. Víctor Hugo Galván Piña, Dr Antonio Corgos López-Prado y Dr. Enrique Godínez Dominguez. 37pp.

DOF (Diario Oficial de la Federación). 2000. *Norma Oficial Mexicana NOM-131-SEMARNAT-1998, Establece lineamientos y especificaciones para el desarrollo de actividades de observacion de ballenas, relativas a suproteccion y la conservacion de su habitat.* Official Gazette. Ministry of Environment, Natural Resources and Fisheries. 8pp.

Fischer, W., Krupp, F., Schneider, W., Somer, C., Carpenter, K.E., and V.H. Niem. 1995. *Guía FAO para la identificación de especies para los fines de la pesca. Pacífico Centro-Oriental.* Vol. I, II, and III. FAO, Rome. 1747pp.

Galvan Piña, V.H., Godínez-Domínguez, E., and F.A. Silva-Bátiz. 2007. *Usos Humanos y Procesos de Cambios en la Laguna Barra de Navidad, Jalisco.* Departamento de Estudios para el Desarrollo Sustentable de Zonas Costeras/Universidad de Guadalajara. Informe Técnico. 23pp.

García, A. and G. Ceballos. 1995. Reproduction and Breeding Success of California Least Terns in Jalisco, Mexico. The Condor. 97: 1084–1087.

Gaviño, G. and Z. Uribe. 1980. Distribución, población y época de la reproducción de las aves de las islas Marietas, Jalisco, México. An. Inst. Biol. U.N.A.M. 51 Ser. Zool. 1: 505–524.

González-Guevara, L.F. 2001. Manifestación de Impacto Ambiental del Desarrollo Turístico "Isla Primavera," Laguna Barra de Navidad, Jalisco. Informational Report. 200pp.

Hendrickx, M.E. 1988. Distribution and bathymetric records of Processidae (Caridea) and Penaeidae (Penaeoidca) in the Gulf of California, México. An. Inst. Cienc. del Mar y Limnol., Univ. Nal. Aut. México 15.

Hernández Vazquez, S. 2005. *Aves acuáticas de la costa de Jalisco, México.* Programa de doctorado en Ciencias Marinas. CICIMAR, México.

Hernández-Cruz, J.F. 2005. Perspectivas de los Recursos Naturales y sus Usos en la Laguna Barra de Navidad, México. Tesis Profesional. Universidad de Guadalajara. 90pp.

Hernández-Vázquez, S. 2000. Aves acuáticas del estero La Manzanilla, Jalisco, México. Acta Zoológica Mexicana, nuevaserie. 80: 143–153.

Hernández-Vazquez, S., De La Cueva-Salcedo, H., and J. Rojo-Vazquez. 2002. Análisis Comparativo De La Avifauna Del Estero Majahuas (Jalisco, México) Entre Un Evento El Niño y Un Año No Niño. Boletín del Centro De Inv. Biol. 36(1): 94–112.

Hutto, R. L. 1989. The effect of habitat alteration on migratory land birds in a west Mexican tropical deciduous forest: a conservation perspective. Conservation Biology. 3: 138–148.

IUCN. 2007. *IUCN Red List of Threatened Species.* (https://www.iucnredlist.org/)

Nelson, J.S. 1994. *Fishes of the World.* 3rd. Edition. John Wiley and Sons, Inc., New York. 600pp.

Pérez, J.J.C. 2001. *Análisis de la pesquería artesanal de tiburones y rayas de Isla Isabel, Nayarit, México.* Tesis de Maestría (Ecología Marina). División de Oceanología. Centro de Investigación Científica y de Educación Superior de Ensenada (CICESE).

Phleger, F.B. 1969. Some general features of coastal lagoons. *Mem. Sim. Intern. Lagunas Costeras.* UNAM-UNESCO, México. p. 5–26.

Rodríguez, S. 1993. Macrofauna de la Laguna de Barra de Navidad, Jalisco (pp. 499–508). In: S. I. Salazar and N. E. González (Eds) *Biodiversidad Marina y Costera de México.* Comisión Nacional para el Conocimiento y Uso de la Biodiversidad (CONABIO) y Centro de Investigaciones de Quintana Roo (CIQRO). México. 865pp.

Ruiz, L.A. 1977. *Algunos aspectose cológicos de la Isla Isabel, Nayarit.* Tesis de Licenciatura (Biología). Facultad de Ciencias, UNAM.

Rzedowski, J. 1978. Vegetación de México. Limusa. México. 432pp.

Toledo, V. M. 1988. La diversidad biológica de México. Ciencia y Desarrollo. 14(81): 7–30.

6 Mexico
States of Colima, Michoacan, and Guerrero

The Ramsar wetlands along the western coast of south-central mainland Mexico, in the states of Colima, Michoacan, and Guerrero, include some of the most unique environments discussed thus far, such as within the Archipiélago de Revillagigedo. The south-central mainland region of wetlands demonstrates a definitive shift that occurs along the Mexican mainland coastal region, from more arid environmental conditions to increasingly tropical environmental conditions, indicative of our continued latitudinal journey south. The criticality of the following nine coastal wetland ecosystems (Figure 6.1), specifically for a multitude of terrestrial and aquatic organisms, as well as for the indigenous people of the region, is evident:

- Laguna de Cuyutlán Vasos III y IV
- Reserva de la Biosfera Archipiélago de Revillagigedo
- Santuario Playa Boca de Apiza-el Chupadero-el Tecuanillo
- Playa de Colola
- Playa de Maruata
- Playón Mexiquillo
- Laguna Costera el Caimán
- Playa Tortuguera Tierra Colorada
- Playa Tortuguera Cahuitán

LAGUNA DE CUYUTLÁN VASOS III Y IV, COLIMA: OVERVIEW

Laguna de Cuyutlán Vasos III y IV (Figures 6.2 and 6.3) is a lagoon, which is the fourth largest coastal wetland in Mexico. The mangrove community at the site is represented by white mangrove (Laguncularia racemosa), red mangrove (Rhizophora mangle), and black mangrove (Avicennia germinans), which all provide important habitat for a variety of species of resident wildlife, migratory birds, fish, and invertebrates that use the lagoon for feeding, resting, reproduction, and/or breeding. The site also supports species under special protection, as well as endemic species, such as the lilac crowned parrot (Amazona finschi), the great horned owl (Bubo virginianus), the banana bat (Musonycteris harrisoni), and the Mexican mud turtle (Kinosternon integrum). The primary human uses of the lagoon include fishing and salt extraction, the latter being the major negative impact to the ecology of the site, along with agriculture.

SITE DETAILS

The site is located in the State of Colima, 8.8 km southeast of the City and Port of Manzanillo. The Cuyutlán Lagoon is located on the coast of the Pacific Ocean, in the coastal plain, bordering the Armería River, to the south. The site is elongated along the coastline, with a sand bar that has a width ranging between approximately 5 km and 6 km. Adjacent to the mangroves, on the margins of those forested areas. are mainly deciduous forest species (particularly to the north and northeast of the lagoon), with water coconut plantations and agriculture throughout. Along the beaches adjacent

DOI: 10.1201/9781003046394-6

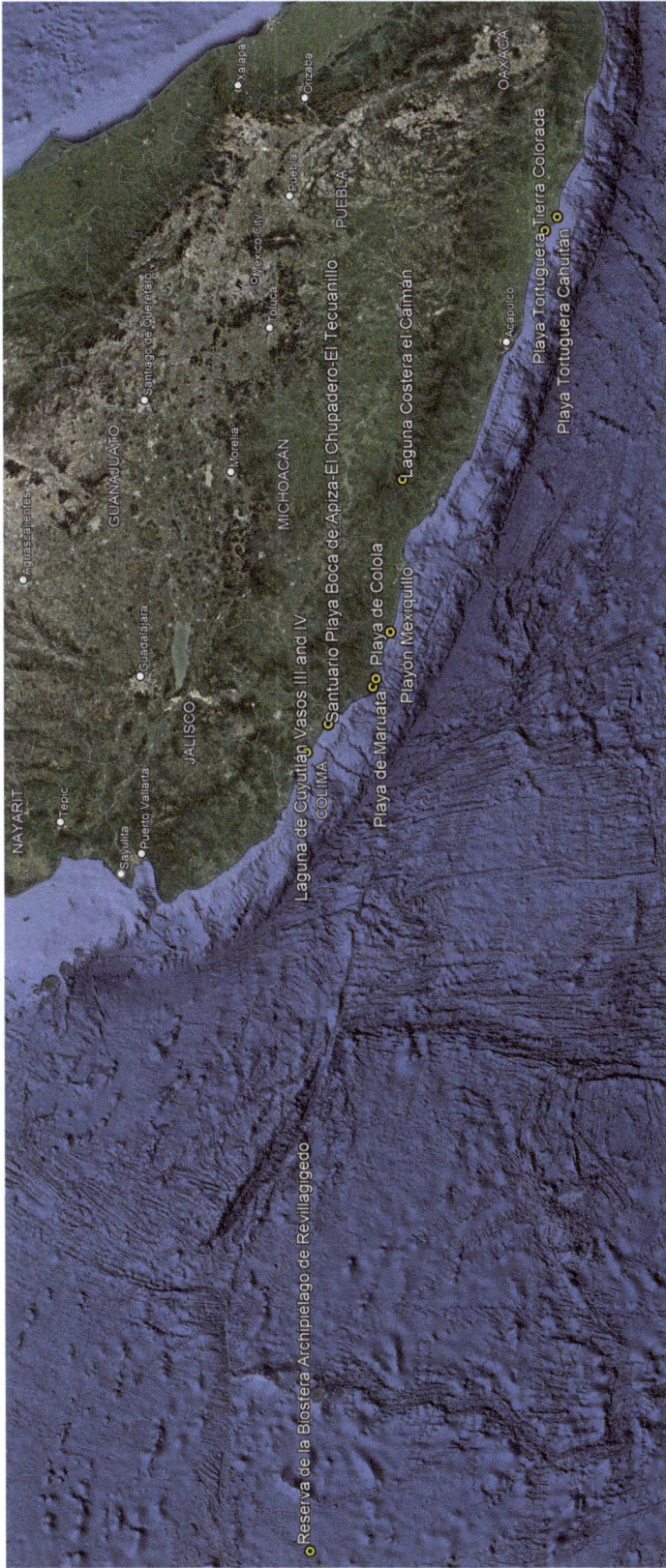

FIGURE 6.1 Referenced Ramsar wetlands generally shown along the western coast of south-central mainland Mexico, in the states Colima, Michoacan, and Guerrero.

to these wetlands, endangered sea turtles, such as olive ridley (Lepidochelys olivacea), leatherbacks (Dermochelys coriacea), and green sea turtles (Chelonia mydas) flourish (Alvarado and Figueroa 1985). Within the Cuyutlan Lagoon itself, other species of special protection are supported, including the reptiles *Ctenosaura pectinata*, *C. similis*, *Iguana iguana*, *Procyon insularis*, *Balaenoptera* sp., *Eschrichtius robustus*, *Crocodylus moreletii*, and *C. acutus* (Alvarado et al. 1996); invertebrates, specifically molluscs, such as *Ancistromesus mexicanus*, *Pinctada mazatlanica*, *Pinna rugosa*, *Pternia sterna*, *Purpura pansa*, and *Noctilio leporinus* (Espino-Barr et al. 1998); crustaceans of commercial importance, such as *Macrobrachium* sp., *Litopenaeus* (white shrimp) and *Callinectes arcuatus* (arched swimming crab) [Andrade 1998]; mammals such as *Leopardus pardalis*, *Leopardus wiedii*, *Felis yagouaroundi*, *Icterus cucullatus*, *Aramides axillaris*, and *Tachybaptus dominicus*; and the birds, such as *Egretta rufescens*, *Mycteria americana*, *Chondrohierax uncinatus*, *Rostrhamus sociabilis*, *Buteo platypterus*, *Buteo albonotatus*, *Micrastur semitorquatus*, *Sternula antillarum*, *Eupsittula canicularis*, and *Glaucidium palmarum*.

Main plant species at the site are red mangrove (Rhizophora mangle), white mangrove (Laguncularia racemosa), and black mangrove (Avicennia germinans), as well as *Astronium graveolens*, *Orbignya guacuyule*, *Guayacum coulteri*, *Cocos nucifera*, *Handroanthus chrysanthus*, and *Sideroxylon capiri*. In the site's semi-deciduous forested areas, the following species are supported: breadnut (Brosimum alicastrum), kapok tree (Ceiba pentandra), trumpet-tree (Cecropia obtusifolia), pink trumpet tree (Tabebuia palmeri), sandbox tree (Hura polyandra), the euphorb *Coccoloba floribunda*, elephant-ear tree (Enterolobium cyclocarpum), gumbo limbo (Bursera simaruba), gliricidia (Gliricidia sepium), Brazilian rose (Cochlospermum vitifolium), *Jacquinia aurantiaca*, stinking toe (Hymenaea courbaril), and the euphorb *Hura polyandra*, among others. Other forest vegetation supported by the site includes: *Cordia elaeagnoides*, *Cercidium praecox*, *Acacia*s pp., *Bursera* spp., *Haematoxylum brasiletto*, and *Cephalocereus* spp. In the coastal dunes of the site, species present include: *Distichlis spicata*, *Ipomoea pes-caprae*, *I. stolonifera*, *Pectis arenaria*, and *P. saturejoides* (Santana et al. 1992).

Faunal species supported at the site that are protected under NOM-059-SEMARNAT-2010 include: masked duck (Nomonyx dominicus), rufous-necked wood rail (Aramides axillaris) [threatened]; least grebe (Tachybaptus dominicus), reddish egret (Egretta rufescens), wood stork (Mycteria americana), hook-billed kite (Chondrohierax uncinatus), snail kite (Rostrhamus sociabilis), broad-winged hawk (Buteo platypterus), zone-tailed hawk (Buteo albonotatus), collared forest falcon (Micrastur semitorquatus), least tern (Sternula antillarum), orange-fronted parakeet (Eupsittula canicularis), Colima pygmy owl (Glaucidium palmarum), and pale-billed woodpecker (Campephilus guatemalensis) [Hernández-Vázquez 2004, SEDUR 2007]. Mammals that are key elements of the ecosystem as seed dispersers include grayish mouse opossum (Marmosa canescens), Virginia opossum (Didelphis virginiana), collared peccary (Pecari tajacu), Mexican gray squirrel (Sciurus aureogaster), Sumichrast's vesper rat (Nyctomys sumichrasti), Nine-banded armadillo (Dasypus novemcinctus), painted spiny pocket mouse (Heteromys pictus), greater bulldog bat (Noctilio leporinus), plains pocket gopher (Geomys bursarius), hooded skunk (Mephitis macroura), marsh rice rat (Oryzomys palustris), coyote (Canis latrans), and ocelot (Leopardus pardalis) [a threatened species]. Most common reptiles at the site include boa (Boa constrictor), arboreal ratsnake (Gonyosoma oxycephalum), the coast patch-nosed snake (Salvadora hexalepis virgultea), milk snake (Lampropeltis triangulum), Sinaloan milk snake (Lampropeltis triangulum sinaloae); Saurians of the order Scuamata, suborder latercilia: iguanas (Iguana iguana and Ctenosaura similis), American crocodile (Crocodylus acutus) [Alvarado et al. 1996], olive ridley sea turtle (Lepidochelys olivacea), green sea turtle (Chelonia mydas), and the leatherback sea turtle (Dermochelys coriacea). Obligate aquatic fauna at the site include: Gerres cinereus, mullet (Mugil curema), Peruvian mojarra (Diapterus peruvianus), Pacific dog snapper (Lutjanus novemfasciatus), Pacific bumper (Chloroscombrus orqueta), longjaw leatherjacket (Oligoplites altus), milkfish (Chanos chanos), snook (Centropomus sp.), spotted-fin sole (Trinectes fonsecensis), tilapia (Sarotherodon), whiteleg

FIGURE 6.2 Composite aerial image of Laguna de Cuyutlán Vasos III y IV. Designated February 02, 2011; Area: 4,051 ha; Coordinates: 18°58′09″N 104°06′42″W; Ramsar Site number: 1,985.

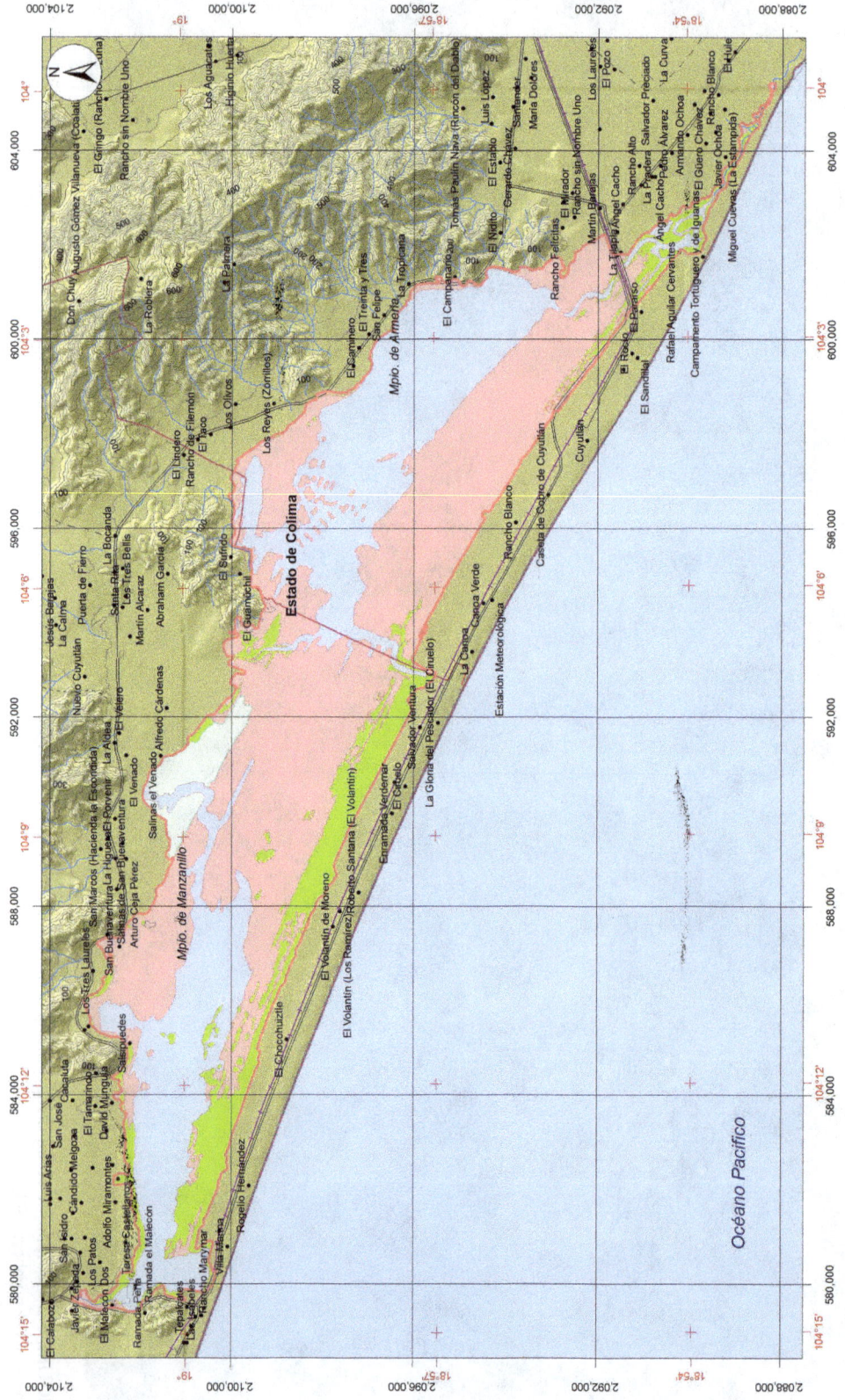

FIGURE 6.3 Map of Laguna de Cuyutlán Vasos III y IV. Designated February 02, 2011; Area: 4,051 ha; Coordinates: 18°58′09″N 104°06′42″W; Ramsar Site number: 1,985.

shrimp (Litopenaeus vannamei), yellowleg shrimp (Farfantepenaeus californiensis), the arched swimming crab (Callinectes arcuatus), and prawn (Macrobrachium sp.).

The history of, and current need for humans to obtain, salt from the sea has been, and is a major economic driver for, human activities in the region. Technology used for salt mining in Cuyutlán is among the oldest in Mexico (known to be in use since the 16th century). "The Museum of Salt" is nearby the site, where the history, process, and the tools used for the production of this precious substance are archived. The site's coastal lagoon is a long-standing source of fishing resources for the various fishing cooperatives in the area, and the irrigated agriculture, rainfed agriculture, grasslands, and small human settlements on the periphery of the site are all of vital importance to the communities of the area, as they have been for many generations.

Human activities that have negative impacts on the ecology of the site include self-consumption fishing, salt extraction, guided boat rides through the mangroves, and agriculture. Maintenance of the Tepalcates canal, which guarantees the necessary flow conditions to guarantee the entrance of salt water into the lagoon, involves routine dredging and removal of silt from the site's channel to the sea, under the criteria set forth by Ecological and Territorial Planning of the Cuyutlán Lagoon Sub-basin (SEDUR 2007); studies of these activities have been conducted so as to ensure that environmental impacts are understood, particularly with regard to crocodile populations, fisheries, birds, and mangroves.

The Cuyutlán Colima Ecological Center (Tortugario) operates in the Palo Verde Estuary, whose objective is to research and better understand the conditions necessary to protect reptiles in danger of extinction in this vicinity of Mexico, for which they have delimited areas where sea turtles come to spawn and incubate hundreds of hatchlings at the beach adjacent to the Cuyutlán Lagoon. This research also involves the study of captive iguanas, crocodiles, and turtles that have been injured either by nets, boat propellers, and/or predators, until they recover. Other associated projects with this captive animal recovery program include education and outreach about the topics of sea turtle, crocodile, green iguana, and bird conservation and general ecological conservation of the Palo Verde Estuary. Other investigations have been carried out to characterize the physical and biotic environment of the site, by the Regional Center for Fisheries Research (CRIP), with headquarters in the city of Manzanillo, Colima. At the Cuyutlán Ecological Turtle Center, public awareness activities are carried out, focusing on the conservation of sea turtles, mangroves, as well as the wetland ecosystem in general. The Ecological Camp of Cuyutlán Colima also conducts awareness and education work by providing tourists with tours of the mangroves, with the highest visitation occurring during the Easter season, as well as on holidays and long weekends.

RESERVA DE LA BIOSFERA ARCHIPIÉLAGO DE REVILLAGIGEDO: OVERVIEW

Reserva de la Biosfera Archipiélago de Revillagigedo (Figures 6.4 and 6.5) is a volcanic archipelago located approximately 400 km offshore of the western coast of Mexico, in the Pacific Ocean, and is a unique home to particular endemic flora and fauna. Isla Socorro (Figures 6.4a and 6.5) is the site's largest island, with the Evermann Volcano peaking at approximately 1050 m above sea level, followed in area by Isla Clarión (Figures 6.4b and 6.5), Isla San Benedicto (Figure 6.4c and 6.5), and Isla Roca Partida (Figure 6.5), respectively. Socorro Island represents vegetation that follows an increasing altitudinal gradient, from coastal halophytes and shrubs *Dodonaea viscosa*, *Guettarda insularis*, and *Croton masonii* to forests of *Ficus cotinifolia* and *Psidium galapageium*. Ten endemic species and subspecies of birds have been recorded on Socorro Island, with three of them extinct in the wild, including the Socorro dove (Zenaida graysoni), with plans of reintroduction from several captive individuals in Germany. The site is well preserved and uninhabited, apart from Mexican Navy activities. Diving and fishing groups visit the islands regularly. Volcanic eruptions, hurricanes, and fires pose risks to the islands' wildlife, and invasive species (sheep, pigs, and rabbits) remain the main threat to the ecology of the site.

SITE DETAILS

The annual average temperatures of the site's archipelago are generally warm, above 22° C, and the temperature of the coldest month is greater than 18° C. In the case of Socorro Island, the warmest month is August, at 27.5° C, the coldest is January, at 22° C, with thermal oscillations between 5° C and 7° C. For the highest area of Socorro Island, the estimated temperature for the coldest month is 11.2° C, on average, while the warmest month is 19.7° C, on average. The average annual precipitation reaches 313.8 mm, with September as the wettest month.

The fluvial network of Isla Socorro is radial centrifugal, typical of volcanic areas. Since river courses lack tributaries at the site, the hydrographic basins are extremely reduced. According to Blásquez (1960), two springs are known to exist on the island, one in the vicinity of Grayson Bay and another in the southeast region along the coast. The spring near Bahía Grayson has a hydraulic flow of 0.05 to 0.1 liters per second and its temperature is 27° C, as it has thermal contact with rocks still hot from prior eruptions. This spring is considered brackish and alkaline. In contrast, the other spring, named "La Tribuna", is sourced from fresh water, with a temperature of 20° C and a flow of 0.3 to 0.5 liters per second during droughts, while in the rainy season it increases approximately ten times this rate.

Only Socorro Island and Clarión Island have flora, with the remaining islands denuded of vegetation as a result of volcanic activity (Johnston 1931). The marine flora that inhabits the archipelago integrate biological elements from the Indo-Pacific, the Gulf of California, and the Mexican Pacific (Verdi et al. 1994). Avendaño and Sotomayor (1986) identified 37 genera of phytoplankton west of Socorro Island, with a predominance of *Bacillariophyta*. Macroalgae of the site include 214 species, of which 43 belong to the Chlorophyta, 32 to the Phaeophyta, and 139 to the Rhodophyta.

Miranda (1960) points out that plant species such as *Forestiera rhamnifolia* were widespread on the continent in ancient times, when they invaded the archipelago. Later, due biogeographic change on the mainland, but not on the archipelago, reduced dominance of this species on the continent occurred, conserving it predominantly on the archipelago, as with the other predominant tree species of the site (amate, cascarillo, copal, guayabillo, and zapotillo). It is possible that plant colonization of the islands occurred at the end of the Tertiary, or beginning of the Quaternary Periods.

The terrestrial fauna of the archipelago are not very diverse, relative to the mainland. The most varied group and the most numerous are the birds. Amphibians lack representatives at the site. Vázquez (1960), who made one of the few descriptions of the entomological fauna of Isla Socorro, refers to the presence of numerous scorpions such as *Vaejovis mexicanus decipiens*. There are also centipedes, or *Scolopendras*, at the site, some of which are poisonous. Recorded entomological observations for the site include a total of 119 species for Isla Socorro, of which 74.9% are insects and 22.7% are arachnids. Jiménez (1991) studied the arachnids of the site and recorded 21 species for the archipelago. The herpetofauna of the archipelago is comprised of four terrestrial species: the lacertilians *Urosaurus clarionensis*, which is endemic to Clarion Island; *Urosaurus auriculatus* (Socorro Island tree lizard), which is endemic to Socorro Island; *Hemidactylus frenatus*, which is introduced; and the snake *Masticophis anthonyi*, which is endemic to Clarión Island. The most picturesque terrestrial species at the site is the Socorro parakeet (Aratinga holochlora brevipes) that gather in large flocks on the branches of the guavaberry tree (Psidium socorrense) as they feed on its fruits, which are similar to the common guava. Likewise, the presence of the Socorro mockingbird (Mimodes graysoni) is notable at the site. Some of the other bird species that are observed among the bushes and wooded areas of the site are: olive verdigris (Dendroica auduboni), Socorro towhee (Pipilo erythrophthalmus socorroensis), Socorro parula (Parula pitiayumi graysoni), Socorro wren (Troglodytes sissonii), Socorro red-tailed hawk (Buteo jamaicensis socorroensis), and barn owl (Tyto alba) [Villa 1960]. The Socorro parula is the most abundant species on the island, is present throughout all plant associations of the site. It feeds mainly on small invertebrates among a wide variety of plants, such as euphorbias of genus *Croton*, shrubs such as *Dodonaea* sp., and trees such as *Guettarda* sp., *Bumelia* sp., *Psidium* sp., *Ficus* sp., and *Oreopanax* sp. The

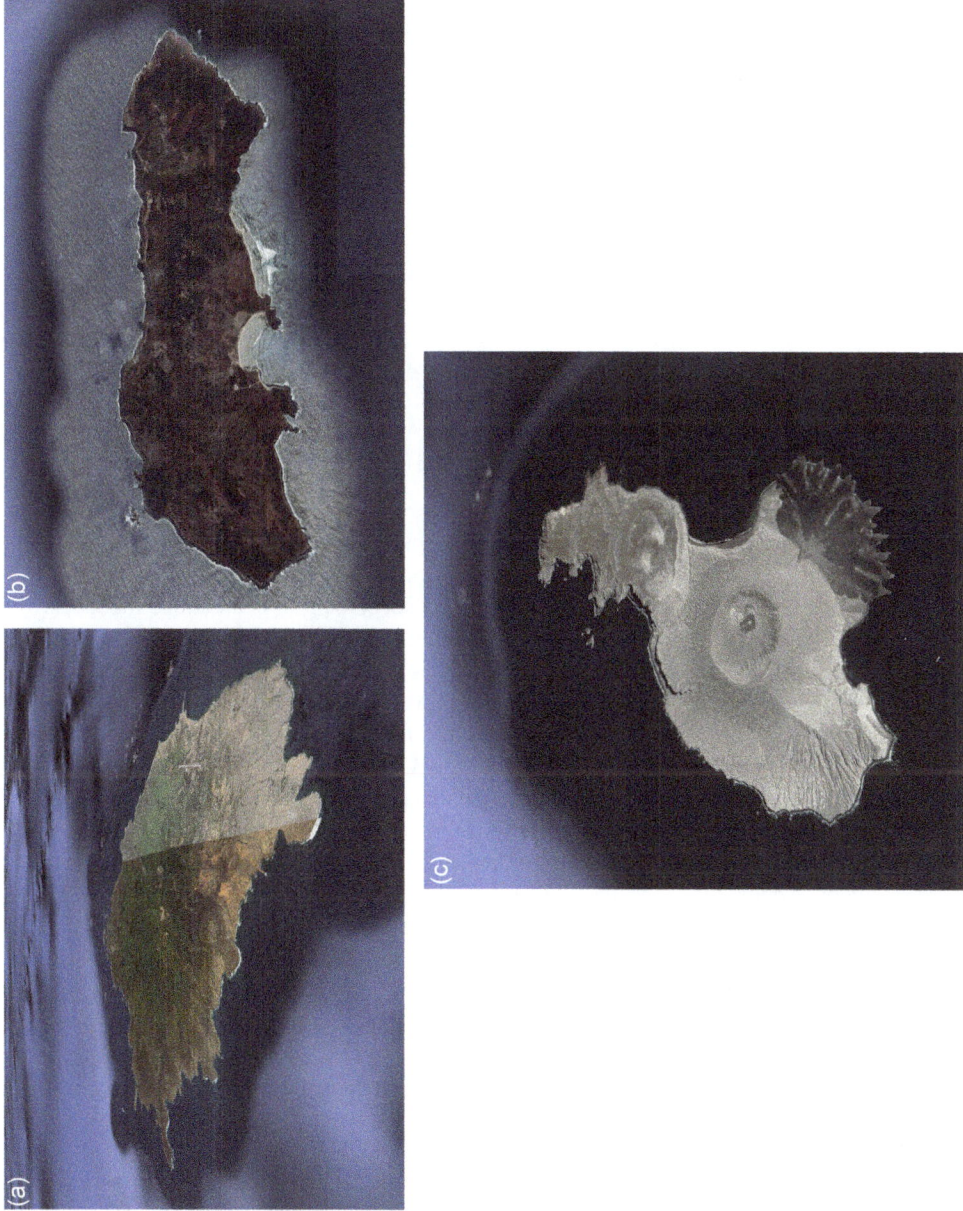

FIGURE 6.4 Composite aerial images of the three main islands of Reserva de la Biosfera Archipiélago de Revillagigedo (a. Isla Socorro; b. Isla Clarión; and c. Isla San Benedicto). Designated February 2, 2004; Area: 636,685 ha; Coordinates: 18°50′N 112°47′W; Ramsar Site number: 1,357.

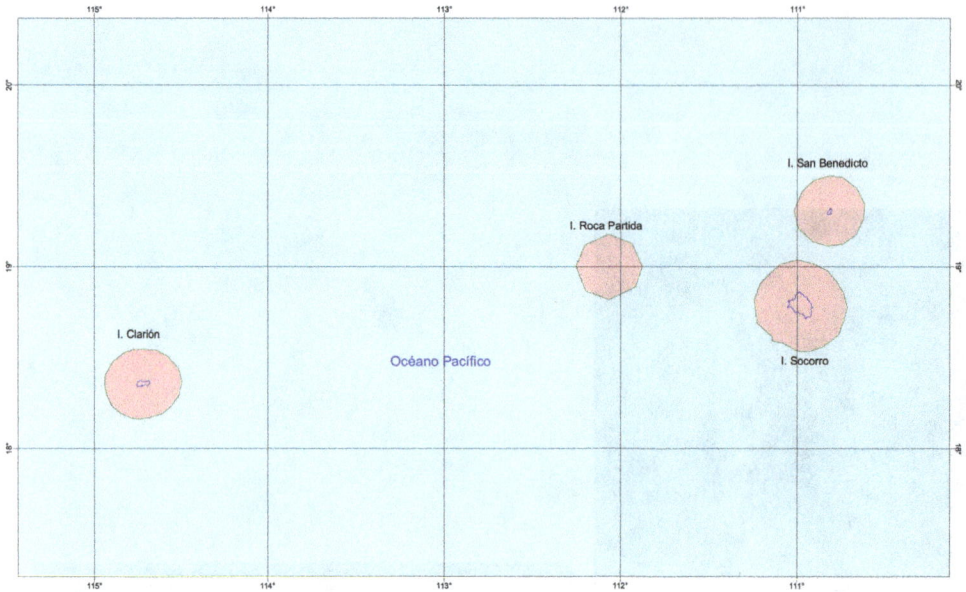

FIGURE 6.5 Map of the islands of Reserva de la Biosfera Archipiélago de Revillagigedo. Designated February 2, 2004; Area: 636,685 ha; Coordinates: 18°50′N 112°47′W; Ramsar Site number: 1,357.

Socorro wren (Troglodytes sissonii) is the second most abundant on Socorro Island, which feeds on small invertebrates and is found along the entire altitudinal gradient and in all plant associations of the site. The Socorro towhee (Pipilo erythrophthalmus socorroensis) is an abundant species in almost all plant associations. The Socorro dove (Columbina passerina socorrensis) is a common species in the lower parts of Socorro Island. This species is declining because of domestic cat predation at the site. Mocking birds (Mimodes graysoni) are distributed throughout Socorro Island, in the forests and scrub. Other species of note at the site are the Revillagigedo shearwater (Puffinus auricularis), the yellow-crowned night-heron (Nyctanassa violacea gravirostris), and the land crab (Gecarcinus planatus). Ichthyofauna at the site are relatively diverse, and include tiger snake-eel (Myrichthys maculosus), moray eels, ray-finned fish (Dermatolepis), grouper (Mycteroperca), giant hawkfish (Cirrhitus rivulatus), scrawled filefish (Alutera scripta), convict tang (Acanthurus triostegus), sergeant major (Abudefduf saxatilis), Cortez damselfish (Eupomacentrus rectifraenum), albacore (Thunnus alalunga), bluefin tuna (Thunnus thynnus), yellowfin tuna (T. albacares), Mexican hogfish (Bodianus diplotaenia), and the spinster wrasse (Halichoeres nicholsi). The site is devoid of known wild mammals, other than some reports of a bat species (Medina 1978). Past attempts to raise donkeys, cattle, goats, chickens, pigs, sheep, and rabbits on Clarión Island, and the presence of domestic cats have had an impact on the ecology of the site. The humpback whale (Megaptera novaeangliae) is the one species studied in greatest detail, as the Revillagigedo Archipelago is a key breeding area for this species, as well as other recorded cetacean species, regarding parts of their life cycles at the site.

Since the discovery of Socorro Island in 1533 by Captain Hernaldo de Grijalva, it was apparent that the islands did not appear to be inhabited, nor contain archaeological features of permanent residency of indigenous people. The site is under federal jurisdiction, and approximately 6 ha of Socorro Island and Clarión Islands is dedicated to national defense forces presence/activities. There is fishing carried out at the site, mainly by commercial companies and the tuna fishing cooperatives of Sonora and Sinaloa, focusing on tuna, billfish, and other commercially important species. Fishing activities occur during all but the winter months.

Major threats to the site include hurricanes (an approximate average of two tropical storms or hurricanes per year, since the 1950s); introduction of exogenous flora and fauna (mainly sheep, pigs, rabbits, cats, and certain flora); volcanic eruptions from the active volcanoes of the site; human-sourced fire, mainly on the southern portion of Socorro Island; lack of freshwater; and overfishing.

Through the Comisión Nacional de Áreas Naturales Protegidas (CNANP) and their program of work, the following general conservation strategies for the site have been defined: guarantee the integrity of the area; promote the protection and conservation of natural resources; promote social participation; promote research and environmental education; and secure funding for the site as a Biosphere Reserve. Scientific exploration at the site began in 1903 and in the 1980s research activity increased, including activities by the UNAM, National Polytechnic Institute (IPN), Secretary of the Navy (SEMAR), Autonomous University of Colima (UCOL), University of Guadalajara (UDG), Autonomous University of Baja California Sur (UABCS), National Fisheries Institute (INP), Biological Research Center of the Northwest SC (CIBNOR, SC), as well as Villanova University, University of California Los Angeles, and the California Academy of Sciences, in close coordination with the Mexico's Ministry of Social Development (SEMAR), the National Council of Science and Technology, and international conservation organizations such as Conservation International. The preceding research endeavors have generated a number of educational and outreach benefits. The main tourist activities at the site are associated with sport fishing, and ship cruises within the archipelago.

SANTUARIO PLAYA BOCA DE APIZA-EL CHUPADERO-EL TECUANILLO, COLIMA: OVERVIEW

Santuario Playa Boca de Apiza-El Chupadero-El Tecuanillo (Figures 6.6 and 6.7) is a coastal wetland with dominant scrub and deciduous forest vegetation, with adjacent coastal dunes, seagrass beds, and mangroves. Three of the seven species of marine turtles recorded in Mexico hatch their eggs on the beaches of the site: *Lepidochelys olivacea*, *Dermochelys coriacea*, and *Chelonia mydas*. The site also provides habitat for resident and migratory, terrestrial and aquatic, birds. Agriculture, livestock, fishing, and tourism are the most common human activities at the site. Among the threats to the ecology of the site are the hunting of female turtles during their hatching period to obtain meat, eggs, and oil; depredation of nests by wild and introduced species; and contamination of nest zones. The site is located in the State of Colima, within the municipality of Tecomán, located on the Pacific coast of Mexico, near the small resort town of El Tecuanillo, and the community of El Ahijadero.

SITE DETAILS

Special protection species at the site include: wood stork (Mycteria americana), reddish egret (Egretta rufescens), great blue heron (Ardea herodias), bare-throated tiger-heron (Tigrisoma mexicanum), and the agami heron (Agamia agami). The gray-headed hawk (Leptodon cayanensis) and Harris' hawk (Parabuteo unicinctus) are at the site and in the CITES. Among the threatened species at the site are the red-billed tropicbird (Phaethon aethereus), least storm-petrel (Oceanodroma microsoma), and the aplomado falcon (Falco femoralis), the latter of which is also in the CITES.

The site is within the "Costa de Jalisco" and "Armería-Coahuayana" Hydrological Regions, which include the mouth of the three main rivers in the state, the Cihuatlán or Marabasco River, the Armería River, and the Coahuayana River, on the border with Michoacán. The main bodies of water adjacent to the sea turtle nesting sites are the Estero El Chupadero, Estero Boca de Apiza, Estero de Bayardo, and Estero El Tecuanillo. The coastline of the site is linear, other than the Las Brisas-Salahua embayment.

FIGURE 6.6 Composite aerial image of Santuario Playa Boca de Apiza-El Chupadero-El Tecuanillo. Designated February 02, 2008; Area: 40 ha; Coordinates: 18°45′N 103°49′W; Ramsar Site number: 1,764.

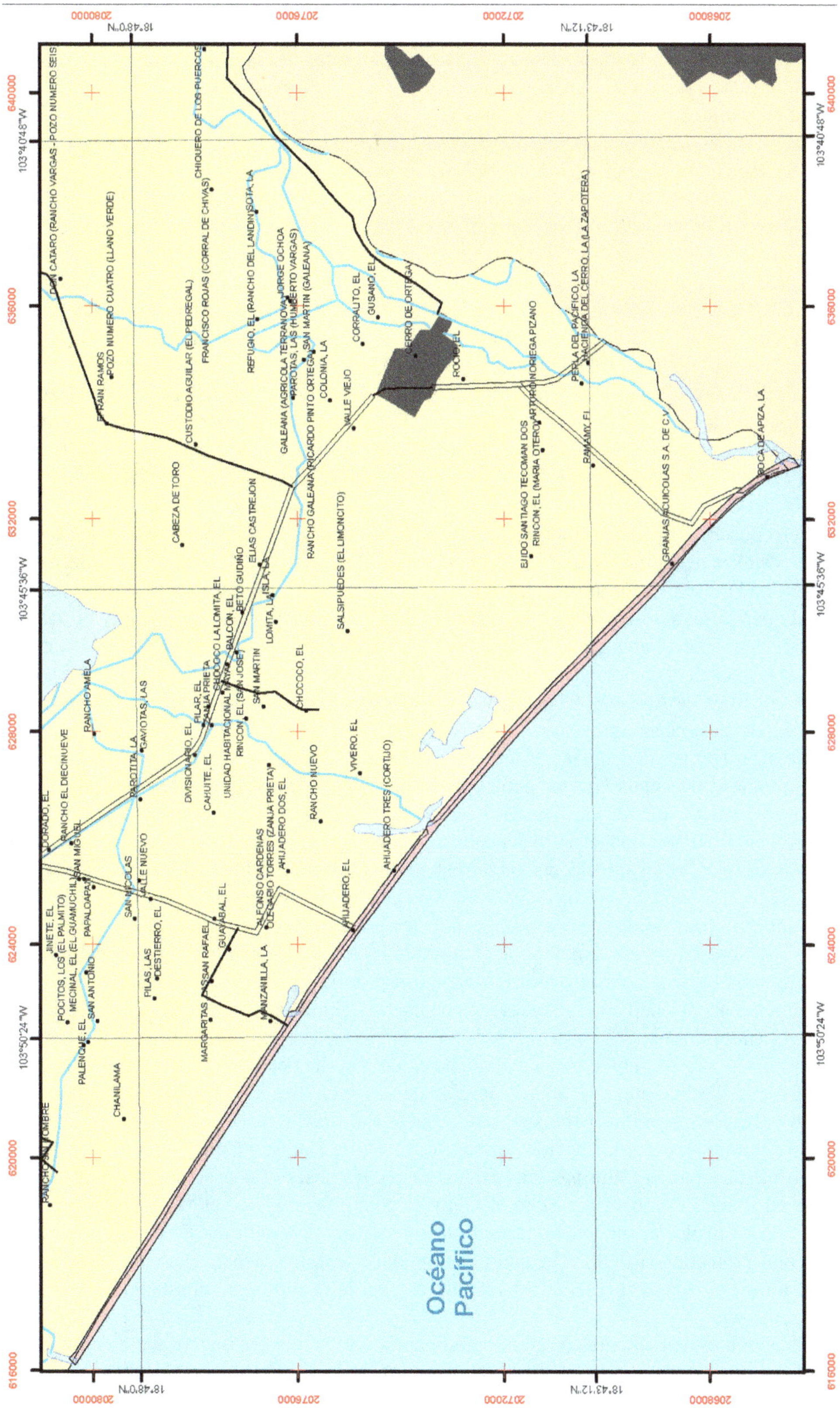

FIGURE 6.7 Map of Santuario Playa Boca de Apiza-El Chupadero-El Tecuanillo. Designated February 02, 2008; Area: 40 ha; Coordinates: 18°45′N 103°49′W; Ramsar Site number: 1,764.

Main flora of the site include *Distichlis spicata* and *Ipomoea pes-caprae*, as well as the legumi-nous tree *Caesalpinia coriaria*, *Acacia* spp., and genus *Prosopis*, *Gliricidia*, *Pithecellobium*, and *Solanum*. In some of the site's adjacent areas, orchards of coconut palms and mangroves (Rhizophora mangle, Laguncularia racemosa, Avicennia germinans, and Conocarpus erectus) are common. The species that determined the creation of the site as a sanctuary, predating the Ramsar designation, are three of the seven species of sea turtles that nest in Mexico. In the El Chupadero Tortuguero Camp, the average number of protected individuals varies in each season, with the annual average of approximately 700 nests. During the late 1990s and early 2000s a total of approximately 700,000 turtle hatchlings were released, recording an average percentage of hatchling emergence of 67%. In addition to the three species of sea turtles that arrive at these beaches to nest (olive ridley, leath-erback, and green sea turtles), other faunal species of wildlife at the site include: squirrel (Sciurus spp.), armadillo (Dasypus novemcinctus), raccoon (Procyon lotor), South American coati (Nasua nasua), common opossum (Didelphis marsupialis), coyote (Canis latrans), gray fox (Urocyon cine-reoargenteus), white-tailed deer (Odocoileus virginianus), American crocodile (Crocodylus acutus), green iguana (Iguana iguana), and Mexican spiny-tailed iguana (Ctenosaura pectinata). The site also provides habitat for a large number of resident and migratory bird species including *Agamia agami* (agami heron), *Leptodon cayanensis* (gray-headed hawk), *Parabuteo unicinctus* (Harris' hawk), and *Falco femoralis* (aplomado falcon). A number of endemic fish are noted at the site (e.g., Lile gracilis and Cynoscion nannus), and rose flower urchin (Toxopneustes roseus).

Chupadero Beach, at the site, is thought to have its name derived by an ancient custom of taking the beasts to drink water at the estuary. These beaches are also an important area for organized fish-ers to gather, and presently, for tourists to relax. The local inhabitants of the areas around the site work mainly in fishing, trade, tourism, and agriculture. Livestock raised in the vicinity of the site include cattle, pigs, sheep, and goats. Turtles that lay eggs are surreptitiously hunted or otherwise used by local residents, to obtain eggs, oil, blood, and meat that is either sold or used for subsistence. The beaches of Boca de Apiza-El Chupadero-El Tecuanillo are also utilized for agriculture, live-stock, fishing, clandestine mangrove forestry, poaching, and tourist activities. Green iguana (Iguana iguana), Mexican spiny-tailed iguana (Ctenosaura pectinata), white-tailed deer (Odocoileus vir-ginianus), armadillo (Dasypus novemcinctus), mourning dove (Zenaida macroura), white-winged dove (Zenaida asiatica), and various species of ducks are all variously utilized by local residents. Aquaculture (mainly tilapia and shrimp) depends on the water resources of the site, and similarly is highly depended upon by local residents for subsistence and income. Additional threats to the ecol-ogy of the site include predation of turtle nests by wild and introduced animals (e.g., dogs and pigs); contamination of nesting areas due river, sea, and beach pollution; clandestine exploitation of man-groves; heavy deforestation and exploitation of aquifers in the middle and lower parts of the basin; population growth; presence of (introduced) tilapia; inappropriate use of fishing nets; poaching and drug cultivation; uncontrolled commercial logging; and an unknown degree of aquifer depletion by urban and agricultural uses.

A consortia has been promoted by the federal delegation of the Secretary of Urban Development and Ecology, and the Secretary of Social Development, which are designed to provide financial resources for the conservation of the site, made up of the State's main companies: Manzanillo Integral Port Administration, Cementos Apasco, Coca Cola, Danisco Cultor Mexicana, and the Benito Juárez Peña Colorada Mining Consortium, for the purposes of conservation and protection of endangered species found in the State of Colima, giving priority to sea turtles. Accordingly, monitoring of sea turtles at the site is carried out in coordination with the Federal Attorney for Environmental Protection (PROFEPA), and in compliance with the General Law of Ecology and Protection of the Environment (LGEEPA), as well as other laws and regulations applicable within the limits of the site.

Research activities and associated infrastructure at the site is focused on morphometry, and the tracking of sea turtle species, nest quality, infertile eggs, as well as other aspects of conservation at the site. In addition, a number of regional education and outreach activities, as well as tourism organizations, utilize the site to share conservation concepts and provide periodic opportunities to engage with local, national, and international visitors.

PLAYA DE COLOLA, MICHOACAN: OVERVIEW

Playa de Colola (Figures 6.8 and 6.9) is located along the western central region of Mexico, in the State of Michoacán, and consists of 4.8 km of sandy beaches, with an approximate width of 150 m, surrounded by a variety of shrublands. The beach is a major nesting site for olive ridley sea turtles (Lepidochelys olivacea), leatherbacks (Dermochelys coriacea), and green sea turtles (Chelonia mydas). It is estimated that about 70% of the population of green sea turtles visit Cocola Beach to nest and breed. The turtles are exposed to the dangers associated with human presence, as nesting females can be easily caught. Sea turtle eggs are extracted by some local residents, for personal consumption, and/or are extracted by animals digging through the nests. Another situation in which the turtles are vulnerable to illegal capture and distribution is when they are transiting onto land, or out to sea from the beach. Colola Beach is bounded by a strip of vegetation approximately 60 m wide, composed of deciduous forest shrubs, mainly of the genera Prosopis, Acacia, Gliricidia, Pithecellobium, and Solanum, which separates the beach from the nearest coastal road. Colola Beach is an open beach with small patches of halophytic grass (Distichlis spicata), with additional areas of Ipomoea pes-caprae (bayhops), Sesuvium portulacastrum, Pectis arenaria, and grasses such as Distichlis spicata, Canavalia maritima, and Okenia hypogaea (beach peanut). Further upland, and inland, are thorny shrubby plants such as boatspine acacia (Acacia cochliacantha), mesquite (Prosopis juliflora), the leguminous Pithecellobium lanceolatum, grey nicker (Caesalpinia bonduc), and other plants that help to retain sand within the dunes. Among these species are some that are important for forage and medicines.

SITE DETAILS

The site is located in the Neotropical Biogeographic Region (Wallace 1876), which corresponds to the Pacific Biogeographical province (Cabrera and Willink 1980). Within Michoacán, the site is located within the Sierra Madre del Sur province and Costa del Sur Sub-province (Garduño 1999). The hydrological region of the coast of Michoacán includes the currents located between the Coahuayana and Mezcala Rivers, which ultimately flow to the Pacific Ocean; the hydrologic region is subdivided into two basins: the Nexpa River and the Cachán or Coalcomán River basins. The site is comprised of hills and alluvial plains, with poorly developed shallow phaeozem, regosol, and cambisol soils.

Because of the particular shape of Colola Beach, it serves as a buffer against sea flooding in the town of Colola, which is located at the end of the beach, on the eastern border, approximately 200 m from the high tide line. Because the grain size of the sand at Colola Bach is relatively large (0.5 mm), seawater percolation occurs quickly, which can reduce seawater flooding on land. At the end of the beach and the strip of vegetation, there are numerous artesian wells that supply water for local human consumption, and for agricultural irrigation of local fields/orchards of papaya, chili, hibiscus, tomato, and lemon. The absence of nearby lotic and lentic water bodies makes it necessary that the water extracted for human consumption in the area come from the groundwater table.

The presence of small vertebrates, such as skunks, raccoons, coatis, armadillos, coyotes, green and black iguanas, and some snakes is noted at the site. Among the birds at the site are magpies, herons, hawks, buzzards, and white-winged doves. The predominant vegetation at the site is within deciduous forest areas (Rzedowski 1978), with additional sub-evergreen forest plants present as well. The deciduous forests of the site include the following flora: gliricidia (Gliricidia sepium), which is used to cure skin irritations; the leguminous tree Caesalpinia eriostachys; copal (Bursera excelsa), with its resin used for religious festivals and masses; palo colorado (Caesalpinia platyloba), from which dye is extracted, and which is also used for construction of houses; vanillo (Cassia

FIGURE 6.8 Composite aerial image of Playa de Colola. Designated February 02, 2008; Area: 286.83 ha; Coordinates: 18°18′N 103°26′W; Ramsar Site number: 1,788.

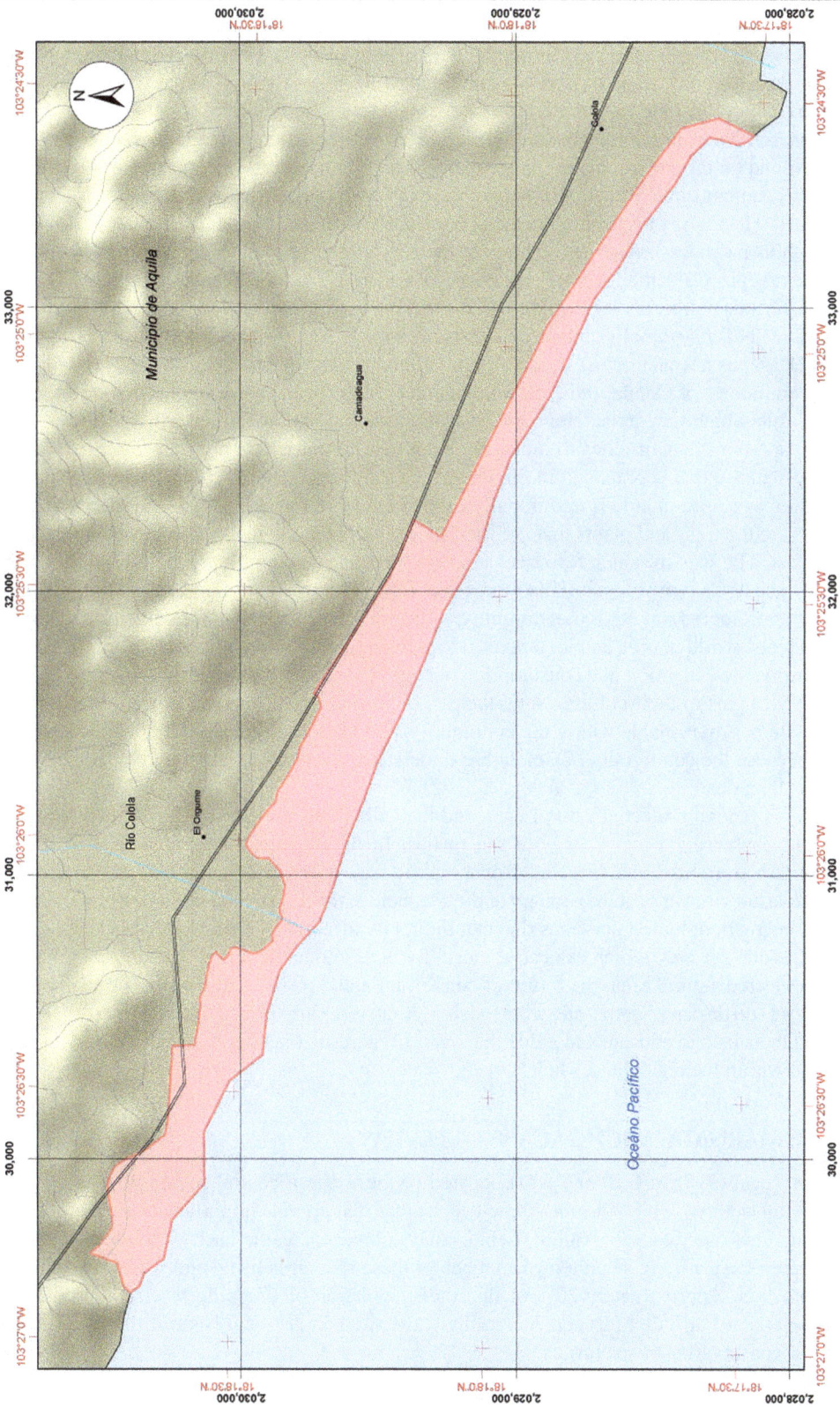

FIGURE 6.9 Map of Playa de Colola. Designated February 02, 2008; Area: 286.83 ha; Coordinates: 18°18′N 103°26′W; Ramsar Site number: 1,788.

tomentosa); bonnet (Jacaratia mexicana); gumbo limbo (Bursera simaruba); bocote (Cordia elaeag-noides), which is used in the construction and furniture manufacturing industries; the cuachalalate (Amphipterygium adstringens), used for medicinal purposes; elephant-ear tree (Enterolobium cyc-locarpum); brazilwood (Haematoxylum brasiletto), used for poles for housing, as dye, and as fuel; and the pink poui (Tabebuia rosea), used for the manufacture of agricultural tools.

Colola Beach represents a key continental site for the reproduction of sea turtles. Other species of reptiles found on the site are the green iguana (Iguana iguana), the Mexican spiny-tailed iguana (Ctenosaura pectinata), both threatened species; the barred whiptail (Ameiva undulata); the Mexican beaded lizard (Heloderma horridum); the boa (Boa constrictor); and the Mexican west coast rattle-snake (Crotalus basiliscus). Among the birds at the site are the yellow-headed amazon (Amazona oratrix), the magpie (Calocitta formosa), the crested caracara (Polyborus plancus), the west Mexican chachalaca (Ortalis poliocephala) [Villaseñor 2000]. Among the mammals supported by the site are opossum (Didelphis marsupialis), the raccoon (Procyon lotor), white-nosed coati (Nasua narica), the armadillo (Dasypus novemcinctus), and northern tamandua (Tamandua mexicana).

In the community of Colola, indigenous inhabitants have an extensive knowledge of the environment within which they exist, and have been immersed within it for generations. The way in which they appropriate nature has traditionally been carried out with a high sense of sustainable use. The agriculture that is practiced in the vicinity of the site is generally for subsistence and the preparation of agricultural land is traditionally carried out by the slash-and-grave system, utilizing burning, where the trees and plants that are used in traditional medicine and/or in the building are extracted first. The way in which resources are exploited in this community is described as only for primary needs, i.e., subsistence. This form of appropriation of nature is one of the fundamental values that is promoted among the community, and it provides cohesion and identity that forms part of residents' worldview as an indigenous group. Even though turtle eggs are part of the diet of the community, the extraction and consumption of eggs of turtles is described as a tradition, which has always been part of their culture. Nevertheless, in present times, such actions are seen as nest poaching, which is punishable within the community, and the slaughter of adult turtles is severely punished because the community of Colola has understood for years that the sea turtle serves all, better alive, than dead.

The site is a federal protected natural area, and the federal maritime-terrestrial zone is reserved to carry out conservation activities with sea turtles. In the surrounding area/watershed, farm-ing occurs as both rainfed and irrigated crops, mainly papaya, chili, sesame, lemon, and tomato. The adverse factors that impact the ecology of the site include the destruction of turtle nesting habitat by urban construction; tourist activities that can indirectly affect the nesting habitat of turtles; and river fishing with gill nets, which can cause mortality in reproductive turtles. Because the site is a federally protected natural area, the future of conservation and research at the site is fairly secure. There is also a 'turtle camp' at the site, where visitor group sizes are limited to fewer than 20 people, where volunteers help to educate and guide the visitors regarding the importance of conservation at the site, and within the region as a whole.

PLAYA DE MARUATA, MICHOACAN: OVERVIEW

Playa de Maruata (Figures 6.10 and 6.11), located in the western central region of Mexico, is a marine-coastal wetland, which includes three beach areas that provide important nesting habitat for three marine turtle species: olive ridley (Lepidochelys olivacea), leatherback (Dermochelys coria-cea), and green sea turtles (Chelonia mydas). Each of these species is listed under different catego-ries of protection. Approximately 20% of the total population of *Chelonia mydas'* reproductive population nests at Playa de Maruata. Accordingly, the site is a National Natural Protected Area, dedicated to conservation of sea turtles.

FIGURE 6.10 Composite aerial image of Playa de Maruata. Designated February 02, 2008; Area: 80 ha; Coordinates: 18°16′N 103°21′W; Ramsar Site number: 1,795.

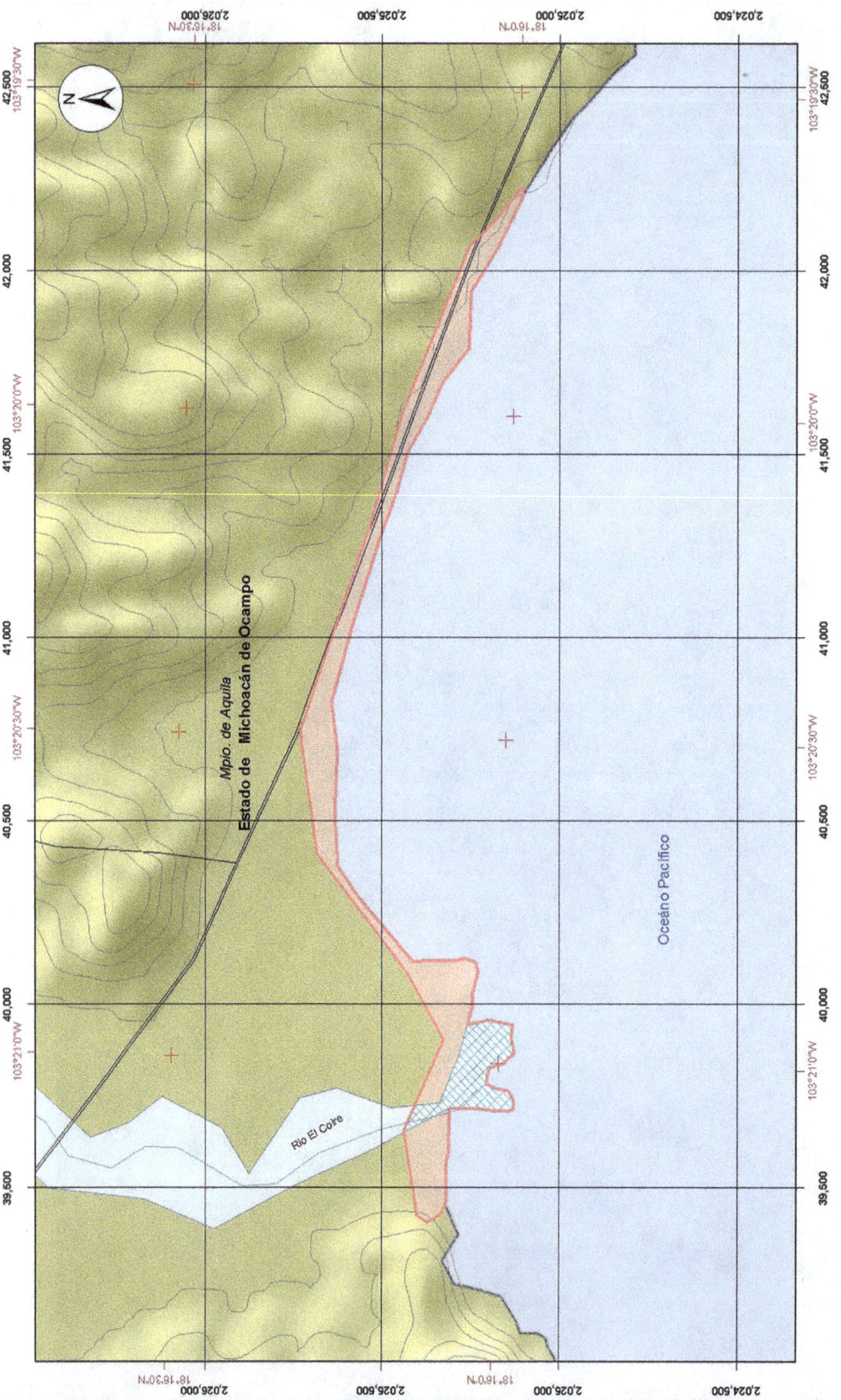

Mpio. de Aquila
Estado de Michoacán de Ocampo

Río El Cote

Oceáno Pacífico

FIGURE 6.11 Map of Playa de Maruata. Designated February 02, 2008; Area: 80 ha; Coordinates: 18°16′N 103°21′W; Ramsar Site number: 1,795.

SITE DETAILS

The site is in the State of Michoacán, within the municipality of Aquila, 150 km from Lázaro Cárdenas, and 62 km from the municipal seat of Aquila. Adjacent to the site is the Nahua de Maruata indigenous community of approximately 1,100 inhabitants, which belongs to the indigenous communal head of Pómaro. Maruata Beach corresponds to the bay of the same name. The main beach (of 3) is located to the east, approximately 2.4 km long, where most of the sea turtles nest; to the center of the coastline of the site is a small beach 90 m long, and at the west end there is a beach approximately 150 m long. The average width of the beaches is 40 m; where the main beach ends, vegetation composed mainly of *Acacia* begins, which marks the limit of the beach habitat and the beginning of rockier terrain. At this edge, deciduous forest shrubs of the genera *Prosopis*, *Acacia*, *Gliricidia*, *Pithecellobium*, and *Solanum* exist.

The sand of Maruata Beach is fine and relatively dark in color due to its high magnetite content (Alvarado and Delgado 2005). The hydrological region of the coast of Michoacán includes the currents located between the Coahuayana and Mezcala Rivers, which flow into the Pacific Ocean. This region in turn is subdivided into two basins: the Nexpa River and the Cachán or Coalcomán River basins. The main input of freshwater to the site, the Maruata River, flows intermittently throughout the year. The site is prone to tropical storms during the rainy season and is periodically affected by El Niño. The landscape of this region is a narrow coastal plain that begins at the border between Michoacán and Colima (specifically at the mouth of the Coahuayana River). The Costa Sur Subprovince is characterized by low mountains of sedimentary origin, with volcanic and metamorphic substrates, where valleys and plains form alluvial materials. The beaches of the site are very dynamic, with significant physiognomic changes occurring over short time periods.

Among the beaches that form the coastline of Maruata, rocky outcroppings exist where cacti and other deciduous forest plants exist. Some of the islets are close to the main beach, where there is also a small jetty for outboard motorboats and where you can see aggregations of seabirds, such as pelicans, cormorants, and seagulls. It is common to see some plantations and home gardens near the houses within the river plain, near the beach, at the west end of the site. It is also common to observe the presence of small- and medium-sized vertebrates, such as skunks, raccoons, coatis, armadillos, coyotes, green and black iguanas, and occasionally snakes at the site. Commonly observed birds at the site include herons, cormorants, magpies, chachalacas, herons, hawks, and vultures. The presence of domestic animals at the site is common as well, such as dogs, horses, donkeys, and pigs, due to the site's proximity to the local community.

The predominant type of vegetation at the site, deciduous forest, is comprised of gliricidia (Gliricidia sepium), which is used to cure skin irritations; the leguminous tree *Caesalpinia eriostachys*; copal (Bursera excelsa), its resins used for religious purposes; palo colorado (Caesalpinia platyloba), from which dye is extracted, as well as use for the construction of houses; vanillo (Cassia tomentosa); bonnet (Jacaratia mexicana); gumbo limbo (Bursera simaruba); bocote (Cordia elaeagnoides), which is used in the manufacturing of furniture; the cuachalalate (Amphipterygium adstringens), used for medicinal purposes; the elephant-ear tree (Enterolobium cyclocarpum); brazilwood (Haematoxylum brasiletto), which is used as poles for housing; and the pink poui (Tabebuia rosea), which is used to make agricultural tools.

Other than turtles, species of reptiles found at the site include the green iguana (Iguana iguana), the Mexican spiny-tailed iguana (Ctenosaura pectinata), both threatened species; the barred whiptail (Ameiva undulata); the Mexican beaded lizard (Heloderma horridum); the boa (Boa constrictor); and the Mexican west coast rattlesnake (Crotalus basiliscus). Birds observed at the site include the yellow-headed amazon (Amazona oratrix), the magpie (Calocitta formosa), the crested caracara (Polyborus plancus), and the west Mexican chachalaca (Ortalis poliocephala) [Villaseñor 2000]. Mammals at the site include opossum (*Didelphis marsupialis*), raccoon (Procyon lotor), white-nosed coati (Nasua narica), armadillo (Dasypus novemcinctus), and northern tamandua (Tamandua mexicana).

In the community of Maruata, the indigenous Nahuas inhabitants are bilingual with the vast majority of the population speaking both Nahuatl and Spanish. Culturally, the indigenous community of Maruata represents one of the most traditional peoples along the coast of Michoacán, and they practice religious ceremonies where Christian and indigenous traditions are combined. The council of elders is the communal authority, found in the community of Pómaro. Maruata fishers, since 1975, have rejected the capture of sea turtles despite having an allowable catch quota. The inhabitants of Maruata have extensive knowledge of the environment, within which they exist and have been immersed for generations. The way in which they appropriate nature is carried out with a high sense of sustainable use. The agriculture that is practiced is generally subsistence and the preparation of agricultural land is traditionally carried out by the slash-and-grave system, utilizing burning, where the trees and plants that are used in traditional medicine and/or in the building of homes/implements are extracted first. The way in which resources are exploited in this community is "naturally and rationally," in that only primary needs are satisfied. This form of appropriation of nature is one of the fundamental values that is promoted among the community and it provides cohesion and identity that forms part of their worldview as an indigenous group. The use of natural resources at the site by the inhabitants of Maruata has allowed the nesting area to persist even while tourism has increased.

There are a number of adverse factors within the site that can potentially affect the ecology of the site, including the destruction of turtle nesting habitat by urban expansion and activity; tourist services; use of all-terrain vehicles on the beach; river fishing in front of the site, since the use of fishing gear such as gill nets can cause mortality to reproductively active turtles.

Ongoing conservation activities at the site are related to the focus on the three species of turtles that nest there, by using semi-natural management techniques (protected nests in incubation pens). Associated research activities are carried out at the site, aimed at monitoring the population of turtles that nest there, the survival rates of hatchlings, and migratory routes of adults. During holiday seasons, environmental education activities are carried out with tourists and school groups from the region that arrive at the site, with basic information about the site and nesting activity of turtles provided. Beach tourism activities occur at the site, however the intensity of tourist use of the beach is limited since it is a location that belongs to the people of Pómaro, who have also implemented alternative tourism activities, such as trail hikes, interpretive tours through the jungle, and river tours.

PLAYÓN MEXIQUILLO, MICHOACAN: OVERVIEW

Playón Mexiquillo (Figures 6.12 and 6.13) is important breeding habitat for three species of marine turtles: leatherbacks (Dermochelys coriacea), olive ridley (Lepidochelys olivacea), and green sea turtles (Chelonia mydas), and among the five most important sites for conservation of sea turtles in the Mexican and Mesoamerican Pacific. The site is relatively well conserved, but an increase in the tourist sector has increased urbanization pressures on the area. The lack of strict protections for such beach areas, and the use of 4-wheel-drive vehicles on the beaches, are major threats to turtle populations, as well as to the fragile vegetation of beach ecosystems, such as is found at the site.

SITE DETAILS

El Playón de Mexiquillo is located in the municipality of Aquila, near the border with the municipality of Lázaro Cárdenas, in the southern part of the coast of the State of Michoacán. The industrial port of Ciudad Lázaro Cárdenas is 80 km away and the nearest town, Caleta de Campos (also known as Bahía Bufadero) is located 9.5 km away, with approximately 2,900 inhabitants.

The site is in the Mexiquillo-Delta del Balsas, with high biodiversity. In general, the beach consists of an inclined plane from the tidal line to the continental shelf, with an average upslope of approximately 10° to the zone of shrubby vegetation that marks the limit of the site with the

FIGURE 6.12 Composite aerial image of Playón Mexiquillo. Designated February 02, 2004; Area: 67 ha; Coordinates: 18°07′N 102°52′W; Ramsar Site number: 1,350.

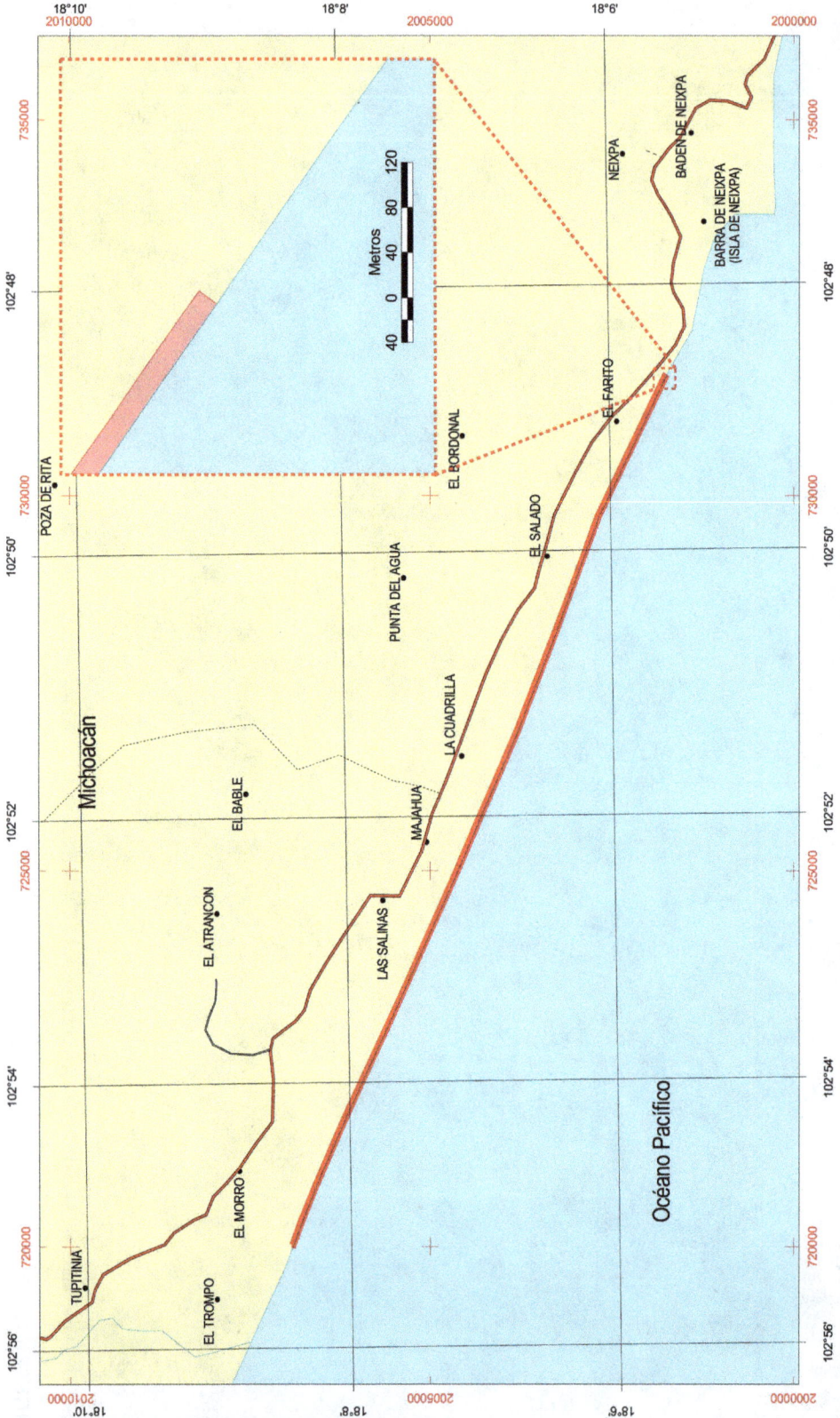

FIGURE 6.13 Map of Playón Mexiquillo. Designated February 02, 2004; Area: 67 ha; Coordinates: 18°07'N 102°52'W; Ramsar Site number: 1,350.

adjoining private coconut and mango orchards (López 1985). Originating in the Sierra Madre del Sur is the Nexpa River, located to the southeast of the site's beach, adjacent to Playón de Mexiquillo, which deposits sediment and other material at the site. Additionally, the Tupitina River is located near the northwest portion of the site. Along the site's entire length there are approximately 10 other channels that flow to the estuary, which support a variety of resident and migratory waterfowl (Correa 1979). The site depends on these fluvial processes, which nourish the entirety of this key coastal ecosystem.

The Michoacán coast is generally characterized as deciduous lowland jungle (Miranda and Hernández 1963), which is highly representative of the forests of Mexico and is generally under threat from land use change (Jaramillo and Villalobos 1994, Janzen 1988), from forest to primarily pasture and other agricultural uses. The relatively natural flora of the site are comprised mainly of creeping plants such as bayhops (Ipomoea pes-caprae), grasses (e.g., Jouvea pilosa), and legumes (e.g., Canavalia maritima). In the upslope areas of the beach, the shrub community predominates with flora such as *Acacia* sp. and mesquite (Prosopis sp.), delimitating the edge of the beach with upland areas. Faunal species at the site include white-tailed deer (Odocoileus virginianus), armadillo (Dasypus novemcinctus), coyote (Canis latrans), skunk (Conepatus mesoleucus and Spilogale gracilis), gray fox (Urocyon cinereoargenteus), lynx (Lynx rufus), jaguarundi (Herpailurus yagouaroundi), Virginia opossum (Didelphis virginiana), white-nosed coati (Nasua narica), northern tamandua (Tamandua mexicana), and squirrel (Sciurus aureogaster). Bird species at the site include wood stork (Mycteria americana), collared plover (Charadrius collaris), cormorant (Phalacrocorax sp.), woodpeckers (Melanerpes chrysogenys, Picoides scalaris, and Dryocopus lineatus), osprey (Pandion haliaetus), herons and egrets (Ardea herodias, Casmerodius albus, and Egretta thula), brown pelicans (Pelecanus occidentalis), vultures (Cathartes aura and Coragyps atratus), common caracara (*Caracara plancus*), falcon (Falco sparverius), hawks (Buteogallus anthracinus, Buteo nitidus, B. albonotatus, and Buteo jamaicensis), and parrots (Eupsittula canicularis and Amazona oratrix). The herpetofauna of the site are well represented, and include boa (Boa constrictor), South American rattlesnake (Crotalus durissus), Daudin's vine snake (Oxybelis aneus), green iguana (Iguana iguana), and Mexican spiny-tailed iguana (Ctenosaura pectinata). American crocodile (Crocodylus acutus) are found at the site, mainly in rivers and in the estuaries of Tupitina, El Chical, El Chico, and Mexiquillo. The site also supports freshwater turtles (Kinosternum integrum and Rhinoclemmys pulcherrima) and the sea snake *Pelamys platurus*. As a result of the establishment of human settlements (i.e., hamlets, ranches, and small communities) in the area, feral dogs, cats, and pigs, and cattle roam the site.

In the town of Caleta de Campos, in close proximity to the site, one of the most important human activities for the community is agriculture, mainly in family orchards of mangoes, coconuts, papaya, and citrus, and to a lesser extent corn, beans, chili, tomato, and watermelon. Another portion of the population is dedicated to artisanal fishing, from which communities obtain shark, sawfish, red snapper, clam, lobster, crab, and sea bass. In the past, oyster and limpet extraction was also very popular with local community members, but the populations of those organisms are now greatly diminished. In more recent years at the site, tourism has become increasingly important to the community, with a number of such activities now organized at beaches to support visitorship, especially during holidays. Broad-scale tourist and agro-industrial uses have been envisioned in the past for the communities of the area, however it is understood that these activities come with high risks for the ecology of the site. Nevertheless, a variety of associated actions have resulted in removal of vegetation along the shrubby edge of the site, during which the beach was dredged to modify its profile, impacting turtle nesting and causing shoreline erosion. Additional threats to the ecology of the site are off-road vehicles used by artisanal fishers as they transport their gear along the shoreline, the use of wood from the site, and the establishment of pasture and crops in the area. An overarching threat to the ecology of the site (and all of the coastal sites of North America and the Pacific) are the impacts from global climate change, which brings local impacts from the associated alterations in precipitation patterns and shoreline erosion.

The site is established as a National Sanctuary for the protection of sea turtles that nest in the Playón. This program is run by the Secretariat of Environment and Natural Resources (SEMARNAT), through the Leatherback Project, under the guidelines of the National Program of Protection, Conservation, and Management of Sea Turtles. In 2003, an Interstate Agreement was signed for the conservation of the leatherback sea turtle, between the States of Michoacán, Guerrero, and Oaxaca. The agreement is intended to establish collaborative work to specifically recover this species. Research at the site includes that of the Center for the Protection and Conservation of Turtles, in the Playón Marinas. Although there is no formal visitor center at the site, when visitors are present there are opportunities for engagement and education that revolves around ongoing monitoring and conservation work related to turtles at the site, so that the public can observe nesting females. Visitation at the site is of medium intensity.

LAGUNA COSTERA EL CAIMÁN, MICHOACAN: OVERVIEW

Laguna Costera El Caimán (Figures 6.14 and 6.15) is a long (approximately 12 km long and 1 km wide) coastal lagoon bordering a relatively steep Pacific shoreline, close to the port of Lázaro Cárdenas. The site is covered by stands of button mangrove (Conocarpus erectus), red mangrove (Rhizophora mangle), and white mangrove (Avicennia germinans). Several endangered species are supported by the site, such as the American crocodile (Crocodylus acutus), the yellow-headed amazon (Amazona oratrix), jaguar (Panthera onca), and the green sea turtle (Chelonia mydas). Coconut production, subsistence fishing, and small-scale tourism are the main human uses of the site. Adverse impacts at the site include the felling of mangroves, sewage and industrial discharges, invasive alien species, dredging and filling, housing development, and improper waste disposal.

SITE DETAILS

The site is located in the State of Michoacán. The eastern end of the estuary/lagoon is located 2 km northwest of the City and the Port of Lázaro Cárdenas. The extreme west of the site is located 2 km southeast of Playa Azul. The local municipality of Lazaro Cardenas has an approximate population of 171,100 inhabitants (INEGI 2000). Because the site is located in a microtidal zone, with an amplitude of less than 1.5 m, the fluctuations of water level are small, which are characteristics that are supportive of the site's halophytes. Accordingly, *Laguncularia recemosa* is the dominant species, occupying most of the flooded area of the site, in dense tree associations, with new individuals branching from near the base of the trees. Less abundant is *Conocarpus erectus*, distributed mainly in the muddy zone. Other tree species in the dry zone (periodically flooded at high tide) are *Hibiscus tiliaceus, Hippomane mancinella, Bravaisia integerrima, Creteva tapia*, and *Pithecellobium lanceolatum*. Among the shrubs at the site are *Mimosa pigra, Prosopis juliflora, Pluchea odorata*, and *Lantana camara*. Common herbaceous species on the sand of the beach are *Ipomoea pescaprae, Canavalia maritima, Distichlis spicata, Batis maritima, Heliotropium curassavicum*, and *Indigofera miniata*.

Main faunal species at the site include American crocodile (Crocodylus acutus); birds: great blue heron (Ardea herodias), white tipped dove (Leptotila verreauxi), grove-billed ani (Crotophaga sulcirostris), black-vented oriole (Icterus wagleri), yellow-rumped warbler (Dendroica coronata), Veracruz wren (Campylorhynchus rufinucha); and fish: seabass, Colorado snapper (Lutjanus colorado), and mullet (Mugil curema).

The site is used mainly for tourism and, to a lesser degree and intensity, for coastal fishing, using artisanal methods, as well as some minor forestry, which supports the construction of local rustic buildings. The site has been studied for potential eco-tourism development, with federal financial support. A number of areas at the site are used to cultivate coconut trees (Orbignya guacuyule). A few palapas have been built at the site to support minimal tourism, built with natural elements (mangrove and palm wood). Any further plans to expand at the site were ceased due to the intervention

FIGURE 6.14 Composite aerial image of Laguna Costera El Caimán. Designated February 02, 2005; Area: 1,125 ha; Coordinates: 17°58'N 101°16'W; Ramsar Site number: 1,448.

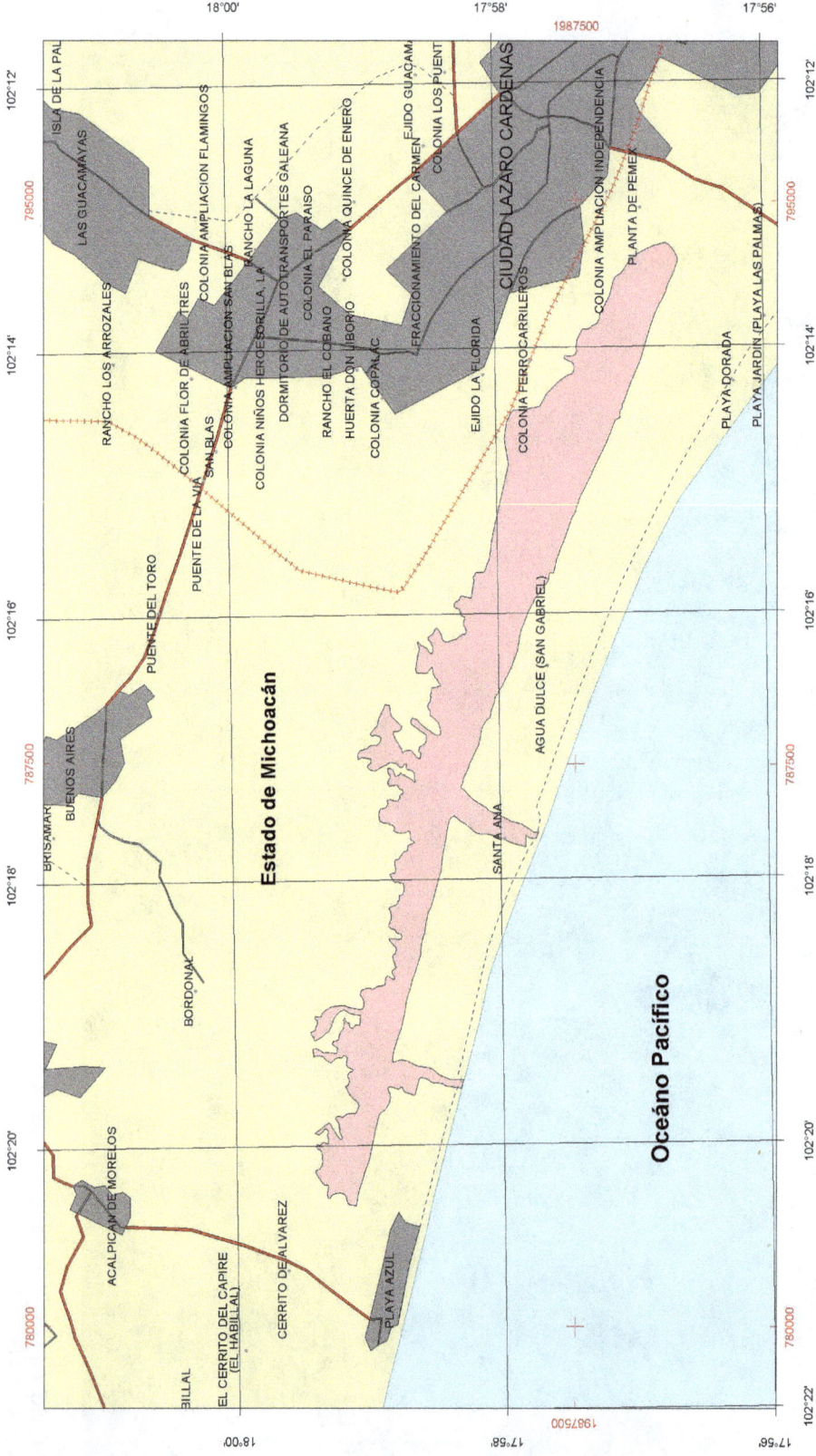

FIGURE 6.15 Map of Laguna Costera El Caimán. Designated February 02, 2005; Area: 1,125 ha; Coordinates: 17°58′N 101°16′W; Ramsar Site number: 1,448.

of governmental conservation initiatives. Within the larger watershed, however, there is continued development of tourism. The Lázaro Cárdenas-Las Truchas Steel Company provides economic opportunities for communities of the area and the industrial and commercial port activities in the area have been in place for several decades. The Government of the State of Michoacán (through the Secretary of Urbanism and Environment) has decreed through a Regional Ecological Ordinance for the Industrial Port Zone of Lázaro Cárdenas, Michoacán, within which the lagoon "El Caimán" is located, that land use is subject to environmental/legal protections.

A number of ecological studies have occurred at the site, including those by Madrigal (2005), who looked at the spatial and temporal distribution of ichthyofauna within the ecosystem, in addition to studies carried out by faculty of the Biology Department of the Michoacán University of San Nicolás de Hidalgo. Coastal birds of Michoacán have also been studied by graduate students at the National Autonomous University of Mexico, as well as other areas of ecological research, such as the influence of environmental gradients on spatial variability of site conditions, and on the distribution and structure of the fish community at the site. A study prepared by CONAGUA and the Fisheries Commission of the State of Michoacán is on the ecological dynamics of mangroves at the site.

PLAYA TORTUGUERA TIERRA COLORADA, GUERRERO: OVERVIEW

Playa Tortuguera Tierra Colorada (Figures 6.16 and 6.17) is an ocean-battered beach on the south-central Pacific Coast of Mexico, a critical nesting area for the endangered leatherback sea turtle (Dermochelys coriacea), olive ridley sea turtle (Lepidochelys olivacea), and green sea turtle (Chelonia mydas). American crocodile (Crocodylus acutus) have also been reported at the site, which is indicative of a relatively functional ecosystem food web. Coastal dune vegetation, mangrove forests, and desert scrub are also present at the site, which although modified by human uses sustains the important biological diversity of the site. Fishing is an important economic activity in the area, and there is considerable exploitation of turtles and other reptiles for sale and consumption. Cattle and dogs on the beaches are a prevalent threat to the ecology of the site, as well as coconut and mango plantations, and tourism. The site is in the State of Guerrero, within the municipality of Cuajinicuilapa. To the northwest, the site begins at the Barra de Tecoanapa Estuary, and ends (southeast) in Punta Maldonado, near the border with the State of Oaxaca.

In addition to turtle habitat, the site supports mangrove forest communities (Rhizophora mangle, Laguncularia racemosa, Avicennia germinans, and Conocarpus erectus), as well as semi-deciduous forest, in addition to coastal dune vegetation and xerophytic scrub, along the beach. In the early 1980s, this area was home to approximately 65% of the world's population of leather back sea turtle (Pritchard 1982). Since then, populations have declined, leading to listing of the species as critically endangered (IUCN 2000). The olive ridley sea turtle (Lepidochelys olivacea) and the green sea turtle (Chelonia mydas) are similarly listed as endangered by both the IUCN and Mexican legislation.

SITE DETAILS

The average annual precipitation at the site is 1110 mm/year, with May–October as the most intense months of the rainy season, with approximately 30 to 59 days of rain during that period (INEGI 1985). During the dry season, the average high temperature is approximately 30° C (SARH 1987). Estuarine circulation results in oligohaline conditions for about 8 months of the year, and mesohaline conditions for the remaining months of the year. Fresh water comes mainly from the Quetzala River and associated streams that provide approximately 48 tons of nutrients per year (e.g., NO_2, NO_3, NH_4, and PO_4), sediment (3.7×10^7 ton/year), and organisms through the site, and on to the ocean ecosystem. The site has a very dynamic and high-energy beach, leading to the redistribution of marine sediments in front of the bay on a periodic basis.

FIGURE 6.16 Composite aerial image of Playa Tortuguera Tierra Colorada. Designated November 27, 2003; Area: 54 ha; 16°25'N 98°38'W; Ramsar Site number: 1,327.

FIGURE 6.17 Map of Playa Tortuguera Tierra Colorada. Designated November 27, 2003; Area: 54 ha; 16°25'N 98°38'W; Ramsar Site number: 1,327.

The site supports a variety of plant communities, with the most representative being the coastal dune, where the vegetation is composed of creeping and stoloniferous plants, pioneers that establish on the sand, and those plants adapted to high temperatures; bayhops (Ipomoea pes-caprae) predominate in the coastal dune areas of the site, and are most abundant in the rainy season (Rzedowski 1978). The tree species *Coccoloba liebmannii* has been recorded in the higher areas of the dunes, where the sand consolidates, forming an ecotone and irregular-shaped patches in those sections of the beach where it is wider. Xerophytic or coastal scrub at the site is characterized by a relative abundance of *Opuntia puberula* and *O. velutina*, mixed with shrubs like *Lantana camara*, among others. The impact of vegetational loss in these areas is the influx and spread of *Prosopis juliflora*, *Acacia riparia*, and *Phragmites australis*.

Local agricultural activities attract raccoon (Procyon lotor), American hogged-nosed skunk (Conepatus mesoleucus), opossum (Didelphis virginiana), rabbit (Sylvilagus floridanus), armadillo (Dasypus novemcinctus), and badger (Nassau narica), among other animals, to the site, which commonly inhabit the mangroves and jungle. Within that jungle area, large numbers of Mexican spiny-tailed iguana (Ctenosaura pectinata) and green iguana (Iguana iguana) feed on the foliage of certain trees, such as the *Ficus mexicana* and *Guazuma ulmifolia*.

Main floral species at the site during the rainy season, in slightly flooded areas, include swamp lily (Crinum erubescens). Only a small portion of the area has remnant species, such as the evergreen tree *Caesalpinia cacalaco* and *Bursera simaruba* (gumbo limbo). Among the locally exploited trees are the elephant-ear tree (Enterolobium cyclocarpum), and primavera (Roseodendron donnell-smithii), among other species (Rzedowsi 1978). Likewise, the surrounding communities of people make extensive use of the mangrove, mainly from *Rhizophora* and *Conocarpus erectus*, as construction materials and fuel (Hernández-Tovilla 1998). There is also constant deforestation by way of the introduction of crops to the adjacent areas, which are sometimes later abandoned.

The first scientific studies of the site found approximately 5,000 nesting females of leatherback sea turtle (Dermochelys coriacea), and an estimated total population of 16,000 individuals (Márquez et al. 1981). Currently, based on reviews of reproduction activities on nesting beaches, a turtle population reduction greater than 80% has been documented for the region during the late 1980s and 1990s (Sarti et al. 1997, 1998, 1999, 2000). Olive ridley sea turtles (Lepidochelys olivacea) have generally fared better during these other downturns in population sizes. Other species of reptiles that have been subject to increased exploitation for food in the area are the green iguana (Iguana iguana), the Mexican spiny-tailed iguana (Ctenosaura pectinata), and the American crocodile (Crocodylus acutus). The belief that snakes are poisonous and dangerous has also led to the indiscriminate killing of boas (Boa constrictor imperator) and sea snakes (Pelamys platurus). Six species of mammals are caught on a daily basis in the areas surrounding the site, an activity carried out by the poorest inhabitants. The most valued species are the armadillo (Dasypus novemcinctus) and raccoon (Procyon lotor), with white-tailed deer (Odocoileus virginianus) more rare, due to declining populations. Habitat destruction in the area has led to negative impacts on coyotes (Canis latrans), jaguarundi (Felis yagouaroundi), northern tamandua (Tamandua mexicana), and gray fox (Urocyon cinereoargenteus), among others (Hernández-Tovilla 1998). The site is known for its carrion birds, such as vultures (*Cathartes aura* and Coragyps atratus), as well as birds of prey, such as the osprey (Pandion haliaetus), the common black hawk (Buteogallus anthracinus), and occasionally the common caracara (Caracara plancus). Important migrations of birds of the genera *Sterna* (terns) and *Anas* (ducks) occur regularly (Hernández-Tovilla 1998). In the estuary, there are abundant populations of cormorants (Phalacrocorax sp.) and the magnificent frigatebird (Fregata magnificens). Black-necked stilts (Himantopus mexicanus) can be seen on the beach, as well as herring gulls (Larus argentatus) and brown pelicans (Pelecanus occidentalis). Fishing at the site is carried out mainly in an artisanal way, but more intensively in Barra de Tecoanapa and Punta Maldonado. The main catch in these areas is estuarine catfish (Galeichthys caerulescens), yellowfin snook (Centropomus robalito and C. nigrescens), and mullet (Mugil curema); these species provide the largest catches in the fishing season, along with crab (Callinectes spp.) and whiteleg shrimp (Litopenaeus vannamei). Among the highly valued species that are transferred to the Port of Acapulco from the site are the spotted nose snapper (Lutjanus guttatus) and sharks (Carcharhinus sp.).

Several ethnic groups come together in the region, the Afro-Mestizo population being very significant, which have displaced the indigenous populations. Over time, black people who were dedicated to tending cattle on estates increased as they fled from slavery and rebuilt their independent lives and settled in various communities on the coast of Guerrero and Oaxaca. Such groups brought a variety of customs such as the characteristic Spanish language spoken in the area, along with traditional medicine attributed to turtle eggs and oil. Although a once-dominant human activity, estuarine artisanal fishing, fishing in recent years has decreased substantially, due to over-exploitation of fishing grounds. The raising of livestock is also among the dominant human land uses in the area. Additionally, agriculture comprises approximately a quarter of human activities in the area, focused on coconut orchards, and cultivation of fruit such as watermelon. Tourism is also very important to the economy of the surrounding communities of Barra de Tecoanapa and Punta Maldonado. As additional people come to the area, though, potable water in the local communities is limited and sewage systems are often impaired, increased use of water from the Quetzala River for irrigation and human consumption is increasingly necessary.

Adverse impacts on the ecology of the site is tourist infrastructure, including boat berths for fishing or any other tourism activities that may impact turtle habitat. Other direct threats to turtles are dogs on the beach, which prey on clutches and kill adult turtles that come out to oviposit; constant accumulation of solid waste on the beach; and looting of sea turtle nests by local people either for consumption or for sale. Deciduous forests in the surrounding areas have been cut down to a large extent to establish temporary crops, which impacts a large number of plant and animal species at the site, and indirectly within the region and along migratory corridors. The lack of strong (and enforced) fishing regulations is a major threat to the estuary.

The site has a management and protection plan for female sea turtles that come to nest, as well as for the collection of clutches and the release of hatchlings, implemented by the Leatherback Project, via the Secretariat of Environment and Natural Resources (SEMARNAT). Since 1996, The Leatherback Project has executed a program for the protection, conservation, investigation, and management of sea turtles in Tierra Colorada, Guerrero. Accordingly, local talks are given on the biology and conservation of sea turtles in schools within the communities close to the site. Likewise, school visits in the region, and occasionally from the Federal District, occur, along with periodic beach cleaning campaigns, often after the Christmas and New Year holidays, with the active participation of the community of Tierra Colorada. There are no formal tourist activities at the site, although there is some pressure and some interest in the local communities to initiate such activities.

PLAYA TORTUGUERA CAHUITÁN, GUERRERO: OVERVIEW

Playa Tortuguera Cahuitán (Figures 6.18 and 6.19) is an important nesting beach for three species of endangered marine turtles in the Pacific Ocean: leatherbacks (Dermochelys coriacea), nesting from October to March; olive ridley (Lepidochelys olivacea), nesting year-round; and green sea turtles (Chelonia mydas), nesting from October to January. The beach is a highly dynamic ecosystem with large tidal surges, which can form sand walls of up to 2 m high. Adjacent areas have been severely deforested, with remnant patches of *Rhizophora mangle*, *Enterolobium cyclocarpum*, and *Roseodendron donnell-smithii* remaining, used locally for house building and woodwork. Leatherback sea turtles have been intensely surveyed since 1996, with a focus on protecting nestlings and releasing newborns into the sea. However, there is an alarming trend of decline for nestlings, attributed to nest looting by locals and the accidental capture of females by fishing operations. Artificial illumination, indiscriminate waste disposal, and the prospects of tourism in the area pose additional threats to the turtles, and the ecology of the site in general.

FIGURE 6.18 Composite aerial image of Playa Tortuguera Cahuitán. Designated February 02, 2004; Area: 65 ha; Coordinates: 16°17′N 98°29′W; Ramsar Site number: 1,347.

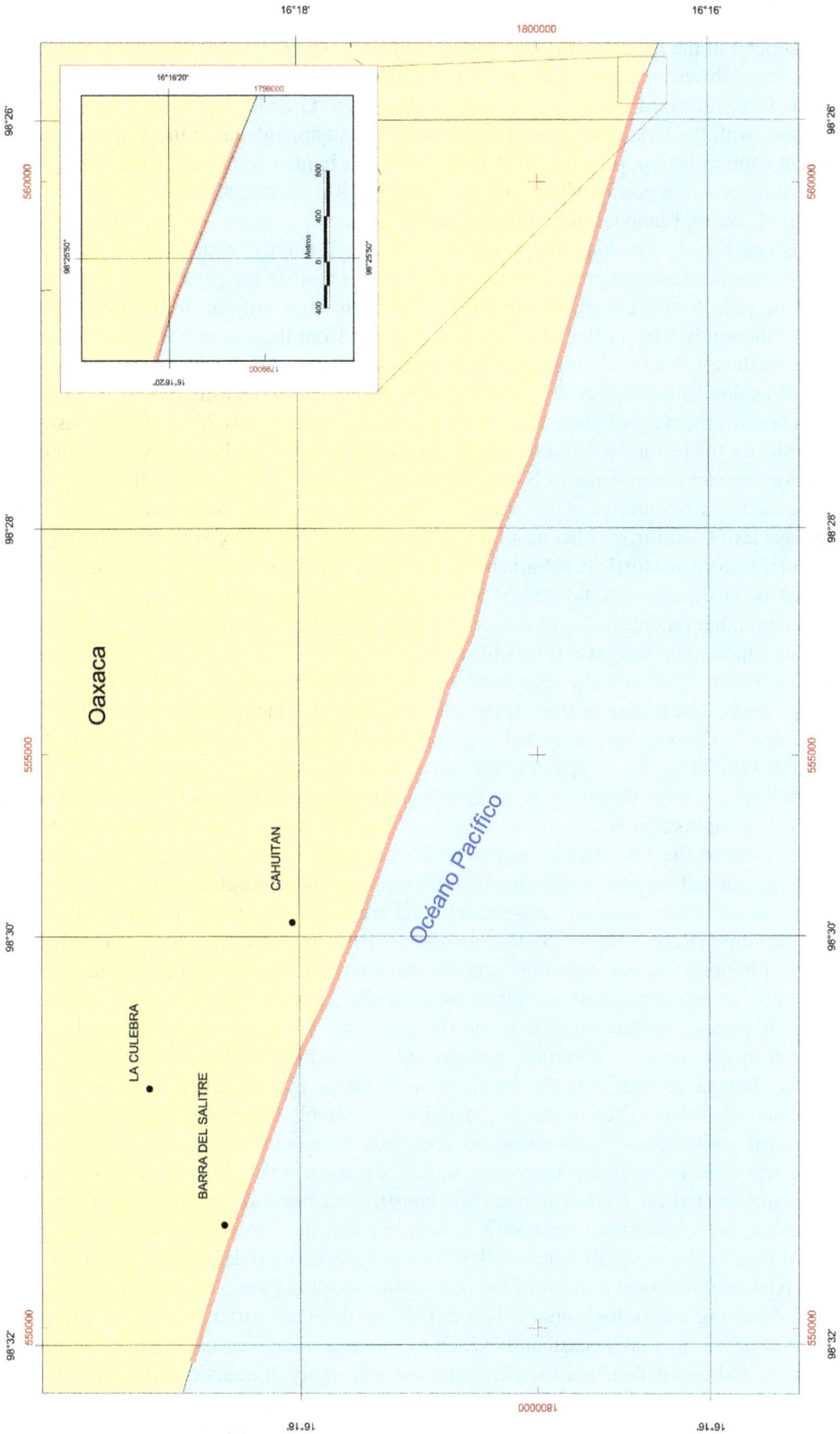

FIGURE 6.19 Map of Playa Tortuguera Cahuitán. Designated February 02, 2004; Area: 65 ha; Coordinates: 16°17′N 98°29′W; Ramsar Site number: 1,347.

SITE DETAILS

The site is located in the Municipality of Santiago Tapextla, District of Jamiltepec, in the State of Oaxaca. To the northwest of the site is the border of the State of Guerrero, and to the south of the site is the Pacific Ocean; toward the northeast of the site is Llano Grande. The nearest town (approximately 25 km), with the largest number of inhabitants, is Cuajinicuilapa, in the State of Guerrero, which has an approximate population of 26,000. There is a human settlement adjacent to the site, with the same name as the beach, which has approximately 100 inhabitants. Other small towns near the site are La Culebra, Llano Grande, Tapextla, and Tecoyame.

The site's beach is 12 km long and is dynamic and high-energy, with intense tidal changes, except in the extreme southeast portion of the site. The tides modify the physiognomy of the beach throughout the year, forming walls of sand up to 2 m high, especially in the extreme northwest. The width of the sandy strip of the site's beach is reduced from the extreme southeast (100 m) to the extreme northwest (50 m), during the months of November to April. The communities bordering the beach exploit its fishing resources, both in the ocean and in the estuaries located along the beach, and consume them or sell them to local towns people. In the 1980s, Mexico was considered a key nesting site for the leatherback sea turtle (Pritchard 1982), however the overexploitation of eggs and the incidental capture of females by the high-seas fisheries, mainly in South America, have caused a decline in the population of this species (Eckert and Sarti 1997, Sarti and Barragan. 2000). Two other species of sea turtles also nest on this beach, the olive ridley sea turtle (Lepidochelys olivacea) and the green sea turtle (Chelonia mydas), both of which are considered endangered in the Red List (IUCN 2003), and in NOM-059-SEMARNAT-2001 and/or -2010.

Site vegetation that predominates in the sandy beach areas are creeping plants, mainly bayhops (Ipomoea pes-caprae) and saltgrass (Distichlis spicata). There are also cacti of genus *Opuntia* on the dunes and within other shrubby vegetation areas of the site; Rubiaceae are predominant in the shrubby edge areas. The felling of trees at the site is common, to facilitate agricultural cultivation and the raising of livestock, which has led to greater development in the shrubby areas of the site as well (Rzedowski 1978). As is typical in the estuaries and seasonal streams of this region, arboreal vegetation that is most dominant is mangrove (Rhizophora mangle), which provides numerous ecological functions and is actively used by the human communities of the area (Rzedowski 1978). Other trees of the site include elephant-ear tree (Enterolobium cyclocarpum), primavera (Roseodendron donnell-smithii), and guapinol (Platymiscium dimorphandrum). These species are exploited locally for the housing construction, and are also used for the manufacture of poles, fences, and furniture (Rzedowski 1978). Red mangrove (Rhizophora mangle) are also exploited for the building of houses, fencing, firewood, and the extraction of the bark to be used in traditional medicines. In the areas surrounding the site, most of the flora are introduced by way of agriculture (i.e., corn, chili, papaya, sesame, sugar cane, and watermelon) and the raising of livestock.

In addition to the hunting of turtles, iguanas are also exploited at the site, specifically the green iguana (Iguana iguana) and the Mexican spiny-tailed iguana (Ctenosaura pectinata). In addition, white-tailed deer (Odocoileus virginianus) is a notable large wild mammal at the site, hunted for local consumption. The armadillo (Dasypus novemcinctus) is another species that, like iguanas, represent an important alternative source of protein in the diet of nearby communities, and is frequently hunted for food. The boa (Boa constrictor imperator), although it is not hunted for consumption, are often killed because it is believed that they are poisonous and dangerous, rather than their actual ecosystem role, which is as a key arboreal predator. The American crocodile (Crocodylus acutus) exists in most of the main estuaries of the area, periodically observed on the beach and moving within the estuary. This population does not suffer much from commercial exploitation; however, they are occasionally killed by humans when they cause losses in livestock. In the estuaries and on the beach of the site, there are a diversity of aquatic and marine bird species, such as the magnificent frigatebird (Fregata magnificens), herring gull (Larus argentatus), white heron (Casmerodius albus), and cormorant (Phalacrocorax sp.). Other winter-month residents observed at the site include the brown pelican (Pelecanus occidentalis), plover (Charadrius sp.), and terns (Sterna sp.). Within semi-deciduous vegetation of the site, the crested caracara (Polyborus plancus), the common black hawk (Buteogallus anthracinus), roadside hawk (Buteo magnirostris), and two species of vultures (Coragyps atratus and Cathartes aura) have been observed. Additionally, the site supports two species of woodpecker (Melanerpes aurifrons and Campephilus guatemalensis), turtle doves (Columbina passerina), and cardinals (Cardinalis cardinalis), among many other species, some of which are captured to be sold as pets in the larger towns of this region, such as is the case for the orange-fronted parakeet (Eupsittula canicularis). Other smaller mammals at the site include white-nosed coati (Nasua narica), raccoon (Procyon lotor), American hog-nosed skunk (Conepatus mesoleucus), gray fox (Urocyon cinereoargenteus), and opossum (Didelphis virginiana). Some of the small mammals at the site are captured for food, medicinal uses (e.g., skunk), or as pets. A number of jaguarundi (Felis yagouaroundi) have also been observed at the site. The site supports groups of marine mammals, such as the common bottlenose dolphin (Tursiops truncatus) and humpback whale (Megaptera novaeangliae). Of note at the site are feral cats and dogs, as well as grazing cattle, horses, and goats, which are left free to feed on the vegetation adjacent to the beach.

The towns adjacent to Playa Tortuguera Cahuitán are Afro-Mestizo communities, typical of the Costa Chica; there are few popular traditions and customs preserved in the area due to growing outside influences. The religions practiced in the nearby communities are Apostolic Catholic and of the Evangelical Church, these being the two strongest religions in the area. A traditional recreational activity at the site is 'mixteca ball,' a sport that has its roots in an ancient Nahuatl (language of the Aztec People) ball game. In addition, each community near the site has at least one soccer field and a basketball field, being the most popular sports in this area. The agricultural production of the region consists mainly of the cultivation of maize and some varieties of chili, as well as coconut, watermelon, mango, banana, and papaya. Water supply for the surrounding communities is from wells that are located in the center of the local towns. In most of the local towns, fishing is mainly for subsistence, and on occasion there is some trade of dried-salted fish. Most farmers in the area of the site cultivate rainfed crops and normally do not use irrigation systems for agriculture, since this method is very expensive and not all farmers in the area have access to consistent fresh water.

The average number of visitors to the site's beach is in the hundreds, with larger numbers during the holidays. Major impacts on the ecology of the site are the clearing of forests for agriculture and cattle grazing; the use of herbicides in agriculture; contamination of groundwater and bodies of water, such as small rivers and estuaries; use of laundry detergents in the area's streams, which flow into the mangroves; indiscriminate dumping of garbage, which reaches the sea; and the over-hunting of species such as armadillo (Dasypus novemcinctus) and iguanas.

The site is active with the work of the Leatherback Conservation Program, supported by the Secretariat of Environment and Natural Resources (SEMARNAT) in collaboration with Conservation International; the work includes the conservation of clutches, nesting females, and hatchlings that emerge from nests, as well as research focused on the biology and reproduction of sea turtles that nest at the site, and in the region. This work has also been used to communicate with the people of La Culebra, Cahuitán, and Llano Grande, sharing sustainable solutions to community needs and site conservation. Economic incentives are granted for children who attend primary and high school, where lessons that include information about the importance of conservation of natural resources occur.

REFERENCES

Alvarado., J. and T. Carlos Delgado. 2005. *Tortugas Marinas de Michoacán: Historia Natural y Conservación.* Morevallado Editores, Morelia, Mich.

Alvarado, J. and A. Figueroa. 1985. *Ecología y Conservación de las tortugas marinas de Michoacán, México.* Cuaderno de investigación La Universidad Michoacana de San Nicolás de Hidalgo. (4) 40pp.

Alvarado Díaz, J. Huacuz, E., and D. del Carmen. 1996. *Guía ilustrada de los anfibios y reptiles más comunes de la reserva Colola-Maruata en la costa de Michoacán, México*. La Universidad Michoacana de San Nicolás de Hidalgo. 122pp.

Andrade, E. 1998. *Analisis de la Pesqueria del camaron de la Laguna de Cuyutlán, Col, México*. Universidad de Colima. Tesis de Maestría, 81pp.

Avendaño-Sánchez, H. and O. Sotomayor-Navarro. 1986. Influencia de la Heterogeneidad Espacialen la Estructura de la Comunidad del Fitoplancton, al oeste de la Isla Socorro, México. *Invest Ocean Bulletin*. 3(1): 1–21.

Cabrera, A.L. and A. Willink. 1980. *Biogeografía de América Latina. Monografía Científica*. Serie de Biología. Secretaría General de la O.E.A., Washington, DC. 177pp.

Correa, G. 1979. *Geografía Física del Estado de Michoacán*. Gobierno del Estado.

Eckert, S. and L. Sarti. 1997. Distant fisheries implicated in the decline of leatherback Pacific populations. *MTN*. 76: 7–9.

Espino-Barr, E., Cruz-Romero, M., Garcia-Boa, A., and A. Sanchez. 1998. *Catalogo de especies de peces marinos con menor valor comercial, capturas en la costa de Colima. México*. SEMARNAPINP-Manzanillo, México. 70pp.

Hernández-Tovilla, C. 1998. Ecología de los bosques de manglar y unos aspectos.

Hernández-Vázquez, S. 2004. Aves acuáticas de la laguna de Agua Dulce y esteros El Ermitaño, Jalisco, México. *Biología Tropical/International Journal of Tropical Biology and Conservation*. 53(1–2): 229–238.

INEGI (Instituto Nacional de Estadística Geografía e Informática). 1985. *Carta de efectos climáticos regionales Mayo-Octubre*. 1:250000. Acapulco. E14–11pp.

INEGI (Instituto Nacional de Estadística Geografía e Informática). 2000. *Censo Nacional de Población y Vivienda*.

Janzen, D. 1988. Tropical dry forest: the most endangered major tropical ecosystem (pp. 130–137). In: E. O. Wilson (Ed.) *BioDiversity*. National Academic Press, Washington.

Jaramillo-Villalobos, V. 1994. *Revegetación y reforestación de las áreas ganaderas en las zonas tropicales de México*. Comisión Técnico Consultiva de Coeficientes de Agostadero, SARH, México DF.

Jiménez, M.L. 1991. Araneofauna de las Islas Revillagigedo, México. *Anales del Instituto de Biología, Universidad Nacional Autónoma de México. Serie Zoología*. 62(3): 417–429.

Johnston, I.M. 1931. The Flora of the Revillagigedo Islands. Proceedings of the California Academy of Sciences. *Fourth Series. USA*. 20(2): 9–104.

López, C. 1985. *Diseño de una reserva para tortugas marinas en el Playón de Mexiquillo, Michoacán. Tesis de Licenciatura*. Facultad de Ciencias, UNAM.

Madrigal, G.X. 2005. *Distribución Espacial y Temporal de la Ictiofauna del Estero de Santa Ana, Michoacán, México. Tesis de Maestría*. UNAM. 80pp.

Márquez, R., Villanueva, O., and S. Peñaflores. 1981. Anidación de la tortuga laud (Dermochelys coriacea schlegelli) en el Pacífico mexicano. *Ciencia Pesquera*. 1(1): 45–51.

Medina, G. M. 1978. Memoria de la Expedición Científica a las Islas Revillagigedo. Abril de 1954. Universidad de Guadalajara, Jal. 333pp.

Miranda, F. 1960. *Vegetación*. (pp. 126–152) In: J. Adem, E. Cobo, L. Blásquez, F. Miranda, A. Villalobos, T. Herrera, B. Villa, and L. Vásquez (Eds.) Isla Socorro, Archipiélago Revillagigedo. Monografías del Instituto de Geofísica. Universidad Nacional Autónoma de México, México, DF.

Miranda, F. and E. Hernández-X. 1963. Los tipos de vegetación en México y su clasificación. *Boletín de la Sociedad Botánica de México*. 28: 29–179.

Pritchard, P. 1982. Nesting of the leatherback sea turtle Dermochelys coriacea in Pacific Mexico, with a new estimate of the world population status. *COPEIA*. 1982(4): 741–747.

Rzedowski, J. 1978. Vegetación de México. Limusa, México. 432pp.

Santana M., F., Lemus, J. S., and M. Vergara S. 1992. *Guía de excursión etnobotánica en el Estado de Colima, México*. Universidad de Colima, 20pp.

SARH (Secretaría de Agricultura y Recursos Hidráulicos). 1987. Volúmenes de *agua transportados por los ríos Nexpa, Quetzalapa, y Verde durante los últimos 10 años. Publ. Esp. Num*. 4: 52pp.

Sarti, M. L. and A.R. Barragán. 2000. Estimación del tamaño de la población anidadora de tortuga laúd Dermochelys coriacea y su distribución en el Pacífico Oriental durante la temporada de anidación 1999-2000. Informe Final de Investigación. Instituto Nacional de la Pesca. SEMARNAP.

Sarti, M. L., Barragán, A., and N. García. 1997. Estimación del tamaño de la población anidadora de tortuga laúd Dermochelys coriacea y su distribución en el Pacífico mexicano durante la temporada de anidación 1996-1997. Informe Final. Laboratorio de Tortugas Marinas, Fac. De Ciencias, UNAM/INP. 39pp.

Sarti, M, L., Barragán, A.R., and S.A. Eckert. 1999. Estimación del tamaño de la población anidadora de tortuga laúd Dermochelys coriacea y su distribución en el Pacífico Oriental durante la temporada de anidación 1998-1999. Informe Final de Investigación. Instituto Nacional de la Pesca. SEMARNAP; Laboratorio de Tortugas Marinas, Fac. De Ciencias-UNAM. 24pp.

Sarti, M, L., Eckert, S., Barragán, A., and N. García. 1998. Estimación del tamaño de la población anidadora de tortuga laúd Dermochelys coriacea y su distribución en el Pacífico mexicano durante la temporada de anidación 1997-1998. Informe Final de Investigación. Instituto Nacional de la Pesca. SEMARNAP; Laboratorio de Tortugas Marinas, Fac. De Ciencias-UNAM. 20pp.

SEDUR (Secretaria Municipal de Desenvolvimento e Urbanismo). 2007. Memoría técnica de la actualización del Programa de Ordenamiento Ecológico del Territorio de la Subcuanca Laguna de Cuyutlán, Estado de Colima. Universidad Autónoma del Estado de Morelos, Red Regional de Recursos Bióticos y Red Mesoamericana de Recursos Bióticos. 495pp.

Vázquez, G. L. 1960. Observaciones sobre las Artrópodos (pp. 217–233). In: J.Adem et al. (Eds.) *La Isla Socorro, Archipiélago de las Revillagigedo*. Monografías del Instituto Geofísica. Monografías del Instituto de Geofísica. Universidad Nacional Autónoma de México. México, DF.

Verdi, L.A., Casteñeda B. E., Contreras, B. G., Aguilera, L. G., García, L., M. de L., Ortiz Gallarza, S.M., and N. Villa A. 1994. El Archipiélago Revillagigedo, Colima, México. Dirección General de Oceanografía Naval. Secretaría de Marina.

Villa, R. 1960. Vertebrados Terrestres (pp. 201–216). In: J. Adem et al. (Eds.) *La Isla Socorro, Archipiélago de las Revillagigedo*. Monografías del Instituto de Geofísica. Universidad Nacional Autónoma de México, México, DF.

Villaseñor, L. E. 2000. *Las aves. Catálogo de Biodiversidad en Michoacán*. Universidad Michoacana de San Nicolás de Hidalgo, Secretaría de Urbanismo y Medio Ambiente (SUMA).

Wallace, A. R. 1876. The Geographical distribution of animals with a study of the relations of living and extinct faunas as elucidating the pas changes of the earth's surface. 1962 (2nd ed.), Hafner publishing Company, New York and London.

7 Mexico
States of Oaxaca and Chiapas

The Ramsar wetlands along this southernmost coastal area of Mexico, in the states Oaxaca and Chiapas, serve as an instructive contrast to those regions further north, and along the migratory pathways taken by both avian and oceanic organisms. This region also epitomizes the definitive tension that exists along the entirety of the coastal region of North America and the Central Pacific, between human use and the conservation imperatives of our time. With some of the most unique wetland environments and organisms present in these eight coastal wetlands, the biological diversity and specialization of organisms of these sites is notable, paired with a notable intensity of human uses. The eight coastal wetlands along this southernmost coastal region of mainland Mexico (Figure 7.1) are:

- Lagunas de Chacahua
- Cuencas and Corales de la Zona Costera de Huatulco
- Playa Barra de la Cruz
- Sistema Estuarino Puerto Arista
- Sistema Estuarino Boca del Cielo
- Reserva de la Biosfera la Encrucijada
- Zona Sujeta a Conservación Ecológica Cabildo-Amatal
- Zona Sujeta a Conservación Ecológica el Gancho-Murillo

LAGUNAS DE CHACAHUA, OAXACA: OVERVIEW

Lagunas de Chacahua (Figures 7.2 and 7.3) is a coastal lagoon ecosystem with a sandy beach facing the Pacific Ocean; distributed lowlands associated with alluvial, fluvial, and deltaic plains; intertidal marshes; and uplands with elevations as high as 140 m above the sea level. The site supports American crocodile (Crocodylus acutus) and a number of notable fish species, such as *Centropomus nigriscens* and *Centropomus robalito*, as well as the leatherback sea turtle (Dermochelys coriacea). The site's mangroves are a source of food and refuge for many birds, mammals, and reptiles. Fishing is the main human activity at the site, with rapidly increasing human populations as the primary ecological threat. The site has been a Natural Protected Area since 1937, and part of the Playa de la Bahía de Chacahua Sanctuary for the protection of marine turtles since 1986. The site is located along a coastal strip in the municipality of Villa de Tututepec de Melchor Ocampo, in the district of Juquila, Oaxaca, 150 km from the town of Juárez, the capital of the State of Oaxaca (approximate population 1,400,000).

SITE DETAILS

Of the 224 species of flora reported at the site, five are under special protection, and two are threatened (DOF 2002, 2010). Notable plant species at the site are red mangrove (Rhizophora mangle), button mangrove (Conocarpus erectus), white mangrove (Laguncularia racemosa), black mangrove (Avicennia germinans), Sonora guaiacum (Guaiacum coulteri), the palm *Cryosophila nana*, and palo de agua (Bravaisia integerrima). Vertebrates at the site include 280 species, including 12 species of amphibians, 26 species of reptiles, 175 species of birds, and 67 species of mammals.

DOI: 10.1201/9781003046394-7

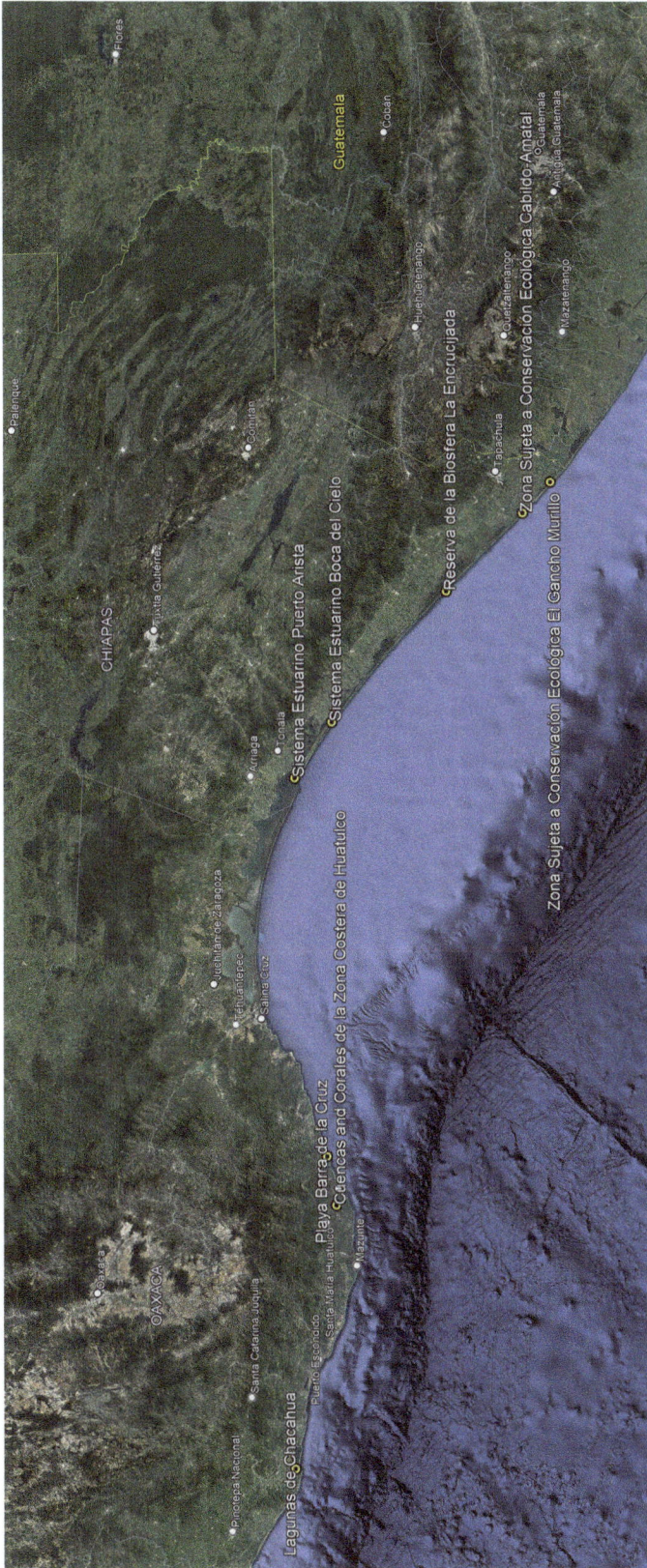

FIGURE 7.1 Referenced Ramsar wetlands generally shown along the western coast of southern mainland Mexico, in the states Oaxaca and Chiapas.

FIGURE 7.2 Composite aerial image of Lagunas de Chacahua. Designated February 02, 2008; Area: 17,424 ha; Coordinates: 16°00′N 97°40′W; Ramsar Site number: 1,819.

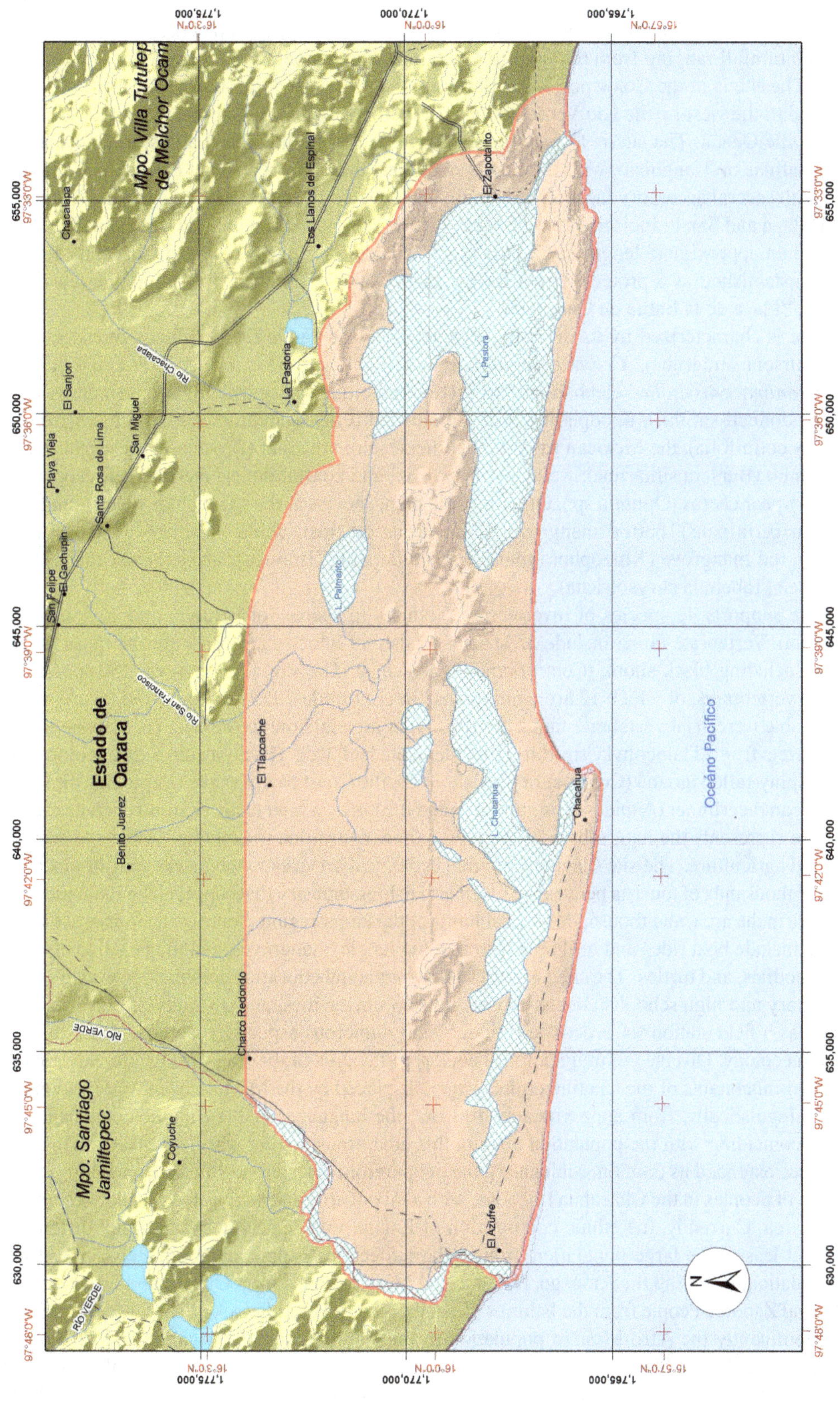

FIGURE 7.3 Map of Lagunas de Chacahua. Designated February 02, 2008; Area: 17,424 ha; Coordinates: 16°00′N 97°40′W; Ramsar Site number: 1,819.

The lagoon ecosystem is located in the subhumid tropics, which includes part of the Mexican Pacific coast, characterized by variable humidity with a considerable period of drought (5–9 months long), with rainfall ranging from 600 mm to 1,500 mm, and temperatures above 20° C throughout the year. The site is in the Colotepec River Basin and San Francisco Sub-basin (660 km²). The site is bordered to the west by the Río Verde, to the east by the Río Grande Sub-basin, and to the south by the Pacific Ocean. The site is formed mainly by three coastal lagoons: La Pastoría, Chacahua, and Las Salinas or Tianguisto, with the first two lagoons linked by a channel called the Canal del Corral, with several accessory lagoons and small channels fed by temporary water courses, such as the Chacalapa and San Francisco Rivers. Beaches of the site (i.e., at Lagunas de Chacahua National Park) have an approximate length of 32 km, which include San Juan Beach and Chacahua Beach, the latter established as a protected area for the turtles, and designated in 2022 as the sea turtle sanctuary, "Playa de la Bahía de Chacahua."

The site is characterized by its diversity of vegetation, such as in the deciduous forest: gumbo limbo (Bursera simaruba), *Ceiba aesculifolia*, and the leguminous *Pithecellobium lanceolatum*, *Byrsonima crassifolia*, elephant-ear tree (Enterolobium cyclocarpum), and primavera tree (Tabebuia donnell-smithii); floodplain forest: lemonwood (Calycophyllum candidissimum), alamo tree (Ficus cotinifolia), the Mexican sabal (Sabal mexicana); breadnut (Brosimum alicastrum), and gumbo limbo (Bursera simaruba); wetlands; *Typha* sp.; and coastal dunes: mesquite (Prosopis sp.) and prickly pear cactus (Opuntia sp.). Other notable plant species at the site include black mangrove (Avicennia germinans), button mangrove (Conocarpus erectus), white mangrove (Laguncularia racemosa), red mangrove (Rhizophora mangle), palo de agua (Bravaisia integerrima), and golden trumpet tree (Tabebuia chrysotricha).

The site supports 42 species of invertebrate molluscs (6 species of Bivalvia and 16 species of Gastropoda). Vertebrate fauna include 52 species of fish, of which 12 species are the most valued for food, including black snook (Centropomus nigrescens). The site also supports 280 species of terrestrial vertebrates, of which 12 are amphibians, 26 are reptiles, 175 are birds, and 67 are mammals. Notable terrestrial vertebrate fauna are those that are endemic to Mexico, such as the dwarf Mexican tree frog (Tlalocohyla smithii), the Mexican leaf frog (Pachymedusa dacnicolor), the Mexican spiny-tailed iguana (Ctenosaura pectinata), Pacific Coast parrot snake (Leptophis diplotropis), Mexican racerunner (Aspidoscelis guttatus), and the Oaxaca mud turtle (Kinosternon oaxacae).

The site represents the nature-base of the regional economy, due to activities at the site such as fishing and agriculture. The site's lagoons provide ecological services to the tourist industry (servicing tens of thousands of tourists per year), as well as a fishing industry that supports the inhabitants of eight towns in the area, and thousands of inhabitants of the larger region. Some of the tourist activities at the site include boat rides and trail tours through the jungle, mangroves, and the beach to observe birds, crocodiles, and turtles. The site also offers environmental education activities, by way of visits from primary and high school students, as well as from universities, and a variety of other visitors. The site has a field station for crocodile research, and numerous aspects of wetland and estuarine ecosystem ecology. Diverse cultural identities coexist in the area of the site, with the descendants of the original inhabitants of the Chatino ethnic group, displaced by the Mixtecs from other regions of Oaxaca. Linguistically, from approximately 400 BC, the languages of the Chatinos and Zapotecs diverged, coinciding with the population peak in this area, transitioning when the Mixtec Kingdom of Tututepec extended its domain, subjugating the people from Guerrero to Huatulco, resulting in the settlement of peoples in the Chacahua Lagoons, with a mix of elite and subjugated peoples occupying the same area. Currently, the ethnic composition of Lagunas de Chacahua is culturally differentiated into at least three large population groups: the indigenous people, approximately 25% of the site's population, including the Amuzgo, Nahuas, and Mixtec groups from Oaxaca and Guerrero; the Chatino and Zapotec People from the Isthmus of Tehuantepec (those of black or Afro-Mestizo ancestry), predominantly the Afro-Mestizo population in the communities of Chacahua and El Azufre; and the Mestizo People. These three groups are linked to that great identity that articulates coastal and river fishers from all over the country, and represents the mix of identities that is part of the rich regional, social, and cultural identity. At present, a high percentage of the site's population is dedicated to artisanal fishing. The communal assets of San Pedro Tututepec is emblematic of the history and tenure of the land in this area, with 90% of the lands administered by the federal and state governments, and the rest by the Community Assembly.

Agriculture plays an important historical and economic role in the communities of the area, with the main crops of coffee, lemon, copra, banana, plum, papaya, avocado, peanuts, sesame, beans, and hibiscus as the primary cash crops. With these activities, the use of agrochemicals and the use of subsurface water resources poses potential threats to the ecology of the site. The accelerated population of the microregion in which the Lagunas de Chacahua National Park is located is one of the main factors that has led to the deterioration, destruction, and overexploitation of its natural resources. Likewise, the irregular growth of human settlements and the lack of provision of basic services to people incentivizes contamination of the soil and water, as well as the accumulation of solid waste, which is another major threat to the ecology of the site. Other threats include land clearing for the establishment of permanent crops and extensive livestock; indiscriminate hunting; illegal exploitation of forest products, wild animals, and turtle eggs; the use of prohibited fishing gear; illegal fishing; and the collection of flora and fauna, specifically the extraction of wood and turtle eggs. Another major impact on the ecology of the site is the Río Verde Dam project, which has severely affected the ecological conditions of the Chacahua Lagoons, particularly due to diversions and infrastructure that have been designed to serve the needs of agricultural irrigation higher in the watershed.

Prior to the establishment of this Ramsar site, Lagunas de Chacahua National Park operated a Program for the protection and conservation of sea turtles on the beaches of San Juan, a program in which residents of the Park participate, with the support of volunteers from various educational institutions. In addition, there is a consolidated surveillance program, in partnership with the San Pedro Tututepec Communal Assets Commission, which represents the agrarian nucleus and legitimate owners of the land, the Municipal Presidency, the Federal Attorney for Environmental Protection (PROFEPA), and the Park; agreements with communities to respect fishing regulations and other conservation measures have been implemented through this consortia. A number of research and monitoring activities, which have involved the construction of some conservation infrastructure, have occurred at the site, primarily by the Faculty of Sciences of the UNAM. The facilities of the Lagunas de Chacahua National Park also provide impetus to ongoing studies and support of the management of natural resources of the site, especially in the management of mangrove forests and turtle camps.

CUENCAS AND CORALES DE LA ZONA COSTERA DE HUATULCO, OAXACA: OVERVIEW

Cuencas and Corales de la Zona Costera de Huatulco (Figures 7.4 and 7.5) features coral reefs associated with shallow inlets along a coast dominated by cliffs. Mangrove forests are found at the outlets of some rivers at the site, associated with several coastal lagoons. Inland, dry forests are intersected with a number of streams, where fauna and flora are supported, including several endangered and endemic species, including amphibians *Bufo marmoreus* and *Hyla sartori*, the Sinaloa wren (Thryophilus sinaloa), the golden-cheeked woodpecker (Melanerpes chrysogenys), and the west Mexican chacalaca (Ortalis poliocephala). Numerous marine mammals transit the coastline of the site, such as the pygmy killer whale (Feresa attenuata), the false killer whale (Pseudorca crassidens), and the long-finned pilot whale (Globicephala macrorhynchus). Shell fishing, agriculture, and tourism are the main human activities of the site, with tourism development as the main threat to the ecology of the site, along with forest clearing, hurricanes, and hunting. The site is located in the municipality of Santa Maria Huatulco, in the district of Pochutla, Oaxaca. The site is located 28 km from the town of Pochutla (approximate population 39,000).

FIGURE 7.4 Composite aerial image of Cuencas and Corales de la Zona Costera de Huatulco. Designated November 27, 2003; Area: 44,400 ha; Coordinates: 15°47′N 96°12′W; Ramsar Site number: 1,321.

FIGURE 7.5 Map of Cuencas and Corales de la Zona Costera de Huatulco. Designated November 27, 2003; Area: 44,400 ha; Coordinates: 15°47'N 96°12'W; Ramsar Site number: 1,321.

SITE DETAILS

The site supports a number of organisms that have a limited distribution, such as the purple snail (Plicopurpura pansa) and the coral species *Pocillopora eydouxi*. Indeed, the site is home to one of the most significant coral reefs of the Mexican Pacific, due to its location at the southern limit of the Mexican biogeographic province, with a unique composition that shares elements of the adjacent Panamica Province (Barrientos and Ramírez 2000). The fauna of the site include 22 threatened species, 58 subject to special protection, and 12 that are in danger of extinction (DOF 2002, 2010), with a high degree of endemism. The dry forests of Huatulco are part of one of the nine highest priority areas for conservation in Central America (i.e., including Southern Mexico, Belize, Guatemala, Honduras, Costa Rica, Nicaragua, Panama, and El Salvador).

The coral communities of the site are made up of 12 species, including genus *Pocillopora*, and these communities support a total of 121 fish species, with the vast majority carnivorous, and a few that feed on plankton and algae, as well as upon coral, such as *Prionurus puncatus*, which helps to regulate the composition and structure of the algal community of reefs and allows them to recover from damage/disturbances (Ramos 2003).

The terrestrial portion of the site supports a wealth of plant species that in turn support a high diversity of reptile, bird, and mammal species. Vegetation associations at the site include deciduous forest, coastal dunes, riparian areas, savannah, mangroves, and wetlands. Notable species include cuachalalate (Amphipterygium adstringens), palo de arco (Apoplanesia paniculata), and gumbo limbo (Bursera simaruba). Among these notable plants, the site also supports a total of 78 families, 289 genera, and 429 species of plants, including 72 legume species, 34 species of the Euphorbiaceae, and 19 species of grass. This diverse plant assemblage supports 282 species of birds, 71 species of reptiles, 15 species of amphibians, and 130 species of mammals. Floral species at the site include the Acanthaceae, Burseraceae, Cactaceae, Combretaceae, Rhizophoraceae, among others, along with special protection species *Pterocereus gaumeri*, *Conocarpus erectus*, *Laguncularia racemosa*, *Rhizophora mangle*, and *Guaiacum coulteri*; and the endemic species *Achatocarpus oaxacanus*, *Manfreda* sp., and *Aeschynomene sousae*.

There are 92 observed fauna species at the site, including Wiegmann's toad (Bufo marmoreus), the yellow tree frog (Hyla sartori), Sinaloa wren (Thryophilus sinaloa), golden-cheeked woodpecker (Melanerpes chrysogenys), west Mexican chachalaca (Ortalis poliocephala), dolphins (Stenella attenuate and S. longirostris), the pygmy killer whale (Feresa attenuata), the false killer whale (Pseudorca crassidens), the gray dolphin (Grampus griseus), the humpback whale (Megaptera novaeangliae), and the long-finned pilot whale (Globicephala macrorhynchus).

The coral species *Pocillopora eydouxi* forms isolated colonies at Playa Violín and Bahía Chachacual, and constitutes one of the only records of this species for the Mexican Pacific. Along the rocky coast of the site, in addition to the purple snail (Plicopurpura pansa), there are a number of species of gastropod molluscs that are endemic to the site (Arene hindsiana, Calliosthoma aequisculptum, Rissoina stricta, Cerithium maculosum, Crucibulum monticulus, Anachis ritteri, and Costoanachis sanfelipensis).

The site resides in a region that historically fell under the Mixtec reign of Tututepec, which after the conquest was the seat of one of the first ports of New Spain in 1539, as an important link for trade with Peru, the rest of South America, China, and the Philippines. Historic use of snail species' locally sourced pigments dates back to pre-Hispanic times, as the pigments are considered to convey fertility and power. The inhabitants of the area also have years of tradition as fishers, in the sea as well as in rivers and lagoons. Fishing is traditionally for self-consumption and local sale. Other agriculture and tourism activities are also historically noted and still practiced at the site.

The site falls under the Ministry of Agriculture, Livestock, and Fisheries (SAGARPA); the Ministry of Communications and Transportation for vessels; and the Directorate of the National Park. Beaches and other areas subject to flooding by seawater exist in a strip that is 20 m wide, from the maximum flood level, which are also federal property and their administration corresponds to the General Directorate of the Federal Maritime Terrestrial Zone and the Secretariat of Environment and Natural Resources (SEMARNAT).

Fishing production cooperatives in the area are organized among the community members, with prawns as a major commodity. Sport fishing is practiced during sport fishing tournaments as well, held annually in the area. Other human activities at the site include light industry concentrated in small workshops where handicrafts are made with clay, wood, and liana; small-scale textile workshops; mezcal distilleries; as well as tortilla shops and coffee processors. The tourist industry is one of the activities with the greatest presence within the site, and neighboring areas, which support restaurant and other tourism services; merchants; and excursion businesses (i.e., for snorkeling, diving, swimming, boat rides, and kayaking). Beginning in the 1980s, the federal government began development of a tourism megaproject called Bahías de Huatulco, which involved, in addition to the expropriation of 20,000 ha of land from the community of Santa Maria Huatulco, a series of modifications to the environmental and socioeconomic conditions of the area, including the change in land use in various areas, the alteration of the natural drainage networks, causing a growing demand for space and resources for the urban and commercial development in the area. Thus, and in general, the construction of infrastructure for tourism is the main threat to site, since estuaries have been converted into marinas, and the streams of the area have been channeled. Docks for cruise ships and golf courses are all part of large tourism infrastructure plans for the site.

Two main research lines at the site have been established: (1) management of non-timber species in the site's dry forests, for artisanal use (Sánchez and Acevedo 2001); and (2) an herbarium, through which a collection of the flora of the region is being built, in collaboration with the Huatulco National Park (Gordon et al. 2003), which further enables partnerships with universities (e.g., the UNAM) and specialized research centers. The activities that are linked to the Huatulco National Park, in coordination with the UNAM, also focus on environmental education and studies of medicinal plants, which enables workshops for authorities, community technicians, and community members on the management of resources at the site, and throughout the region.

Due to the tourism developments since the 1980s, as mentioned above, tourism at the site is in the hundreds of thousands of visitors per year, with hundreds of associated vessels, multiple diving companies, and an indeterminate number of restaurants located on the main beaches and bays of the area. Similarly, in the terrestrial areas of the site, tours are carried out on foot, on all-terrain vehicles, and on bicycles, promoted by companies and individuals for the purpose of observing the local landscape, flora, and fauna.

PLAYA BARRA DE LA CRUZ, OAXACA: OVERVIEW

Playa Barra de la Cruz (Figures 7.6 and 7.7) consists of sandy beaches that are key habitat for three marine turtle species: *Dermochelys coriacea*, *Lepidochelys olivacea*, and *Chelonia mydas*. The site also supports a diversity of vertebrates, including migratory birds such as *Pelecanus occidentalis*, *Charadrius* sp., and *Sterna* sp.; resident bird species such as *Fregata magnificens*, *Larus argentatus*, *Casmerodius albus*, and *Phalacrocorax* sp.; a number of small mammals such as coati (Nasua), racoon (Procyon lotor), and American hog-nosed skunk (Conepatus mesoleucus); as well as numerous reptiles, amphibians, and fish species. Common bottlenose dolphins (Tursiops truncatus) are regularly observed close to the coast and groups of humpback whales (Megaptera novaeangliae) can be seen during December through January, and then again in March through April. The beach of Barra de la Cruz is located in the southern portion of the State of Oaxaca, bordering the small towns of Barra de la Cruz and Playa Grande, a sandy and pebble beach type ideal for turtle nesting.

SITE DETAILS

Given the global trends in the decline of sea turtle numbers, the site is critical habitat for the leatherback sea turtle (Dermochelys coriacea), which is considered a critically endangered species by the IUCN. The other two species of sea turtles also nesting at this site, the olive ridley sea turtle (Lepidochelys olivacea) and the green sea turtle (Chelonia mydas), are both considered endangered in the IUCN Red List, and in NOM-059-ECOL-1994. Vegetation at the site that is listed in NOM-059-ECOL-1994, classified as endangered, are red mangrove (Rhizophora mangle), black mangrove (Avicennia germinans), and button mangrove (Conocarpus erectus) [Rzedowski 1978]. The beach at the site is very dynamic and high-energy, with substantial changes in the physiognomy, a result of strong winds and tides (Alvarado 1991).

The predominant vegetation communities at the site are jungle, cultivated forest, and coastal dune; clearing in the jungle areas has resulted in secondary vegetation that is typically thorny forest, primarily mesquite (Prosopis juliflora) with cacti of the genus *Stenocereus* (Rzedowski 1978). The predominant vegetation along the sandy beach area of the site includes creeping plants, mainly bayhops (Ipomoea pes-caprae) and patches of saltgrass (Distichlis spicata). Cacti of genus *Opuntia* are

FIGURE 7.6 Composite aerial image of Playa Barra de la Cruz. February 02, 2008; Area: 18 ha; Coordinates: 15°50′N 95°56′W; Ramsar Site number: 1,821.

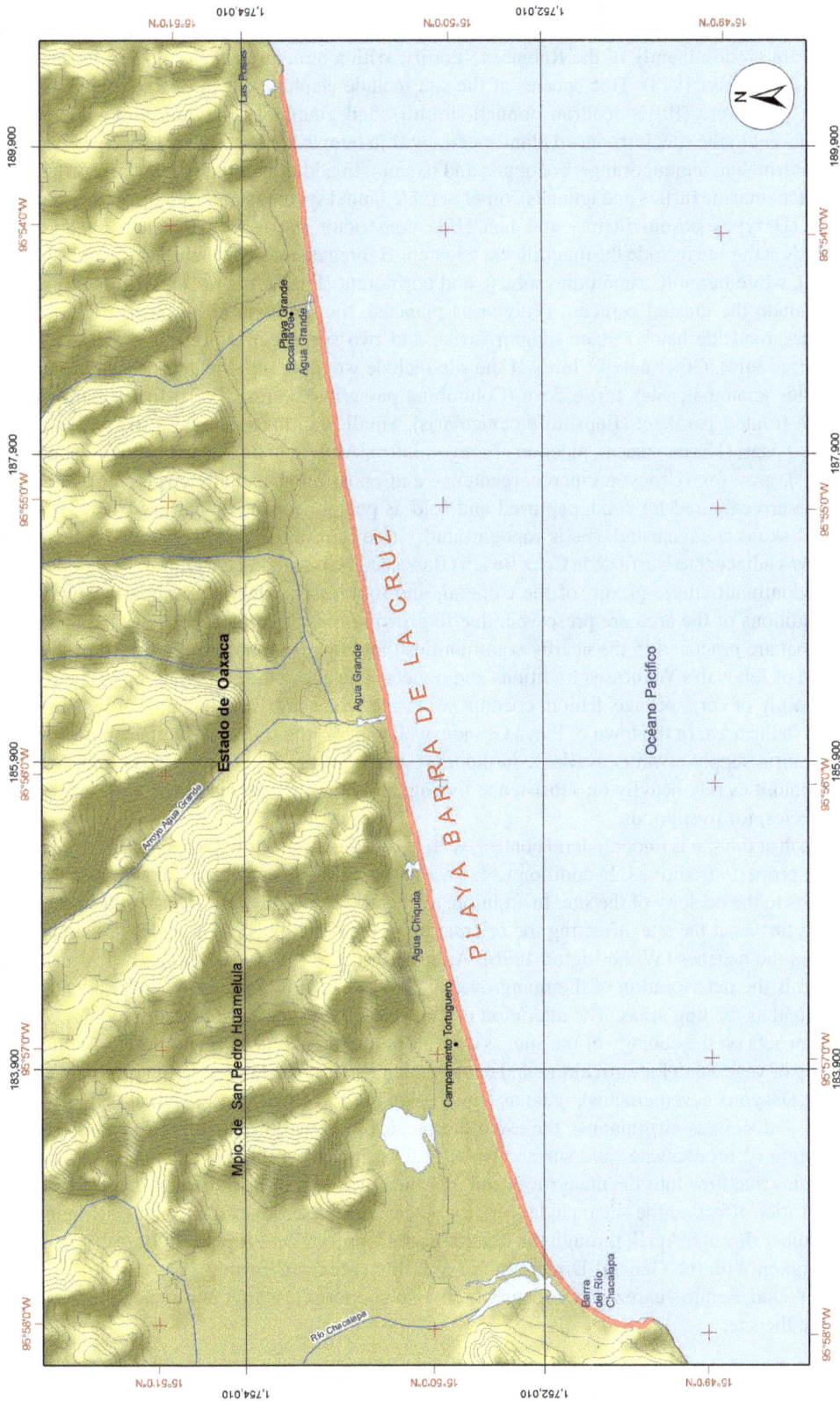

FIGURE 7.7 Map of Playa Barra de la Cruz. February 02, 2008; Area: 18 ha; Coordinates: 15°50′N 95°56′W; Ramsar Site number: 1,821.

also present in the dune areas of the site, with the upland extents of the site populated by endemic plant species such as *Melocactus delessertianus* and *Diospyros oaxacana*. Shrubs in the dune areas of the site are predominantly of the Rubiaceae family, with a number of areas cleared for grazing livestock (Rzedowski 1978). Tree species at the site include elephant-ear tree (Enterolobium cyclocarpum), primavera (Roseodendron donnell-smithii), and guapinol (Platymiscium dimorphandrum). Adjacent to the site, introduced plant species within human settlements include corn, papaya, mango, watermelon, lemon, orange, coconuts, and banana. In addition to the species of reptiles that use the beach (marine turtles and iguanas), other notable faunal species supported by the site include armadillo (Dasypus novemcinctus) and boa (Boa constrictor imperator). Resident aquatic and marine birds at the site include the magnificent frigatebird (Fregata magnificens), herring gull (Larus argentatus), white heron (Casmerodius albus), and cormorant (Phalacrocorax sp.). Birds of prey at the site include the crested caracara (Polyborus plancus), the common black hawk (Buteogallus anthracinus), roadside hawk (Buteo magnirostris), and two species of vulture (Coragyps atratus and Cathartes aura). Other notable birds at the site include woodpecker (Melanerpes aurifrons and Campephilus guatemalensis), turtle dove (Columbina passerina), cardinal (Cardinalis cardinalis), and orange-fronted parakeet (Eupsittula canicularis). Small mammals of note at the site include white-nosed coati (Nasua narica), raccoon (Procyon lotor), American hog-nosed skunk (Conepatus mesoleucus), gray fox (Urocyon cinereoargenteus), and opossum (Didelphis virginiana). Some of these fauna are captured for food, captured and sold as pets, or for medicinal use (such as is the case for the skunks). Jaguarundi (Felis yagouaroundi) have occasionally been observed at the site.

The towns adjacent to Barra de la Cruz Beach (Barra de la Cruz and Playa Grande) are populated by the predominant ethnic groups of the Chontal, and to a lesser extent the Zapotec. Few of the ancient traditions of the area are preserved, due to growing outside influences; however, there are religions that are practiced in the nearby communities that draw from both Catholic Apostolic and the Church of Jehovah's Witnesses traditions and beliefs. The agricultural production of the region consists mainly of corn, orange, lemon, coconut, watermelon, papaya, and mango; cattle and goats are grazed in the area. In the town of Playa Grande, well water is the source of human consumption, with the central supply of water available in the largest town of the area, Barra de la Cruz. Both of these communities rely heavily on subsistence fishing, and hunting of some of the aforementioned animal species, for livelihoods.

The beach at the site is under federal control, with land tenure in the towns of the area based upon communal property traditions. In addition to extraction of sea turtle eggs, tourism activities are the main threats to the ecology of the site. In addition, artificial (i.e., urban) light impacts the success of nesting sea turtles at the site, affecting the orientation of the hatchlings and females that come out to spawn on the beaches (Witherington 1996). Another threat to the site, along the beach of Barra de la Cruz, is the deterioration of the mangroves, which are used by migratory and resident birds as refugia and as feeding areas. The alteration of mangrove habitat endangers many species. Other negative impacts on the ecology of the site, as a result of human activities surrounding the site, are the clearing of vegetation for agriculture and cattle grazing; overhunting of some species such as the armadillo (Dasypus novemcinctus), iguanas (Iguana iguana and Ctenosaura pectinata), and white-tailed deer (Odocoileus virginianus); the use of herbicides in agriculture that impacts soil fertility; contamination of groundwater and surface water bodies, including the use of laundry detergents in the streams that flow into the mangroves and into the sea; and indiscriminate garbage dumping.

Conservation efforts at the site include work to protect and conserve sea turtles that lay their eggs during October through April, through the Leatherback Conservation Program of Barra de la Cruz, in collaboration with the General Directorate of Wildlife (Mexican Turtle Center), SEMARNAT, and Universidad Benito Juárez of Oaxaca, which also supports research and education/outreach activities at the site.

SISTEMA ESTUARINO PUERTO ARISTA, CHIAPAS: OVERVIEW

Sistema Estuarino Puerto Arista (Figures 7.8 and 7.9) is an estuarine coastal wetland. The main floral species at the site include *Pithecellobium dulce*, *Coccoloba caracasana*, *Acanthocereus pentagonus*, and *Bursera excelsa*. Mangrove species at the site include *Rhizophora mangle*, *Laguncularia racemosa*, *Avicennia germinans*, and *Conocarpus erectus*. The site's beaches are used as nesting grounds by endangered marine turtles *Lepidochelys olivacea*, *Dermochelys coriacea*, and *Chelonia mydas*. The site also supports many resident and migratory waterfowl, such as *Recurvirostra americana*, *Tringa flavipes*, *Tringa melanoleuca*, and *Calidris minutilla*. Subsistence fishing, tourism, and exploitation of marine turtles are current threats to the ecology of the site.

SITE DETAILS

The site is located in the Istmo Costa Region of the State of Chiapas, in the municipality of Tonalá. All of the communities in the area are accessed by highway, 19 km from Puerto Arista. There are various species of notable vertebrates at the site, such as raccoon (Procyon lotor), northern tamandua (Tamandua mexicana), ocelot (Leopardus pardalis), jaguarundi (Felis yagouaroundi), and *Caiman crocodilus fuscus*, the latter of which is vulnerable per the IUCN and the CITES; the American crocodile (Crocodylus acutus) is under special protection per NOM-059-SEMARNAT-2010. Other species of concern at the site are green iguana (Iguana iguana), listed in the CITES and under special protection per NOM-059-SEMARNAT-2010; and the Mexican spiny-tailed iguana (Ctenosaura pectinata), which is threatened per NOM-059-SEMARNAT-2010. Freshwater turtles like *Trachemys scripta grayi* and the giant musk turtle (Staurotypus salvinii) are also under special protection per NOM-059-SEMARNAT-2010. The red-cheeked mud turtle (Kinosternon scorpioides cruentatum) is under special protection per NOM-059-SEMARNAT-2010 and are declining in the area, likely as a consequence of poaching. The site supports a large number of resident and migratory bird species, both terrestrial and aquatic, which are in some category of risk per NOM-059-SEMARNAT-2010, among which are: American avocet (Recurvirostra americana), lesser yellowlegs (Tringa flavipes), greater yellowlegs (Tringa melanoleuca), least sandpiper (Calidris minutilla), short-billed dowitcher (Limnodromus griseus), marbled godwit (Limosa fedoa), and Wilson's phalarope (Steganopus tricolor). The beaches of Puerto Arista known also as "Puerto Arista Tortuguero Camp or Zone for the Protection and Conservation of Sea Turtles" (DOF Decree dated October 29, 1986), are considered to be among the 17 most important nesting beaches for the olive ridley sea turtle (Lepidochelys olivacea) in the Mexican Pacific.

Key migratory birds at the site include the wood stork (Mycteria americana), reddish egret (Egretta rufescens), great blue heron (Ardea herodias), bare-throated tiger-heron (Tigrisoma mexicanum), red-billed tropicbird (Phaethon aethereus), least storm-petrel (Oceanodroma microsoma), agami heron (Agamia agami), gray-headed hawk (Leptodon cayanensis), Harris' hawk (Parabuteo unicinctus), and the aplomado falcon (Falco femoralis). The mangrove and lagoon ecosystem of the site support the main species of the coastal fishery of the area, such as long-arm river prawn (Macrobrachium tenellum), cauque river prawn (Macrobrachium americanum), guavina (Guavina guavina), common snook (Centropomus undecimalis), mullet (Mugil curema and M. cephalus), rhomboid mojarra (Diapterus rhombeus), skipjack (Katsuwonus pelamos), dorado (Coryphaena hippurus), tuna (Thunnus thynnus), blacktip shark (Carcharhinus limbatus), blue shark (Galeocerdo cuvier), dogfish (Charchahrinus porosus), sierra (Scomberomorus maculatus), hammer or horned shark (Sphyrna lewini), whiteleg shrimp (Litopenaeus vannamei), and blue shrimp (L. stylirostris).

Key floral species of the site include madras thorn (Pithecellobium dulce), papaturro (Coccoloba caracasana), chaco (*Acanthocereus pentagonus*), copal (Bursera excelsa), indigoberry (Randia armata), Mexican alvaradoa (Alvaradoa amorphoides), the euphorb *Coccoloba floribunda*, mesquite (Prosopis juliflora), bayhops (Ipomoea pes-caprae), croton draco (Coccoloba barbadensis),

FIGURE 7.8 Composite aerial image of Sistema Estuarino Puerto Arista. Designated February 02, 2008; Area: 62,138 ha; Coordinates: 16°00′05.8″N 93°53′06″W; Ramsar Site number: 1,823.

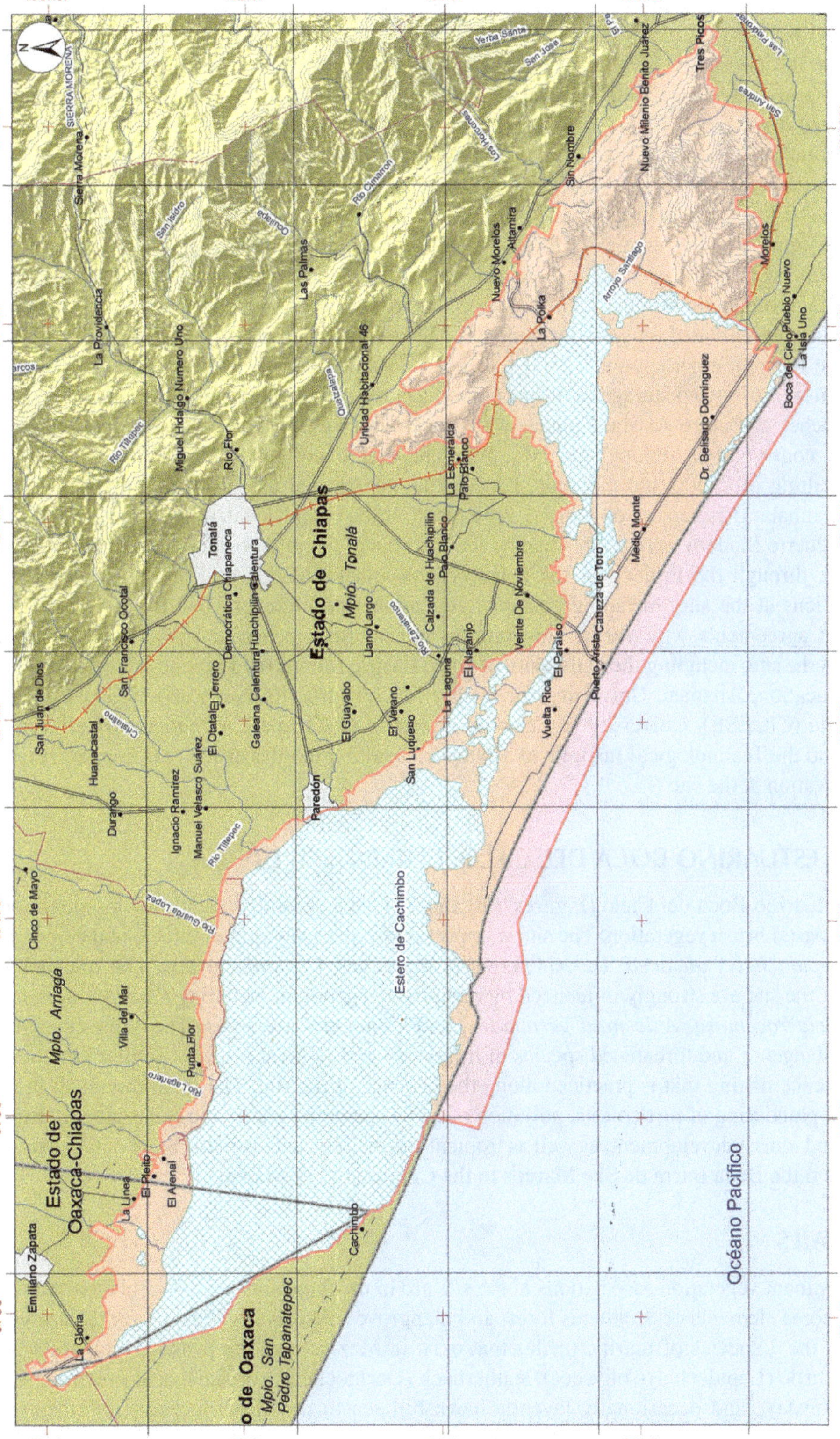

FIGURE 7.9 Map of Sistema Estuarino Puerto Arista. Designated February 02, 2008; Area: 62,138 ha; Coordinates: 16°00'05.8"N 93°53'06"W; Ramsar Site number: 1,823.

and the grasses *Jouvea pilosa* and *Sporobolus domingensis*. The four mangrove species present at the site are *Rhizophora mangle*, *Laguncularia racemosa*, *Avicennia germinans*, and *Conocarpus erectus* (IHNE 2004).

Resident and migratory bird species at the site, both terrestrial and aquatic, include the wood stork (Mycteria americana), reddish egret (Egretta rufescens), great blue heron (Ardea herodias), bare-throated tiger-heron (Tigrisoma mexicanum), red-billed tropicbird (Phaethon aethereus), least storm-petrel (Oceanodroma microsoma), agami heron (Agamia agami), gray-headed hawk (Leptodon cayanensis), Harris' hawk (Parabuteo unicinctus), aplomado falcon (Falco femoralis), as well as an assortment of amphibians and reptiles.

Adverse impacts on the ecology of the site include looting of the nests that sea turtles deposit on the beach; killing of nesting females on the beaches to obtain meat, eggs, and oil; predation of nests by wild and introduced animals (mainly dogs); fishing bycatch on trammel nets; poorly planned infrastructure; unregulated tourism development; the harvesting of mangroves and palms; deforestation of the surrounding areas and the site for the expansion of livestock and agriculture; poaching, and the transportation and storage of drugs; as well as unregulated logging (IHNE 2004).

The beaches of Puerto Arista, Chiapas were declared a Reserve Zone and Refuge Site for the protection, conservation, repopulation, development, and control of the various species of sea turtles, resulting in a total and indefinite ban on the commercial use of the seven species of sea turtles that inhabit these coasts of Mexico. As a result, three camps located in Puerto Arista, Costa Azul, and Puerto Madero were established by the Institute of Natural History and Ecology (IHNE). The IHNE, through the Project for the Conservation and Protection of Sea Turtles, carries out various actions at the site, including research, education, and outreach (IHNE 2004). IHNE has also signed agreements with various institutions and public and private universities to develop research on the site, including the University of Guadalajara, Monterrey Institute of Technology and Higher Education, Anahuac University, the Center for Scientific Research and Higher Education of Ensenada (CICESE), University of Sciences and Arts of Chiapas, Autonomous University of Chiapas, and the Technological Institute of Tuxtepec, resulting in infrastructure to support research and conservation at the site.

SISTEMA ESTUARINO BOCA DEL CIELO, CHIAPAS: OVERVIEW

Sistema Estuarino Boca del Cielo (Figures 7.10 and 7.11) is a coastal wetland with predominantly dune and coastal brush vegetation. The site is important for the spawning of three endangered turtle species: *Lepidochelys olivacea*, *Dermochelys coriacea*, and *Chelonia mydas*. The marshes and channels of the site are strongly influenced by mangrove vegetation, including *Rhizophora mangle*, *Laguncularia racemosa*, *Avicennia germinans*, and *Conocarpus erectus*. The site's ecosystems support endangered and threatened species of migratory and resident birds, as well as commercial and subsistence fishing that is practiced along the beaches of the site. The main threats to the site include the plundering of turtle nests, pollution of turtle spawning zones, the construction of infrastructure and tourist development, as well as tropical storms. The area is in the State of Chiapas, and extends from the Boca Barra de San Marcos to the Community of Majahual.

SITE DETAILS

The predominant vegetation associations at the site are in the dune and coastal scrub communities, as well as some elements of deciduous forest and mangroves. At this key location for the spawning of three of the 7 species of marine turtles that exist in Mexico, the site primarily supports olive ridley sea turtle (Lepidochelys olivacea), leatherback (Dermochelys coriacea), and green sea turtle (Chelonia mydas), and occasionally juvenile hawksbill sea turtles (Eretmochelys imbricata) that forage inside the site's lagoon. All of these species are protected per the CITES and NOM-059-SEMARNAT-2010. The beaches of Boca del Cielo are located within what is referred to as a Priority

FIGURE 7.10 Composite aerial image of Sistema Estuarino Boca del Cielo. Designated February 02, 2008; Area: 8,931 ha; Coordinates: 15°48'N 93°35'W; Ramsar Site number: 1,770.

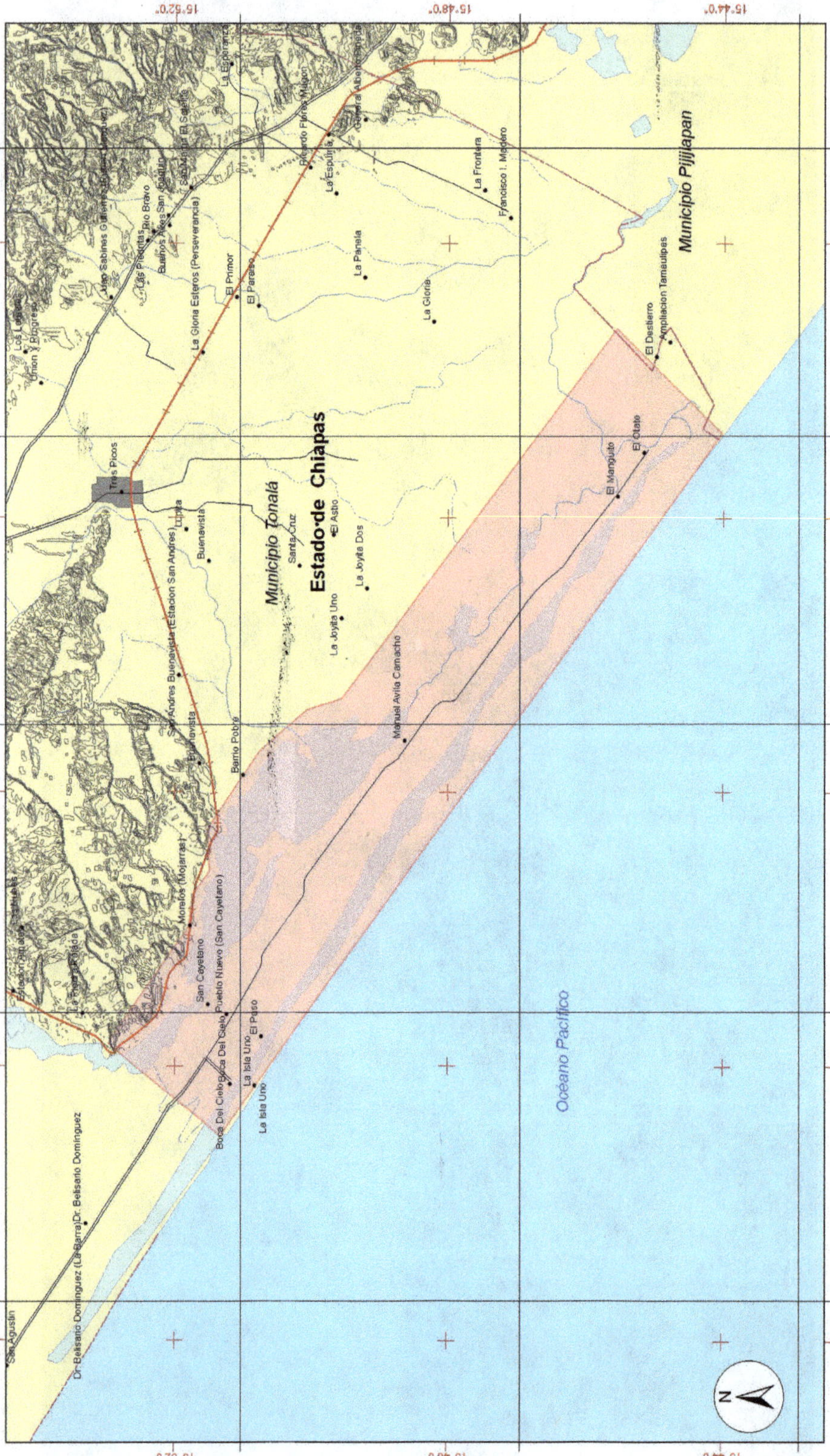

FIGURE 7.11   Map of Sistema Estuarino Boca del Cielo. Designated February 02, 2008; Area: 8,931 ha; Coordinates: 15°48'N 93°35'W; Ramsar Site number: 1,770.

Marine Region, called Port Arista, established by the National Commission for the Knowledge and Use of Biodiversity (Arriaga et al. 2002). There are a number of other key species at the site, including raccoon (Procyon lotor); northern tamandua (Tamandua mexicana); ocelot (Leopardus pardalis); jaguarundi (Felis yagouaroundi); *Caiman crocodilus fuscus*, which is vulnerable per the IUCN and the CITES; and the American crocodile (Crocodylus acutus), which is under special protection per NOM-059-SEMARNAT-2010. Additionally, the site supports the green iguana (Iguana iguana) and the Mexican spiny-tailed iguana (Ctenosaura pectinata), which are under special protection and threatened, respectively, per the IUCN, the CITES, and NOM-059-SEMARNAT-2010. Freshwater turtles like *Trachemys scripta grayi*, giant musk turtle (Staurotypus salvinii), and the red-cheeked mud turtle (Kinosternon scorpioides cruentatum), which are all under special protection per NOM-059-SEMARNAT-2010, are less frequent at the site, likely due to poaching and other predation. Seabirds of note at the site are the American white pelican (Pelecanus erythrorhynchos), gulls (Larus atricilla, L. pipixcan, and L. argentatus), and the magnificent frigatebird (Fregata magnificens). Other site species at risk per NOM-059-SEMARNAT-2010 are: agami heron (Agamia agami) [special protection], Muscovy duck (Cairina moschata) [endangered], and wood stork (Mycteria americana) [special protection]. Additional resident and migratory birds at the site are reddish egret (Egretta rufescens), great blue heron (Ardea herodias), bare-throated tiger-heron (Tigrisoma mexicanum), red-billed tropicbird (Phaethon aethereus), least storm-petrel (Oceanodroma microsoma), gray-headed hawk (Leptodon cayanensis), Harris' hawk (Parabuteo unicinctus), and the aplomado falcon (Falco femoralis).

The lagoon mangrove ecosystems at the site support feeding, reproduction, and breeding habitat for the main aquatic species that support the coastal fishery of the region, including the long-arm river prawn (Macrobrachium tenellum), cauque river prawn (Macrobrachium americanum), guavina (Guavina guavina), common snook (Centropomus undecimalis), mullets (Mugil curema and M. cephalus), rhomboid mojarra (Diapterus rhombeus), skipjack (Katsuwonus pelamos), dorado (Coryphaena hippurus), tuna (Thunnus thynnus), blacktip shark (Carcharhinus limbatus), blue shark (Galeocerdo cuvier), dogfish (Charchahrinus porosus), sierra (Scomberomorus maculatus), hammer or horned shark (Sphyrna lewini), whiteleg shrimp (Litopenaeus vannamei), and blue shrimp (L. stylirostris).

The Boca del Cielo Estuary is located within two basins: La Joya Basin and its main Basin that contains the Agua Dulce, Yerba Santa, El Pedregal, and Los Horcones Rivers. The Agua Dulce River is made up of four tributaries from the upper part of the basin, which flow into the La Joya Lagoon. There is a significant contribution of surface fresh water during the rainy season, through rivers and lagoon ecosystems, with monthly average precipitation at site that exceeds 750 mm (SEMARNAP 2000).

A number of invasive plant species affect the site, including bayhops (Ipomoea pes-caprae), croton draco (Coccoloba barbadensis), and *Sporobolus domingensis*. The estuaries and canals that run through and interconnect the towns of Majahual, El Manguito, Ponte Duro, Buenavista, La Barra, Boca del Cielo, and Isla San Marcos are strongly influenced by their mangroves' biological diversity, with 4 species of mangrove present there: red mangrove (Rhizophora mangle), white mangrove (Laguncularia racemosa), black mangrove (Avicennia germinans), and button mangrove (Conocarpus erectus). Within the deciduous forests of the site, noted species are monkey pod (Pithecellobium sp.), papaturro (Coccoloba caracasana), chaco (Acanthocereus pentagonus), copal (Bursera excelsa), indigoberry (Randia armata), Mexican alvaradoa (Alvaradoa amorphoides), the euphorb *Coccoloba floribunda*, and mesquite (Prosopis juliflora). Much of this vegetation has been disturbed by agricultural activities in the area, which has converted it into areas for livestock, coconut palm, cane, and banana crops. Within the aquatic open water areas of the site, there exist floating aquatic vegetation, where water lettuce (Pistia stratiotes) and gibbous duckweed (Lemna gibba) predominate.

Other predominant floral species supported by the site include southern cattail (Typha domingensis), nanche (Byrsonima crassifolia), Mexican calabash (Crescentia alata), Guiana chestnut (Pachira aquatica), stinking toe (Hymenaea courbaril), trumpet-tree (Cecropia obtusifolia), strawberry tree

(Muntingia calabura), common water hyacinth (Eichhornia crassipes), *Ficus glabrata*, primavera (Roseodendron donnell-smithii), elephant-ear tree (Enterolobium cyclocarpum), kapok tree (Ceiba pentandra), coconut (Cocos nucifera), the euphorb *Jatropha curcas*, cucumber tree (Parmenteria aculeata), sweet acacia (Acacia farnesiana), west Indian elm (Guazuma ulmifolia), gumbo limbo (Bursera simaruba), *Lonchocarpus rugosus*, and *Eysenhardtia adenostyles*.

Notable faunal species supported by the site that are not mentioned above include the red-billed tropicbird (Phaethon aethereus), American avocet (Recurvirostra americana), lesser yellowlegs (Tringa flavipes), greater yellow legs (Tringa melanoleuca), least sandpiper (Calidris minutilla), short-billed dowitcher (Limnodromus griseus), marbled godwit (Limosa fedoa), Wilson's phalarope (Steganopus tricolor), neotropical cormorant (Phalacrocorax brasilianus), magnificent frigatebird (Fregata magnificens), American coot (Fulica americana), black-necked stilt (Himantopus mexicanus), American white pelican (Pelecanus erythrorhynchos), laughing gull (Larus atricilla), and blue-winged teal (Spatula discors) [NAWCC, IHNE, CONANP, IDESMAC 2002a, 2002b], as well as a number of amphibians and reptiles.

The communities surrounding the beaches of Boca del Cielo are known to engage in both marine and freshwater subsistence fishing, with small commercial fishing enterprises existing in the area; rainfed agriculture; extensive livestock ranching; clandestine logging of mangroves; poaching; and tourist services (e.g., restaurants and palapas). Some of the adverse impacts on the site include poaching or hunting of turtles and their eggs; bycatch of turtles on trammel nets; poorly planned infrastructure; and transport/storage of narcotics (IHNE 2006).

As outlined for the previous site, since 1994 the Institute of Natural History and Ecology has led the efforts to develop operations and infrastructure in this region to protect and conserve turtles along a large proportion of the coast of the State of Chiapas. Accordingly, the Institute of Natural History and Ecology carries out a number of important monitoring, research, education, and outreach activities in the region, positively impacting the ecology of this and several other coastal wetland sites. Communities in which the Institute of Natural History and Ecology has done education/outreach work include Boca del Cielo, Belisario Domínguez (known as La Barra), and Miguel Ávila Camacho. Tourist and recreational activities at this and other sites of the region are substantial because they are traditional beach destinations, which provide for a large influx of visitors during Holy Week festivities, as well as during summer and winter holidays.

RESERVA DE LA BIOSFERA LA ENCRUCIJADA, CHIAPAS: OVERVIEW

Reserva de la Biosfera La Encrucijada (Figures 7.12 and 7.13) is composed of coastal lagoons, swamps, and marshes forming the largest area of mangrove forest in the North American Pacific; the only Guiana chestnut (Pachira aquatica) woodland in Mesoamerica. The site supports a large variety of wildlife that is threatened, endemic, rare, or in danger of extinction. A total of 183 bird species have been reported at the site, and it is a temporary and seasonal habitat for numerous migratory species. Nine of 19 vegetative associations recorded for Chiapas are present at the site. Archaeological remains from 5,500 years ago are found here, where the present human activities include an important commercial fishery, slash-and-burn agriculture, extensive cattle ranching, and various exploitation of the area's fauna. Sedimentation due to poorly planned water projects, deforestation, and slash-and-burn agriculture impact the ecology of the lagoons and lakes of the area.

SITE DETAILS

Seventeen rivers discharge into the site's wetlands, which consist of an extensive area of swamps. The site supports 210 resident bird species, in addition to 94 migratory bird species from the United States and Canada, during the winter. Located in the south of the State of Chiapas, Municipalities of Pijijiapan, Mapastepec, Acapetahua, Huixtla, Villa Comaltitlán, and Mazatlán, the site is comprised of coastal lagoons, estuaries, sandbars, and marshes. Coastal lagoons at the site are an average

FIGURE 7.12 Composite aerial image of Reserva de la Biosfera La Encrucijada. March 20, 1996; Area: 144,868 ha; Coordinates: 15°11′N 092°53′W; Ramsar Site number: 815.

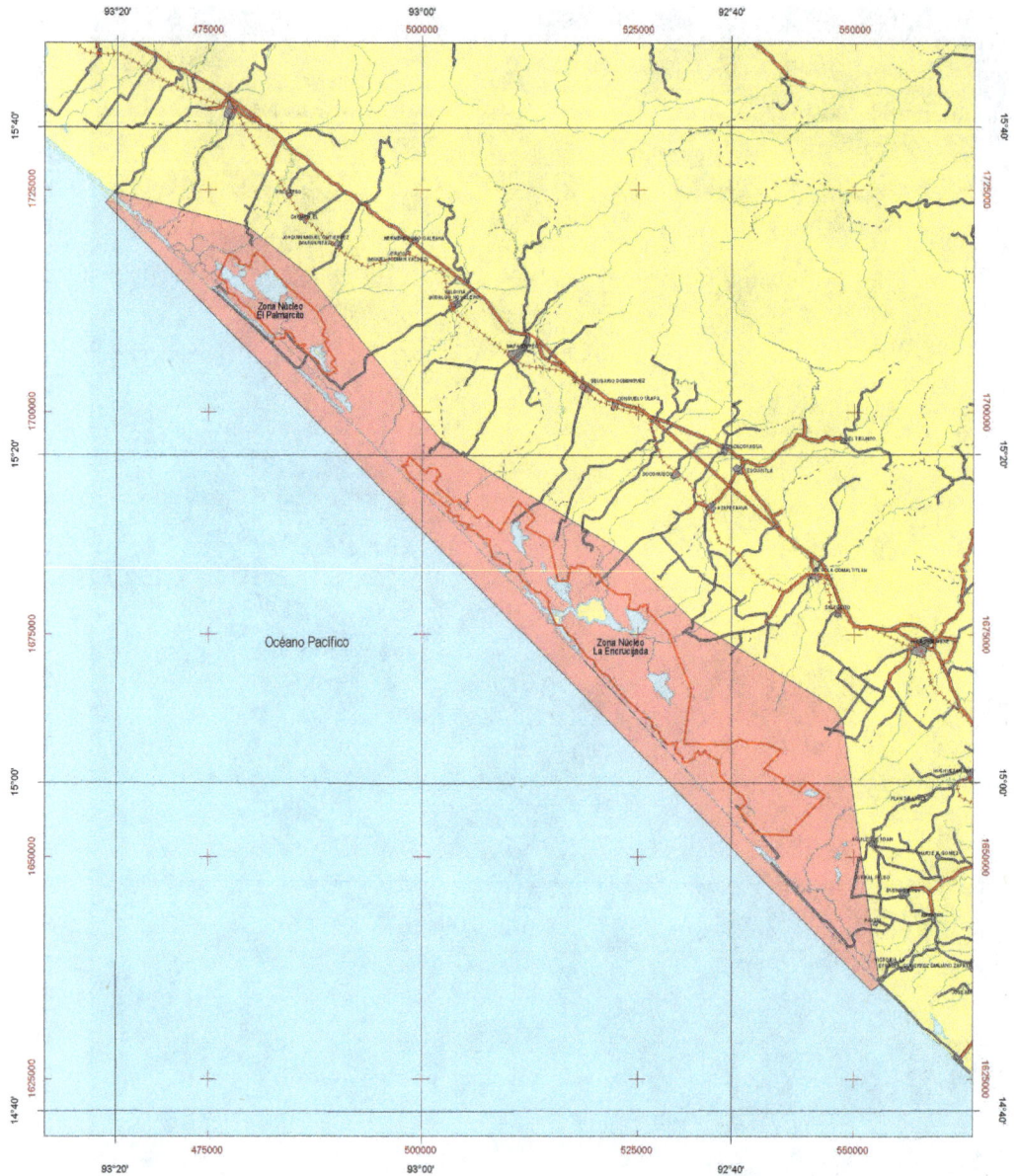

FIGURE 7.13   Map of Reserva de la Biosfera La Encrucijada. March 20, 1996; Area: 144,868 ha; Coordinates: 15°11'N 092°53'W; Ramsar Site number: 815.

depth of between 50 cm and 1.5 m, seasonally fluctuating. Minimum annual rainfall at the site is 1,300 mm, with a maximum of 3,000 mm, distributed between 100 and 200 rainy days per year.

Main floral species of the site are coyol (Acrocomia mexicana), white mangrove (Laguncularia racemosa), button mangrove (Conocarpus erectus), red mangrove (Rhizophora mangle), black mangrove (Avicennia germinans), and the hybrid mangrove *Rhizophora* x *harrisonii*. Main faunal species at the site are jaguar (Panthera onca), spider monkey (Ateles geoffroyi), the agami heron (Agamia agami), giant wren (Campylorhynchus chiapensis), boa (Boa constrictor), leatherback sea turtle (Dermochelys coriacea), green sea turtle (Chelonia mydas), olive ridley sea turtle (Lepidochelys olivacea), scorpion mud turtle (Kinosternon scorpioides), giant musk turtle (Staurotypus salvinii), giant green anole (Anolis biporcatus), Guatemalan helmeted basilisk (Corytophanes percarinatus), and the green iguana (Iguana iguana).

Human activities at the site include fishing, agriculture, and the raising of livestock. Primary crops in the area are corn, beans, watermelon, cucumber, banana, cashew, African palm, mango, sugar cane, and papaya. Fishing within the Carretas Pereyra, Chantuto Panzacola, and Maragato la Cantileña allow for organized groups, as fishing cooperatives, for the extraction of whiteleg shrimp (Litopenaeus vannamei) and scale fish. Fishers use temporarily installed tapos, or rustic enclosures, in the coastal lagoons to facilitate production of shrimp and fish species of high commercial value. More than 95% of communities in the area have basic services (water, telephone, and electricity), with the exception being the municipality of Mapastepec, where electricity is less reliable. Land tenure in the area is a mix of federal, state, and local/private-ownership.

Historical use of precious woods and the pre-Hispanic cultivation of cocoa gave way to agriculture and livestock activities, however natural flooding of the region necessitated a shift to the production of bovine meat and milk. Further south, in the region known as Soconusco, newly cleared areas were developed into banana farms. In recent years, there has been some diversification of crops, with mango, sugar cane, papaya, African palm, and to a lesser extent cashew plantations. Negative impacts on the ecology of the site include those related to different social, cultural, economic, and, to a certain extent, political drivers. The impacts are primarily from the transformation of forests, mangroves, and tules for the establishment of cattle and agricultural lands; the felling of trees to obtain wood for the construction of houses or use as firewood; forest fires caused by the extraction of wildlife; slash-and-burn clearing; unregulated human settlements; overexploitation of animal species, many of them threatened or endangered; contamination with organic and inorganic waste dumped into the channels of estuaries and coastal lagoons; construction of hydraulic engineering projects that result in diverting rivers, opening channels, constructing drains, and draining of wetlands; dredging of rivers and canals; increased/unsustainable fishing; and increased/unsustainable shrimp farming in the area.

The site is within an established Biosphere Reserve. Subsequently, in 1988, the Secretary of Development and Ecology, and the State Government's Institute of Natural History and Ecology (IHNE) began efforts to expand the site, and with that came additional conservation focus. Additionally, the State government reinforced legal measures at the site, as a zone subject to ecological protection, in addition to participating in its administration, management, and operational aspects. The site has since established collaborative agreements with Universidad Autónoma Metropolitana, Colegio del Frontera Sur, Universidad Autónoma de Chapingo, and the University of Science and Arts of Chiapas, for joint development of studies on water quality, community regulations, agrochemical contamination, land use, and socioeconomics. The site also has a visitor services program, where students, researchers, and national and foreign tourists are received. The site also has an interpretive trail, which has a bird observatory. There are also a number of informative brochures and posters that endeavor to promote awareness of conservation at the site.

ZONA SUJETA A CONSERVACIÓN ECOLÓGICA CABILDO-AMATAL, CHIAPAS: OVERVIEW

Zona Sujeta a Conservación Ecológica Cabildo-Amatal (Figures 7.14 and 7.15) is located in the coastal plains of the Pacific coast of Chiapas. The site supports a number of flora and fauna, including the olive ridley sea turtle (Lepidochelys olivacea), boa (Boa constrictor), Muscovy duck (Cairina moschata), piping plover (Charadrius melodus), wood stork (Mycteria Americana), snail kite (Rostrhamus sociabilis), northern tamandua (Tamandua mexicana), and the margay (Leopardus wiedii), as well as the mangrove species *Rhizophora mangle* and *Laguncularia racemosa*. The main human activities at the site are agriculture, raising of livestock, and fishing. The main threats to the ecology of the site include the use of agrochemicals, deforestation, flora and fauna trafficking, furtive hunting, new human settlements, and open dumpsters.

FIGURE 7.14 Composite aerial image of 'Zona Sujeta a Conservación Ecológica Cabildo-Amatal. February 02, 2008; Area: 2,832 ha; Coordinates: 14°46'N 092°28'W; Ramsar Site number: 1,771.

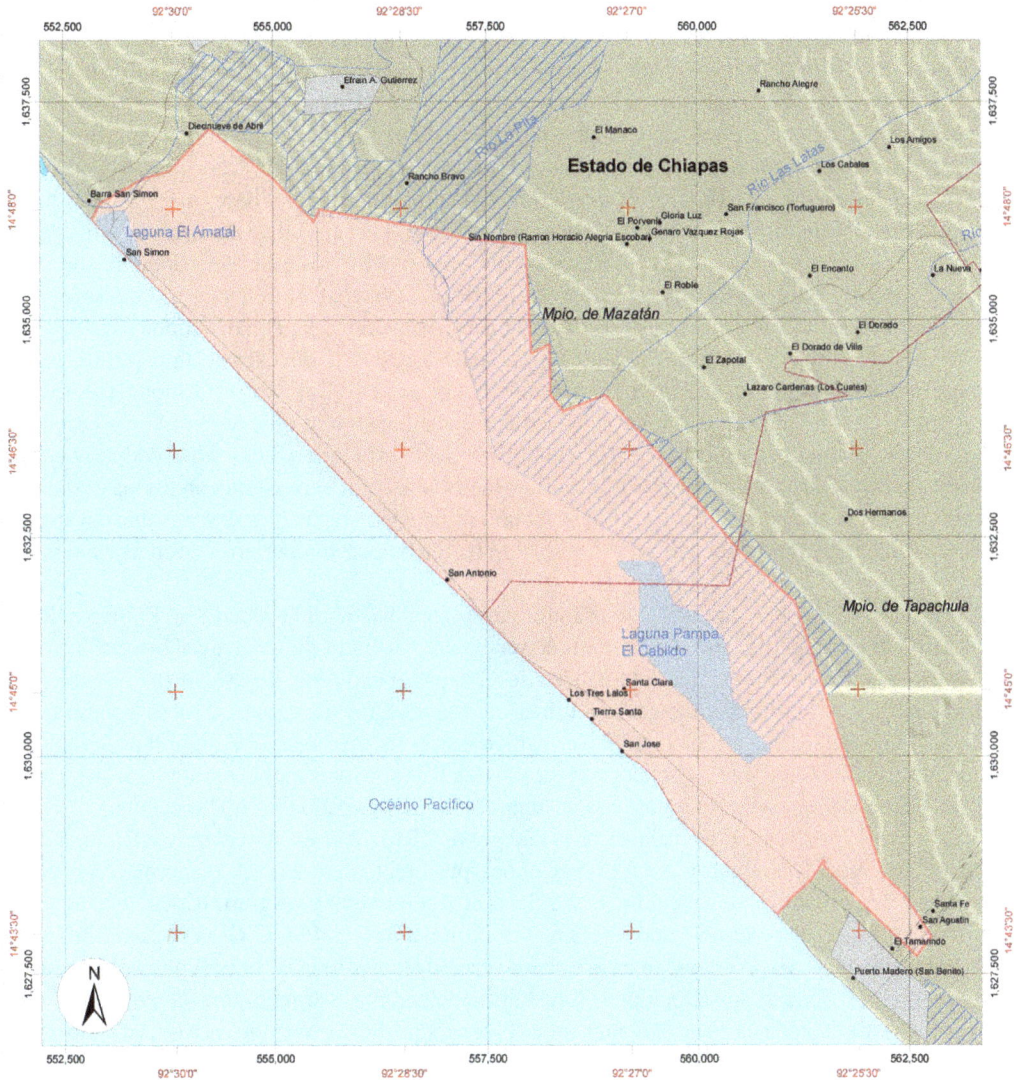

FIGURE 7.15 Map of Zona Sujeta a Conservación Ecológica Cabildo-Amatal. February 02, 2008; Area: 2,832 ha; Coordinates: 14°46'N 092°28'W; Ramsar Site number: 1,771.

SITE DETAILS

The site's wetlands cover the region called Laguna Pampa El Cabildo, Los Mangroves from Efraín to Laguna El Amatal, in the municipalities of Tapachula and Mazatán, Chiapas. The site is within the physiographic province of the Pacific Coastal Plain, which extends along 280 km of the Pacific coast, from the State of Oaxaca in the lagoon known as the Dead Sea, to the neighboring country of Guatemala.

Notable flora and fauna species that are listed as endangered (per NOM-059-SEMARNAT-2001 and/or -2010 and the CITES) include the ocelot (Leopardus pardalis), margay (Leopardus wiedii), jaguarundi (Herpailurus yagouaroundi), and river otter (Lontra longicaudis). Also noteworthy at the site are the olive ridley sea turtle (Lepidochelys olivacea), boa (Boa constrictor), northern tamandua (Tamandua mexicana), red mangrove (Rhizophora mangle), white mangrove (Laguncularia racemosa), *Caiman crocodilus fuscus*, black mangrove (Avicennia germinans), and button mangrove (Conocarpus erectus). Twenty-two bird species of note at the site are listed in NOM-059-SEMARNAT-2001 and/or -2010, including red-crowned amazon (Amazona viridigenalis), yellow-naped parrot (Amazona auropalliata), peregrine falcon (Falco peregrinus), and agami heron (Agamia agami). The site also supports the presence of the yellow-winged parakeet (Brotogeris jugularis), Veracruz wren (Campylorhynchus rufinucha), Muscovy duck (Cairina moschata), piping plover (Charadrius melodus), least grebe (Tachybaptus dominicus), bare-throated tiger-heron (Tigrisoma mexicanum), wood stork (Mycteria americana), snail kite (Rostrhamus sociabilis), Mississippi kite (Ictinia mississippiensis), black-collared hawk (Busarellus nigricollis), the common black hawk (Buteogallus anthracinus), white-bellied chachalaca (Ortalis leucogastra), sungrebe (Heliornis fulica), least tern (Sternula antillarum), pale-vented pigeon (Columba cayennensis), orange-fronted parakeet (Eupsittula canicularis), pale-billed woodpecker (Campephilus guatemalensis), and the giant wren (Campylorhynchus chiapensis) [IHNE 2007]. Among the species that winter at the site are the white heron (Casmerodius albus), great blue heron (Ardea herodias), hooked egret (Eudocimus albus), and the wood stork (Mycteria americana). The productivity and the intricate root system of mangroves, as well as the rich detrital ecosystem conditions at the site, provide support to spawning, mating, and protection of juvenile stages of numerous marine species of commercial importance, such as whiteleg shrimp (Litopenaeus vannamei), as well as oysters and clams (Pitar spp.).

The coast of Chiapas is subdivided into four hydrological basins: The Dead Sea, Pijijiapan River Basin, River Basin Huixtla, and the Suchiate River Basin. Rivers in the area are short-reach, steepdescent surface flows from the Sierra Madre de Chiapas, becoming less steep in the mountain foothills as they flow through the site, and then out into the Pacific Ocean. The two main rivers in the Suchiate Basin are the Suchiate River, which serves as the border between the Republic of Guatemala and the Republic of Mexico, and the Coatan River, which supplies water to a large number of coffee farms, as well as for human consumption in the municipality of Tapachula.

Floral species at the site include red mangrove (Rhizophora mangle), white mangrove (Laguncularia racemosa), button mangrove (Conocarpus erectus), and black mangrove (Avicennia germinans). The coastal dune vegetation at the site is characterized by seagrape (Coccoloba uvifera), bayhops (Ipomoea pes-caprae), and the grasses *Canavalia maritima*, *Croton punctatus*, and *Sporobolus domingensis*. Wetland areas are characterized by southern cattail (Typha domingensis), *Cyperus* spp., and *Scirpus* spp. Deciduous forest flora include the evergreen tree *Stemmadenia mollis*, papaturro (Coccoloba caracasana), mesquite (Prosopis juliflora), triangle cactus (Acanthocereus tetragonus), madras thorn (Pithecellobium dulce), and copal (Bursera excelsa). Other plant species of note at the site are Mexican sabal (Sabal mexicana), and manaca palm (Scheelea preussii). Aquatic vegetation at the site includes dot leaf waterlily (Nymphaea ampla), water lettuce (Pistia stratiotes), common water hyacinth (Eichhornia crassipes), lesser duckweed (Lemna aequinoctialis), and Mexican paspalum (Paspalum convexum).

The site supports numerous bird species, including 22 species listed in NOM-059-SEMARNAT-2001 and/or -2010, including red-crowned amazon (Amazona viridigenalis), yellow-naped parrot (Amazona auropalliata), peregrine falcon (Falco peregrinus), and agami heron (Agamia agami). Additional bird species at the site include the yellow-winged parakeet (Brotogeris jugularis), Veracruz wren (Campylorhynchus rufinucha), Muscovy duck (Cairina moschata), piping plover (Charadrius melodus), least grebe (Tachybaptus dominicus), bare-throated tiger-heron (Tigrisoma mexicanum), wood stork (Mycteria americana), snail kite (Rostrhamus sociabilis), Mississippi kite (Ictinia mississippiensis), black-collared hawk (Busarellus nigricollis), the common black hawk (Buteogallus anthracinus), white-bellied chachalaca (Ortalis leucogastra), least tern (Sternula antillarum), orange-fronted parakeet (Eupsittula canicularis), pale-billed woodpecker (Campephilus guatemalensis), and the giant wren (Campylorhynchus chiapensis) [IHNE 2007]. Mammals at the site that are listed in NOM-059-SEMARNAT-2001 and/or -2010 include ocelot (Leopardus pardalis), margay (Leopardus wiedii), and northern tamandua (Tamandua mexicana) [endangered], as well as the threatened jaguarundi (Herpailurus yagouaroundi) and river otter (Lontra longicaudis). Reptiles at the site include the endangered olive ridley sea turtle (Lepidochelys olivacea), and those under special protection (i.e., the caiman, *Caiman crocodilus fuscus*, and the green iguana, *Iguana iguana*). The black iguana (Ctenosaura similis) is also present at the site and listed as threatened per NOM-059-SEMARNAT-2001 and/or -2010.

Soconusco (the region in the southwest corner of the State of Chiapas, along its border with Guatemala) was one of the most densely populated territories, both before and during conquests of the area by foreigners. Between the years 1901 and 1908, the 355 km railway, following the old trajectory from Arriaga to Ciudad Hidalgo, on the banks of the Suchiate River, opened along the entire coastal region of Chiapas, also establishing communication with the narrow-gauge railway network of Guatemala over the international bridge of the Suchiate River. This opening of a transport route allowed an increase in coffee cultivation and general economic advancements for the region, as well as the mixture of the diversity of peoples of the Soconusco (Helbig 1964). The people who inhabit the coast, known as "La pampa de agua" have and do take advantage of the plant and animal species present. During the 1960s, "La pampa de agua" ceased to have the importance it had before (the Center for Higher Studies of Mexico and the Central America-University of Sciences and Arts of Chiapas 1996), and yet the area is rich in archaeological remains of past cultures that existed at the site, long before the Aztecs came to conquer the Soconusco region, in approximately 1486. According to the Center for Higher Studies of Mexico and Central America (1995), there is archaeological evidence that the estuarine zone of the site was inhabited long before the post-classic period (900 AD), during which times the communities of the area utilized the flora, fauna, and lands of the site much like they have, until modern day.

Current land use includes scale fishing and harvesting of whiteleg shrimp (Litopenaeus vannamei), mullet (Mugil spp.), snook (Centropomus spp.), berrugata (Menticirrhus sp.), dorado (Coryphaena hippurus), blue sea catfish (Arius guatemalensis), roughback shrimp (Trachypenaeus similis), black skipjack (Euthynnus lineatus), crevalle jack (Caranx hippos), and crab (Callinectes spp.). Cooperatives operate in the region, all of them with permits for their catches. Agricultural activities in the vicinity of the site include cropping of corn, cucumber, melon, sesame, banana, cashew, watermelon, chili, coconut, tomato, oil palm, coffee, cocoa, sorghum, and cassava. Livestock activities in the area of the site involve raising horses, cattle, sheep, pigs, chickens, ducks, and turkeys. The main commercial marine species from the site include fin shark (Family Carcharhinidae), bull shark (Carcharhinus leucas), crappie (family Gerreidae), whiteleg shrimp (Litopenaeus vannamei), Indo-Pacific sailfish (Istiophorus platypterus), and rays (Dasyatis spp.). To capture these species, fishers use nets, lines, longlines, cast nets, trammel nets, seine nets, mesh, and hooks.

One of the major environmental problems for the site is the use of various agrochemicals for crops, which cause soil contamination, affect groundwater as well as coastal waters, and directly affect biological diversity. Other negative impacts on the ecology of the site include clandestine logging, looting and trafficking of flora and fauna, poaching, unregulated new human settlements, clandestine open-air garbage dumps, and the expansion of the agricultural frontier. A number of these negative impacts are monitored and addressed in coordination with governmental agencies, and the Institute of Natural History and Ecology (IHNE) [Crocker et al. 2007].

ZONA SUJETA A CONSERVACIÓN ECOLÓGICA EL GANCHO-MURILLO, CHIAPAS: OVERVIEW

Zona Sujeta a Conservación Ecológica El Gancho Murillo (Figures 7.16 and 7.17) is located in the plains of coastal Chiapas. The site supports a diversity of organisms within mangroves, palm forests, deciduous forests, and secondary vegetation. The site is mainly composed of mangroves, with a predominancy of red (Rhizophora mangle) and white (Laguncularia racemosa) mangrove. The site supports a number of faunal species, most notably *Leopardus pardalis*, *Leopardus wiedii*,

FIGURE 7.16 Composite aerial image of Zona Sujeta a Conservación Ecológica El Gancho Murillo. February 02, 2008; Area: 4,643 ha; Coordinates: 14°37′N 92°18′W; Ramsar Site number: 1,772.

FIGURE 7.17 Map of Zona Sujeta a Conservación Ecológica El Gancho Murillo. February 02, 2008; Area: 4,643 ha; Coordinates: 14°37'N 92°18'W; Ramsar Site number: 1,772.

and *Lepidochelys olivacea* (all endangered). Threatened species at the site include *Tamandua mexicana*, *Ardea herodias*, and *Caiman crocodilus fuscus* (special protection). Predominant human activities at the site include agriculture, river fishing, and prawn capture, as well as the raising of livestock. The beaches of the site are a tourist attraction, where sport fishing, swimming, boat rides, and camping are the main human activities.

SITE DETAILS

The areas of the site called "El Gancho," "Barra de Cahoacán," and "Murillo" are located in the Pacific Coastal Plain, which is approximately 280 km long, adjacent to the Pacific coast, from the State of Oaxaca in the lagoon known as the "Dead Sea," to the neighboring country of Guatemala. Throughout, the site has a few hills, and the general slope is only about 0.1%, especially in places prone to permanent and temporary flooding (García 1969). The coast of Chiapas has around 258 km of coastline, where coastal lagoons and mangrove ecosystems predominate. The Area Subject to Ecological Conservation, "El Gancho Murillo," is shared by four hydrological sub-basins. The most important input of surface water to the site is the Suchiate River, being a large part of its channel and the natural border between Mexico and Guatemala, borne in Guatemalan territory and following a southwestern direction, passing between the Tacaná and Tlajocomulco Volcanoes.

The dominant plant species at the site are southern cattail (Typha domingensis), *Cyperus* spp., and *Scirpus* spp. Floating aquatic vegetation at the site includes water lettuce (Pistia stratiotes), swamp lily (Crinum erubescens), found in the marshy and lake areas, as well as on the shores of ditches, canals, and the backwaters of rivers. Other plant species of note at the site are Guiana chestnut (Pachira aquatica), huilihuiste (Karwinskia calderonii), button mangrove (Conocarpus erectus), and black mangrove (Avicennia germinans). In areas of palm, Mexican sabal (Sabal mexicana) and manaca palm (Scheelea preussii) are present, which are often cleared by burning the areas to make new land available for cattle ranching. Aquatic vegetation at the site includes dot leaf waterlily (Nymphaea ampla), water lettuce (Pistia stratiotes), common water hyacinth (Eichhornia crassipes), gibbous duckweed (Lemna gibba), and Mexican paspalum (Paspalum convexum). The site's coastal dune areas support seagrape (Coccoloba uvifera), bayhops (Ipomoea pes-caprae), *Canavalia maritima*, *Croton punctatus*, and *Sporobolus domingensis*. Deciduous forests at the site support diverse flora, such as the evergreen tree *Stemmadenia mollis*, papaturro (Coccoloba caracasana), mesquite (Prosopis juliflora), triangle cactus (Acanthocereus tetragonus), madras thorn (Pithecellobium dulce), and red elephant tree (Bursera hindsiana) [SERNyP 1998].

The site supports a host of bird species, including the great egret (Ardea alba), snowy egret (Egretta thula), cattle egret (Bubulcus ibis), seagulls and terns (Larus atricilla, L. pipixcan, L. argentatus, Sterna maxima, and S. hirundo), American white pelican (Pelecanus erythrorhynchos), magnificent frigatebirds (Fregata magnificens), turkey vultures (Cathartes aura), black-necked stilts (Himantopus mexicanus), northern jacana (Jacana spinosa), scissor-tailed flycatcher (Tyrannus forficatus), roadside hawk (Buteo magnirostris), the black-bellied whistling duck (Dendrocygna autumnalis), and neotropic cormorants (Phalacrocorax brasilianus). Mammals supported by the site include paca (Agouti paca), armadillo (Dasypus novemcinctus), raccoon (Procyon lotor), white-nosed coati (Nasua narica), gray fox (Urocyon cinereoargenteus), eastern cottontail (Sylvilagus floridanus), Virginia opossum (Didelphis virginiana), and greater bulldog bat (Noctilio leporinus). Notable reptiles at the site include the green iguana (Iguana iguana), the black iguana (Ctenosaura similis), the brown basilisk (Basiliscus vittatus), long-tail spiny lizard (Sceloporus siniferus), and the silky anole (Anolis sericeus). Crustaceans noted at the site include blue shrimp (Penaeus stylirostris), several species of crabs of the genus *Uca*, crabs (Callinectes bellicosus, C. toxotes, and C. arcuatus), and cinnamon river shrimp (Macrobrachium acanthurus and M. tenellum). The fish of the site include tropical gar (Atractosteus tropicus), black snook (Centropomus nigrescens), mullet (Mugil cephalus), southern blue catfish (Ictalurus meridionalis), and Pacific four-eyed fish (Anableps dowei) [SERNyP 1998].

Records reveal that along the Chiapas coast, between 7000 BC and 4500 BC, humans had already cultivated corn and made their first ceramic pieces. Also, during that time there were the first commercial exchanges between and among the people of the site, and areas to the south, what is now Central America. It was during this period that the Olmec Culture flourished (Helbig 1964). Subsequently, the Aztecs raided and conquered the region during the 15th and 16th centuries. There is a knowledge gap in the history during the colonial period at the site. Current land use in the vicinity of the site is based mainly on agriculture, riverside fishing, and the capture of estuarine shrimp, as well as the raising of cattle. Among the main agricultural crops are corn, sesame, banana, soybean, corn, cotton, and sorghum. Livestock activities are dominated by cattle and pigs. It should be noted that human activities in the area of the site have overexploited resources, causing ecological deterioration of the site, including the impacts from overuse of agrochemical products; general changes in land use (including development projects); and discharges from coffee farms and cattle ranches. Discharge of untreated wastewater and garbage from towns and cities in the vicinity of the site, into rivers that reach the lagoons, estuaries, and marine zone; clandestine logging, looting, and trafficking of flora and fauna; hunting; and clandestine open air dumps are all causing marked deterioration of the ecology of the site and the environs of the associated coastal areas.

On June 16, 1999, a decree declaring the site a protected natural area was published, which has resulted in a number of monitoring, research, education, and outreach activities to occur, improving awareness and causing actions that have improved conservation initiatives in the region, and the site (Estrada-Croker et al. 2007). Among the most relevant of these actions are tours of illegal activities, organized by the IHNE; talks on the prevention of fire and formation of Community Brigades for firefighting; implementation of a local committee for clean beaches of Tapachula; implementation of actions with the Municipal Council for Environmental Protection of Tapachula to directly address solid waste clean-up; technical visits to address siltation and water use management; turtle conservation projects; environmental education for students and religious groups in the town of Playa Linda and Puerto Madero; restoration of altered areas in mangrove ecosystems in the El Cabildo Lagoon and in the Pozuelos Lagoon; and monitoring of seawater quality in Playa Linda town. The El Colegio de la Frontera Sur (ECOSUR), Tapachula Unit, has also conducted a number of monitoring and research projects at the site, among which there were opportunities for environmental education in mangrove forests, carried out by Tovilla et al. (2004). Bello Mendoza et al. (2003) carried out an evaluation of the water quality in the coastal lagoon ecosystem (Pozuelos-Murillo) and Coutiño and Tovilla Hernández (2003) has conducted faunal studies at the site, and has also published a bird guide for local use.

The beaches of the area are a tourist attraction, near Suchiate, Metapa de Domínguez, Tuxtla Chico, and other areas of Suchiate. The site's beach is also a very popular site for visitation, especially during Easter and Christmas holiday seasons. The Municipal Council of Suchiate supports activities, such as a music festival, as one more attraction, in addition to sport fishing, swimming, boat rides, and camping. During such events, efforts are made to incorporate opportunities for visitors to observe the mangroves, where visitors have an opportunity to better understand the importance of the diversity of the flora and fauna at the site.

REFERENCES

Alvarado, P.J.C. 1991. *Características Físicas de la Playa de Barra de la Cruz, Oaxaca, Como Factores Que Influyen en la Incidencia de Hembras Anidadoras de Tortuga Laúd (Dermochelys coriacea)*. Universidad Autónoma Metropolitana Unidad Xochimilco Informe Final de Servicio Social. 25pp.

Arriaga, L., Aguilar, V., and J. Alcocer. 2002. *Aguas continentales y diversidad biológica de México*. Comisión Nacional para el Conocimiento y Uso de la Biodiversidad, México.

Barrientos Luján, N.A., and S. Ramírez Luna. 2000. *Moluscos asociados a coral en La Mixteca y La Montosa, Bahías de Huatulco*, Oaxaca, México. XII Congreso Nacional de Oceanografía, Huatulco, Oaxaca.

Bello Mendoza, R., Hernández Romero, A.H., Malo Rivera, E. A., and C. Tovilla Hernández. 2003. *Evaluación de la calidad del aguaen el sistema lagunar costero "Pozuelos-Murillo" en el Soconusco, Chiapas*. Fundación Produce Chiapas, A.C. 30pp.

Coutiño Barrios, R. and C. Tovilla Hernández. 2003. *Guía de aves de la Laguna Pampa "El Cabildo"*. ECOSUR, Tapachula, Chiapas, México. 106pp.

Estrada-Croker, J.C., Alvarez-Vilchis, C., González-García, Y., and M. Hernández-López. 2007. Informe de Acciones Realizadas en los Humedales Costeros "El Cabildo Amatal" y "El Gancho Murillo," Chiapas, México. Resultados 2003-2007. Delegación Regional Tapachula. Instituto de Historia Natural y Ecología. Documento Interno. Tapachula de Córdova y Ordóñez, Chiapas.

García, E. 1969. *Geografía General de Chiapas*. México. 375pp.

Gordon, J., González, M.A., Vázquez, H.J., Ortega, L.R., and A. Reyes García. 2003. *Guaiacum coulteri: an overexploited Mexican dry forest tree with potential as an indicator species*. Department of Geography, University of Durham, UK; Grupo Autónomo para la Investigación Ambiental, A.C.; and Herbario Nacional, Instituto de Biología de la UNAM, México.

Helbig, C. M. 1964. *El Soconusco y su Zona Cafetalera en Chiapas. Instituto de Ciencias y Artes de Chiapas.* Tuxtla Gutiérrez, Chiapas, México. 133pp.

IHNE (Instituto de Historia Natural y Ecología). 2004. *Modelo de Ordenamiento Ecológico Territorial Subcuenca del Río Zanatenco, Tonalá Chiapas.* Gobierno del Estado de Chiapas, México.

IHNE (Instituto de Historia Natural y Ecología). 2006. "Informe Final Proyecto de Protección y Conservación de la Tortuga Marina en Chiapas." Documento interno.

IHNE (Instituto de Historia Natural y Ecología). 2007. *Proyecto Monitoreo Biológico en ANPs Chiapas.* Instituto de Historia Natural y Ecología. Gobierno del Estado de Chiapas. Informe Final Interno.

NAWCC, IHNE, CONANP, IDESMAC. 2002a. Informe Final del Proyecto Conservación y Manejo Integral de Cuencas en el complejo Reserva de la Biosfera La Sepultura - Sistema Lagunario Mar Muerto - La Joya Buena Vista, Chiapas. México. 148pp.

NAWCC, IHNE, CONANP, IDESMAC. 2002b. Monitoreo de aves acuáticas y terrestres del complejo REBISE - Humedales Costeros. 34pp.

Ramos Santiago, E. 2003. *Estructura de la comunidad de peces de la Bahía La Entrega, Huatulco, Oaxaca.* Tesis de Maestríaen Biología, División de Ciencias Biológicas y de la Salud, Universidad Autónoma Metropolitana-Iztapalapa, México.

Rzedowski, J. 1978. Vegetación de México. Limusa, México. 432pp.

Sánchez, A. and M. Acevedo. 2001. Caracterización preliminar de plantas ornamentales, melíferas y artesanales con mayor potencial de aprovechamiento en Santa Ma. Huatulco, Oax. Memoria de residencia profesional. Lic. en Biología. Inst. Tecnol. Agropecuario de Oaxaca. No. 23.

(SERNyP) Secretariade Ecologia Recursos Naturales y Pesca. 1998. Estudio Técnico Justificativo de los humedales "El Gancho Murillo" para proponerlos como una Área Natural Protegida. Secretaría de Ecología, Recursos Naturales y Pesca. Subsecretaría de Ecología, Recursos Naturales y Pesca. Dirección de Ecología y Protección Ambiental. Tuxtla Gutiérrez, Chiapas. 46pp.

SEMARNAP. 2000. Estudio Especializado de Acuacultura y Ordenamiento Ecológico en el Estado de Chiapas. 92pp.

Tovilla-Hernández, C., Román-Salazar, A.V., Simuta-Morales, G.M., and R.M. Linares-Mazariegos. 2004. Recuperación del manglar en la barra del río Cahoacán, en la costa de Chiapas. *Madera y Bosques.* 2: 77–91.

Witherington, B., Crady, C., and L. Bolen. 1996. *A "Hatchling Orientation Index" for assessing orientation disruption from artificial lighting.* Proceedings of the Fifteenth Annual Symposium on Sea Turtle Biology and Conservation. NOAA Tech. Mem. NMFS-SEFSC-387. p. 344–347.

Spanish to English Translations/ Definitions for Terms Used

asamblea:	assembly
bahía:	bay
Baja California:	Baja California (a State in Mexico)
Baja California Sur:	Baja California South (a State in Mexico)
canal:	channel
complejo:	complex
estero:	estuary
humedal:	wetland
humedales:	wetlands
isla:	island
laguna:	lagoon
lagunar:	lagoon
Marismas Nacionales:	National Marshlands
oasis:	oasis
Parque Nacional:	National Park
Pesca:	fishing
playa:	beach
playón:	land which is formed by accumulation of sediment on the banks of rivers after floods
reserva de la biosfera:	biosphere reserve
río:	river
santuario:	sanctuary
sistema:	(eco)system
sistema estuarino:	estuarine (eco)system
sistema lagunar estuarino:	estuary lagoon (eco)system
sistema ripario:	riparian (eco)system
tortuguera:	turtle
y:	and
Zona Sujeta a Conservación Ecológica:	Area Subject to Ecological Conservation

Annotated Acronyms Used

ACCP	Amigos para la Conservación de Cabo Pulmo
AD	Anno Domini
BC	Before Christ
BC	British Columbia
BCE	Before Common Era
BI	BirdLife International
BLA	Bahía de Los Angeles
CBD	Convention on Biological Diversity
CCLASF	Corredor Costero La Asamblea-San Francisquito
CECARENA	Centro de Conservación y Aprovechamiento de los Recursos Naturales
CESUES	Centro de Estudios Superiores del Estado de Sonora
CI	Conservation International
CIAD	Centro de Investigación en Alimentación y Desarrollo
CIBNOR	Centro de Investigaciones Biológicas del Noroeste, S. C.
CICESE	Center for Scientific Research and Higher Education of Ensenada
CICIMAR	Centro Interdisciplinario de Ciencias Marinas
CICIMAR-IPN	Centro Interdisciplinario de Ciencias Marinas del Instituto Politécnico Nacional
CIIDIR	Centro Interdisciplinario de Investigación para el Desarrollo Integral Regional
CIPACTLI	The Agency for Forest and Wildlife Restoration
CIQRO	Centro de Investigaciones de Quintana Roo
CITES	Convention on International Trade in Endangered Species
CMS	Convention on Migratory Species
CNANP	Comisión Nacional de Áreas Naturales Protegidas
CONABIO	La Comisión Nacional para el Conocimiento y Uso de la Biodiversidad
CONACYT	Consejo Nacional de Humanidades, Ciencias y Tecnologías
CONAFOR	Comisión Nacional Forestal
CONAGUA	Comisión Nacional del Agua
CONANP	La Comisión Nacional de Áreas Naturales Protegidas
CONAPESCA	Comision Nacional de Acuacultura y Pesca
COP	Conference of the Contracting Parties
CRIP	Regional Center for Fisheries Research
DDT	Dichlorodiphenyltrichloroethane
DEDSZC	Departamento para el Desarrollo Sustentable de Zonas Costeras
DOF	Diario Oficial de la Federación (Official Journal of the Federation of Mexico)
ECOSUR	El Colegio de la Frontera Sur
FONATUR	Fondo Nacional de Fomento al Turismo
GEA	Grassland Ecological Area
IAIA	International Association for Impact Assessment
IBA	BirdLife's Important Bird and Biodiversity Area
IDESMAC	El Instituto para el Desarrollo Sustentable en Mesoamérica, A.C.
IHNE	Instituto de Historia Natural y Ecología (English translation: Institute of Natural History and Ecology

INEGI	Instituto Nacional de Estadística, Geografía e Informática (National Institute of Geography and Informatics Statistics)
INP	(Mexico's) National Fisheries Institute
IPN	(Mexico's) National Polytechnic Institute
ISPM	San Pedro Mártir Island
ITESM-CG	Instituto Tecnológico y de Estudios Superiores de Monterrey-Campus Guaymas
ITMAR	(University of Arizona) Technological Institute of the Sea
IUCN	International Union for the Conservation of Nature
IWMI	International Water Management Institute
LGEEPA	General Law of Ecology and Protection of the Environment
MAB-UNESCO	Man and the Biosphere Programme, United Nations Educational, Scientific, and Cultural Organization
NAWCA	North American Wetlands Conservation Act
NMBCA	Neotropical Migratory Bird Conservation Act
NOAA	National Oceanographic and Atmospheric Administration
NOM	Normas Oficiales Mexicanas (Official Mexican Standards), which is the name of each of a series of official, compulsory standards and regulations, commonly referred to as 'normas' or the 'Mexican normative.'
NOM-059-SEMARNAT-2001	(Mexico's) primary document that establishes a list of wildlife species classified as either endangered, threatened, under special protection, or probably extinct in the wild.
NOM-059-SEMARNAT-2010	(Mexico's) primary document that establishes a list of wildlife species classified as either endangered, threatened, under special protection, or probably extinct in the wild.
NOM-131-ECOL-1998	(Mexico's) primary document that establishes regulations that applies to ecotourism in Mexico, specifically regarding the observation of whales.
NPS	National Park Service
NWR	National Wildlife Refuge
POECO	Ecological Management Program of the Oaxaca Coast
PROFEPA	Federal Attorney for Environmental Protection
RBCC	Chamela-Cuixmala Biosphere Reserve
REBISA	La Sepultura Biosphere Reserve
SAGARPA	Ministry of Agriculture, Livestock and Fisheries
SARH	Secretariat of Agriculture and Hydraulic Resources
SEMAR	Secretaría de la Marina (Mexican Secretariat of the Navy)
SEMARNAP	Secretariat of the Environment, Natural Resources and Fisheries
SEMARNAT	Secretariat of Environment and Natural Resources
SEPESCA	Secretariat of Fisheries
SERNyP	Secretaría de Ecología, Recursos Naturales y Pesca
SFBE	San Francisco Bay/Estuary
SRA	Agrarian Reform Secretariat
STRP	Scientific and Technical Review Panel
SWS	Society of Wetland Scientists
TNC	The Nature Conservancy
TRNERR	Tijuana River National Estuarine Research Reserve
UABCS	Autonomous University of Baja California Sur
UCOL	Autonomous University of Colima
UDG	University of Guadalajara

UNAM	Universidad Nacional Autónoma de México (English translation: National Autonomous University of Mexico)
UNCCD	United Nations Convention to Combat Desertification
UNDP	United Nations Development Program
UNECE	United Nations Economic Commission for Europe
UNEP	United Nations Environmental Program
UNESCO	United Nations Educational, Scientific, and Cultural Organization
US	United States (of America)
USA	United States of America
USACOE	United States Army Corps of Engineers
USFWS	United States Fish and Wildlife Service
USGS	United States Geological Survey
WHC	World Heritage Convention
WWF	World Wildlife Fund
WWT	Wildfowl and Wetlands Trust

Appendix A (Taxa Cited)

Scientific Nomenclature	Common Descriptor(s) [Not Exhaustive]
Abraliopsis affinis	species of enoploteuthid cephalopod
Abronia gracilis	graceful sand verbena
Abronia maritima	red sand verbena
Abudefduf saxatilis	sergeant major
Acacia cochliacantha	boat thorn acacia
Acacia farnesiana	sweet acacia
Acacia pennatula	feather acacia
Acacia riparia	box-katsin
Acacia spp.	acacia species
Acanthemblemaria crockeri	browncheek blenny
Acanthocereus occidentalis	sword pear
Acanthocereus pentagonus	chaco
Acanthocereus tetragonus	triangle cactus
Acanthurus triostegus	convict tang
Accipiter cooperii	Cooper's hawk
Accipiter striatus	bird hawk
Accipiter striatus velox	sharp-shinned hawk
Achatocarpus oaxacanus	tree in the family Achatocarpaceae
Achirus mazatlanus	Mazatlán sun
Acipenser medirostris	green sturgeon
Acipenser transmontanus	white sturgeon
Acrocomia mexicana	coyol
Acropora	crustose coralline algae
Acrostichum danaeifolium	giant leather fern
Actitis macularia	spotted sandpiper
Adiantum concinnum	maidenhair
Aeschynomene sousae	flowering plant in the family Fabaceae
Agamia agami	agami heron
Agave angustifolia	Caribbean agave
Agelaius phoeniceus	red-winged blackbird
Agelaius tricolor	tri-colored blackbird
Agkistrodon bilineatus	pit viper
Agouti paca	paca
Agropecten circularis	ladybug clam
Allenrolfea occidentalis	iodine bush
Alutera scripta	scrawled filefish
Alvaradoa amorphoides	Mexican alvaradoa

(Continued)

Scientific Nomenclature	Common Descriptor(s) [Not Exhaustive]
Amaranthus palmeri	bledo
Amaranthus watsonii	Watson's amaranth
Amazilia violiceps	violet-crowned hummingbird
Amazona auropalliata	yellow-naped parrot
Amazona finschi	lilac crowned parrot
Amazona oratrix	yellow-headed amazon
Amazona viridigenalis	red-crowned amazon
Ambrosia ambrosioides	chicura
Ambrosia chamissonis	silver burr ragweed
Ambystoma californiense	California tiger salamander
Ameiva undulata	barred whiptail
Ammospermophilus leucurus	white-tailed antelope squirrel
Amoreuxia palmatifida	Mexican yellowshow
Amphipterygium adstringens	cuachalalate
Amphiroa beauvoisii	species of thalloid red algae
Amphiroa misakiensis	species of red algae
Amphiroa rigida	sturdy needleweed
Amphiroa valonioides	thalloid red algae under the family Corallinaceae
Amphiroa vanbosseae	a species of red algae in the family Corallinaceae
Amphispiza bilineata	black-throated sparrow
Amphora sp.	diatom species
Anabaena	genus of division of cyanobacteria
Anabaenopsis	genus of filamentous, heterocystous cyanobacteria
Anableps dowei	Pacific four-eyed fish
Anachis ritteri	sea snail in the family Columbellidae
Anadara	genus of saltwater bivalves, ark clams, in the family Arcidae
Anadara perlabiata	ark clam
Anadara tuberculosa	mangrove cockle
Anas acuta	northern pintail
Anas americana	American widgeon
Anas clypeata	northern shoveler
Anas collaris	ring-necked duck
Anas crecca	green-winged teal
Anas cyanoptera	chestnut teal
Anas platyrhynchos	mallard
Anas spp.	ducks
Anas strepera	gadwall
Anchoa mundeola	false Panama anchovy
Anchoa mundeoloides	northern gulf anchovy
Anchoa panamiensis	anchovy
Anchovia macrolepidota	large scale anchovy

(Continued)

Scientific Nomenclature	Common Descriptor(s) [Not Exhaustive]
Ancistromesus mexicanus	giant Mexican limpet
Anhinga anhinga	anhinga
Ankistrodesmus	genus of a green algae
Anniella geronimensis	legless lizard
Anolis biporcatus	giant green anole
Anolis nebulosus	the clouded anole
Anolis schmidti	anole
Anolis sericeus	silky anole
Anona glabra	pond apple
Anous minutus	black noddies
Anous stolidus	brown swallow
Anous stolidus	brown noddy
Anser albifrons	greater white-fronted goose
Antrozous pallidus	pallid bat
Antrozous pallidus minor	pallid bat
Aphriza virgata	surfbird
Apoplanesia paniculata	palo de arco
Aquila chrysaetos	golden eagle
Aquila chrysaetos canadensis	golden eagle
Ara militaris	green macaw
Aramides axillaris	rufous-necked wood rail
Aratinga holochlora brevipes	Socorro parakeet
Archilochus colubris	ruby-throated hummingbird
Arctocephalus townsendi	Guadalupe fur seal
Ardea alba	great egret
Ardea herodias	great blue heron
Ardea herodias herodias	great blue heron
Ardea herodias occidentalis	great blue heron
Arenaria interpres	ruddy turnstone
Arene hindsiana	species of sea snail
Argopecten circularis	ladybug clams
Ariidae	catfish or chihuiles
Aristida ternipes	spidergrass
Arius guatemalensis	blue sea catfish
Arius planiceps	flathead sea catfish
Arius platypogon	slender-spined catfish
Ariopsis seemanni	tete sea catfish
Artemisia	genus of aromatic herbs and shrubs in the Asteraceae family
Artemisia californica	California sagebrush
Arthrocnemum subterminale	Parish's glasswort
Artibeus intermedius	great fruit-eating bat

(Continued)

Scientific Nomenclature	Common Descriptor(s) [Not Exhaustive]
Arundo donax	giant reed
Asclepias subulata	rush milkweed
Asio flammeus	short-eared owl
Aspidoscelis guttatus	Mexican racerunner
Aspidoscelis hyperythrus	orange-throated whiptail
Aspidoscelis tigris	western whiptail
Asplenium nidus	bird's nest fern
Asthianus viminalis	species of riparian aquatic vegetation
Astragalus harrisonii	Harrison's milk-vetch
Astrangia	genus of stony corals in the family Rhizangiidae
Astrea turbanica	rockpile turban
Astrea undosa	panocha snails
Astreopora	genus of stony corals in the family Acroporidae
Astronium graveolens	species of flowering tree in the family Anacardiaceae
Astropecten armatus	spiny sand star
Ateles geoffroyi	spider monkey
Athene cunicularia	burrowing owl
Atherinella crystallina	blackfin silverside
Atherinops affinis	topsmelt silverside
Atherinops affinis affinis	topsmelt silverside
Atractosteus tropicus	tropical gar
Atrina maura	Maura pen shell
Atriplex barclayana	dwarf saltbush
Atriplex canescens	four wing saltbush
Atriplex spp.	species of saltbush
Atyoida bisculata	shrimp
Auriparus flaviceps	verdin
Avicennia germinans	black mangrove
Awaous guamensis	'o'opu nākea
Awaous tajasica	sand fish
Awaous transandeanus	Pacific river goby
Axoclinus nigricaudus	Cortez triplefin
Aythya affinis	lesser scaup
Aythya americana	redhead
Aythya marila	greater scaup
Aythya valisineria	canvasback
Baccharis consanguinea	coyote brush
Baccharis salicifolia	mule fat
Baccharis sarothroides	desertbroom
Baccharis viminea	mule fat
Bacillariophyta	the phylum of diatom
Bacopa monnieri	water hyssop

(Continued)

Scientific Nomenclature	Common Descriptor(s) [Not Exhaustive]
Baiomys musculus	southern pygmy mouse
Bairdiella icistia	ronco croaker
Balaenoptera acutorostrata	minke whale
Balaenoptera borealis	sei whale
Balaenoptera brydei	Bryde's whale
Balaenoptera edeni	Eden's whale
Balaenoptera musculus	blue whale
Balaenoptera physalus	fin whale
Basiliscus vittatus	brown basilisk
Bassariscus astutus	northern cacomixtle
Bassariscus astutus palmarius	palm grove ringtail
Batis maritima	saltwort
Bebbia juncea	bebbia
Bipes biporus	worm lizard
Blennosperma bakeri	Sonoma sunshine
Boa constrictor	boa
Boa constrictor imperator	boa imperator
Bodianus diplotaenia	Mexican hogfish
Bogertophis rosaliae	Baja California rat snake
Bolboschoenus maritimus	saltmarsh tuber-bulrush
Bombycilla cedrorum	cedar waxwing
Bostrychia radicans	species of red algae in the family Rhodomelaceae
Bourreria sonorae	Sonoran strongbark
Brachiaria mutica	California grass
Branchinecta conservatio	fairy shrimp
Branchinecta longiantenna	longhorn fairy shrimp
Branchinecta lynchi	vernal pool fairy shrimp
Branchinecta sandiegonensis	San Diego fairy shrimp
Branta bernicla	brant
Branta bernicla nigricans	black brant
Branta canadensis	Canada goose
Branta canadensis leucopareia	Aleutian Canada goose
Branta canadensis taverni	Taverner's Canada goose
Branta sandvicensis	Hawaiian goose or nēnē
Bravaisia integerrima	palo de agua
Bromelia pinguin	wild pineapple
Brosimum alicastrum	breadnut
Brotogeris jugularis	yellow-winged parakeet
Bubo virginianus	great horned owl
Bubulcus ibis	cattle egret
Bucephala albeola	bufflehead
Bucephala clangula	common goldeneye

(Continued)

Scientific Nomenclature	Common Descriptor(s) [Not Exhaustive]
Bufo marmoreus	Wiegmann's toad
Bufo punctatus	red-spotted toad
Bulbostylis	genus of plants in the sedge family
Bumelia sp.	bumelias buckthorn
Bursera epinnata	southern elephant tree
Bursera excelsa	copal
Bursera hindsiana	red elephant tree
Bursera laxiflora	torote prieto
Bursera microphylla	elephant tree
Bursera simaruba	gumbo limbo
Busarellus nigricollis	black-collared hawk
Buteo albonotatus	zone-tailed hawk
Buteo jamaicensis	red-tailed hawk
Buteo jamaicensis calurus	western red-tailed hawk
Buteo jamaicensis socorroensis	Socorro red-tailed hawk
Buteo magnirostris	roadside hawk
Buteo nitidus	gray-lined hawk
Buteo platypterus	broad-winged hawk
Buteo regalis	golden hawk
Buteo swainsoni	Swainson's hawk
Buteogallus anthracinus	common black hawk
Butorides striatus	striated heron
Butorides virescens	green heron
Byrsonima crassifolia	nanche
Caecidotea tomalensis	Tomales isopod
Caesalpinia arenosa	a species of tree in the family Fabaceae
Caesalpinia bonduc	grey nicker
Caesalpinia cacalaco	a species of tree in the family Fabaceae
Caesalpinia coriaria	leguminous tree
Caesalpinia eriostachys	a species of tree in the family Fabaceae
Caesalpinia platyloba	palo colorado
Caesalpinia sp.	cascalote
Caiman crocodilus fuscus	spectacled caiman
Cairina moschata	muscovy duck
Cakile maritima	marine rocket
Calamagrostis canadensis	bluejoint grass
Calidris alba	sanderling
Calidris alpina	dunlin
Calidris canutus	short-billed sandpiper
Calidris mauri	western sandpiper
Calidris minutilla	least sandpiper

(*Continued*)

Scientific Nomenclature	Common Descriptor(s) [Not Exhaustive]
Calidris ptilocnemis	rock sandpiper
Calliandra magdalenae	plant species in the family Fabaceae
Callinectes	genus of crabs, containing 16 extant species in the family Portunidae
Callinectes arcuatus	arched swimming crab
Callinectes bellicosus	warrior swimming crab
Callinectes crissum	esser blue crab
Callinectes toxotes	giant swimcrab
Calliosthoma aequisculptum	species of sea snail, a marine gastropod mollusk in the family Calliostomatidae
Callipepla californica	California quail
Callipepla californica achrustera	California quail (achrustera)
Callisaurus draconoides	zebra-tailed lizard
Callisaurus draconoides inusitatus	Sonoran zebra-tailed lizard
Calocitta formosa	Magpie
Calycophyllum candidissimum	lemonwood
Calypte costae	Costa's hummingbird
Campanulaceae	the bellflower family, of the order Asterales
Campephilus guatemalensis	pale-billed woodpecker
Campylorhynchus brunneicapillus affinis	cactus wren
Campylorhynchus chiapensis	giant wren
Campylorhynchus rufinucha	Veracruz wren
Canavalia maritima	Grasses
Cancer spp.	rock crab
Canis latrans	coyote
Canis lupus	gray wolf
Caracara cheriway	northern crested caracara
Caracara plancus	crested caracara
Caranx sp.	genus of tropical to subtropical marine fishes in the jack family Carangidae
Caranx hippos	crevalle jack
Carcharhinidae	a family of sharks known as the requiem sharks
Carcharhinus	genus of the family Carcharhinidae, the gray sharks
Carcharhinus leucas	bull shark
Carcharhinus limbatus	blacktip shark
Cardinalis cardinalis	northern cardinal
Cardinalis cardinalis seftoni	northern cardinal
Cardinalis sinuatus peninsulae	desert cardinal
Cardita laticostata	wide-ribbed cardita
Carduelis psaltria hesperophilus	lesser goldfinch
Caretta caretta	loggerhead sea turtle
Carex	genus of sedges

(Continued)

Scientific Nomenclature	Common Descriptor(s) [Not Exhaustive]
Carica papaya	papaya
Carpobrotus aequilaterus	sea fig
Carpobrotus chilensis	Chilean sea fig
Carpobrotus edulis	iceplant
Carpodacus mexicanus	house finch
Carpodacus mexicanus ruberrimus	house finch (ruberrimus)
Casmerodius albus	white heron
Cassia tomentosa	woolly senna
Castilla elastica	Panama rubber tree
Castilleja exserta	purple owls-clover
Castor canadensis	American beaver
Cathartes aura	turkey vulture
Catharus guttatus slevini	hermit thrush
Catherpes mexicanus	canyon wren
Catoptrophorus semipalmatus	curlew
Caulerpa sp.	genus of seaweeds in the family Caulerpaceae (among the green algae)
Caulerpa sertulariodes	willet
Cecropia obtusifolia	trumpet-tree
Cecropia peltata	trumpet-tree
Ceiba aesculifolia	kapok tree
Ceiba pentandra	kapok tree
Cenchrus ciliaris	buffelgrass
Cenchrus echinatus	southern sandbur
Centropomidae	sea bass
Centropomus	genus of predominantly marine fish comprising the family Centropomidae
Centropomus nigrescens	black snook
Centropomus pectinatus	tarpon snook
Centropomus robalito	tellofin snook
Centropomus undecimalis	common snook
Centurus uropygialis	Gila woodpecker
Cephalocereus spp.	genus of slow-growing, columnar-shaped, blue-green cacti
Cercidium floridum	blue palo brea
Cercidium floridum peninsulare	palo verde
Cercidium microphyllum	yellow paloverde
Cercidium peninsulare	blue palo verde
Cercidium praecox	palo brea
Cercidium sonorae	palo verde
Cerithium maculosum	species of the family Cerithiidae; gastropod mollusks that have small to medium sized shells
Cerithium stercusmuscarum	species of snails in the family Cerithiidae

(Continued)

Scientific Nomenclature	Common Descriptor(s) [Not Exhaustive]
Ceryle alcyon	belted kingfisher
Chaenopsis alepidota	orangethroat pikeblenny
Chaetoceros	a genus of diatom
Chaetodipus arenarius	little desert pocket mouse
Chaetodipus baileyi	Bailey's pocket mouse
Chaetodipus dalquesti	dalquesti mouse
Chaetodipus rudinoris	Baja pocket mouse
Chaetodipus spinatus	spiny pocket mouse
Chamaedorea pochutlensis	pochutla bamboo palm
Chamaesyce thymifolia	gulf sandmat
Chanos chanos	milkfish
Charadrius	genus of plovers, a group of wading birds in the family Charadriidae
Charadrius alexandrinus	kentish plover
Charadrius alexandrinus nivosus	western snowy plover
Charadrius collaris	collared plover
Charadrius melodus	piping plover
Charadrius semipalmatus	semipaleated tildillo
Charadrius vociferus	Killdeer
Charadrius wilsonia	Wilson's plover
Charadrius wilsonia beldingi	Wilson's plover (Belding's)
Charchahrinus porosus	dogfish
Chelonia mydas	green sea turtle
Chen caerulescens	snow goose
Chen caerulescens caerulescens	lesser snow geese
Chen canagica	emperor geese
Chilomeniscus cinctus	variable san snake
Chilomeniscus stramineus	spotted sandbow
Chione californiensis	California venus
Chione compta	small banded venus
Chione subrugosa	semi-rough chione
Chione succinta	clam species
Chione undatella	frilled venus
Chiroptera	genus of the bats
Chlamydomonas	genus of green alga
Chloris inflata	swollen fingergrass
Chloropyron molle hispidus	hispid bird's beak
Chloroscombrus orqueta	Pacific bumper
Choeronycteris	genus of bat in the family Phyllostomatidae
Choeronycteris mexicana	Mexican bat
Chondestes grammacus strigatus	lark sparrow
Chondracanthus squarrulosus	a species of red algae in the family Gigartinaceae

(Continued)

Scientific Nomenclature	Common Descriptor(s) [Not Exhaustive]
Chondrohierax uncinatus	hook-billed kite
Chordeiles acutipennis	lesser nighthawk
Chorizanthe chaetophora	prostrate spineflower
Chorizanthe interposita	a species in the family Polygonaceae
Chorizanthe jonesiana	a species in the family Polygonaceae
Chorizanthe turbinata	a species in the family Polygonaceae
Chroroceryle americana	herons kingfishers
Circus cyaneus	hen harrier
Cirrhitus rivulatus	giant hawkfish
Cistothorus palustris plesius	marsh wren
Citarichthys gilberti	tongue cover
Citrus spp.	citrus species
Cladium leptostachyum	sawgrass
Cladonia spp.	reindeer mosses
Cladophora	genus of reticulated filamentous green algae in the family Ulvophyceae
Clangula hyemalis	oldsquaw
Clarias fuscus	Chinese catfish
Clemmys marmorata	western pond turtle
Clevelandia ios	arrow goby
Clupea harengus pallasi	Pacific herring
Clupea pallasi	herring
Cnemidophorus martyris	San Pedro Martir whiptail
Cnemidophorus maximus	cape giant whiptail
Cnemidophorus tigres	San Diegan tiger whiptail
Coccoloba barbadensis	croton draco
Coccoloba caracasana	papaturro
Coccoloba floribunda	a woody euphorb
Coccoloba goldmanii	Sonoran sea grape
Coccoloba liebmannii	a species of sea grape
Coccoloba uvifera	seagrape
Cocconeis	a genus of diatom
Cochlearius cochlearius	boat-billed heron
Cochlospermum vitifolium	Brazilian rose
Cocos nucifera	coconut
Colaptes auratus	gilded flicker
Coleonyx switaki	Switak's banded gecko
Coleonyx variegatus	western banded gecko
Colpichthys regis	charal silverside
Coluber aurigulus	Baja California striped whip snake
Coluber flagellum	coachwhip
Coluber lateralis	California whipsnake

(*Continued*)

Scientific Nomenclature	Common Descriptor(s) [Not Exhaustive]
Columba cayennensis	pale-vented pigeon
Columbina inca	Mexican dove
Columbina passerina	turtle dove
Columbina passerina pallescens	common ground dove
Columbina passerina socorrensis	Socorro dove
Commelina diffusa	honohono
Commicarpus scandens	climbing wartclub
Conepatus mesoleucus	American hogged-nosed skunk
Conocarpus erectus	button mangrove
Contopus sordidulus	western wood-pewee
Conus brunneus	Wood's brown cone
Conus princeps	prince cone
Coragyps atratus	vulture
Corallina vancouveriensis	graceful coral seaweed
Cordia alliodora	salmwood
Cordia elaeagnoides	bocote
Cordylanthus maritimus maritimus	salt marsh bird's beak
Cordylanthus palmatus	palmate-bracted bird's beak
Corvus corax	common raven
Corvus corax clarionensis	Clarion Island raven
Corynorhinus townsendii	Townsend's big-eared bat
Coryphaena hippurus	dorado
Corytophanes percarinatus	Guatemalan helmeted basilisk
Costoanachis sanfelipensis	a species of sea snails in the family Columbellidae
Cottidae	sculpin
Cottus asper	prickly sculpin
Crassostrea corteziensis	Cortez oyster
Crassostrea gigas	Pacific Oyster
Crassostrea iridescens	a species of rock oyster
Crassostrea virginica	eastern Oyster
Crassulaceae	also known as the stonecrop family or the orpine family, are a diverse family of dicotyledon flowering plants characterized by succulent leaves
Crescentia alata	Mexican calabash
Cressa truxillensis	spreading alkaliweed
Creteva tapia	spider flower tree
Cricetidae	family of rodents
Crinum	genus of about 180 species of perennial plants that have large showy flowers on leafless stems, and develop from bulbs
Crinum erubescens	swamp lily
Crocodylus acutus	American crocodile
Crocodylus moreletii	Morelet´s crocodile
Crotalus atrox	pink rattlesnake

(Continued)

Scientific Nomenclature	Common Descriptor(s) [Not Exhaustive]
Crotalus atrox atrox	western diamondback rattlesnake
Crotalus basiliscus	Mexican west coast rattlesnake
Crotalus cerastes	little horned viper
Crotalus durissus	South American rattlesnake
Crotalus enyo	Baja California rattlesnake
Crotalus mitchellii	speckled rattlesnake
Crotalus ruber	red diamond rattlesnake
Crotalus viridis	prairie rattlesnake
Crotaphytus collaris	eastern collared lizard
Crotaphytus vestigium	Baja California collared lizard
Croton californicus	California croton
Croton masonii	species in the spurge family, Euphorbiaceae
Croton panamensis	species in the spurge family, Euphorbiaceae
Croton punctatus	beach-tea
Crotophaga sulcirostris	groove-billed ani
Crucibulum monticulus	a species in the family Calyptraeidae
Crucibulum spinosum	a species in the family Calyptraeidae
Cryosophila nana	palm
Cryptostegia grandiflora	rubber vine
Ctenosaura hemilopha	Sonoran woodpecker
Ctenosaura pectinata	Mexican spiny-tailed iguana
Ctenosaura similis	black iguana
Cyanocorax beecheii	purple-backed jay
Cyclotella	a genus of diatom
Cygnus buccinator	trumpeter swans
Cygnus columbianus	tundra swan
Cylindropuntia alcahes var. burrageana	a species of cholla cactus
Cylindropuntia alcahes var. gigantensis	a species in the family Cactaceae
Cylindropuntia bigelovii var. ciribe	teddy bear cholla
Cylindropuntia spp.	cholla
Cynodon sp.	bermudagrass
Cynoscion	genus of marine ray-finned fishes belonging to the family, Sciaenidae
Cynoscion macdonaldi	totoaba
Cynoscion nannus	dwarf weakfish
Cynoscion reticulatus	corvina
Cynoscion xanthulus	yellow-mouthed corvin
Cyperaceae	the sedge family
Cyperus	a genus of sedges
Cyperus dentoniae	hairy flatsedge
Cyperus ligularis	swamp flatsedge
Cyperus sanguineo-ater	species of sedge that is native to Central America

(Continued)

Scientific Nomenclature	Common Descriptor(s) [Not Exhaustive]
Cyprinodon macularius	desert pup
Cyprinus carpio	carp
Cyrtocarpa edulis	desert plum
Dalea maritima	a species of prairie clover
Dasyatis	genus of stingray in the family Dasyatidae
Dasyatis depterura	diamond stingray
Dasyatis spp.	ray species
Dasypus novemcinctus	Armadillo
Delphinus capensis	long-beaked common dolphin
Delphinus delphis	common dolphin
Dendrocygna autumnalis	arboreal ducks
Dendroica auduboni	olive verdigris
Dendroica coronata	yellow-rumped warbler
Dendroica coronata auduboni	Audubon's warbler
Dendroica nigrescens	black-throated gray warbler
Dendroica pensylvanica	chestnut-sided warbler
Dendroica petechia	American yellow warbler
Dendroica towsendi	Townsend's warbler
Denticula subtilis	a species of diatom in family Bacillariaceae
Dermatolepis	ray-finned fish
Dermochelys coriacea	leatherback turtle
Diapterus peruvianus	Peruvian mojarra
Diapterus rhombeus	rhomboid mojarra
Diclidurus albus	white bat
Didelphis marsupialis	opossum
Didelphis virginiana	Virginia opossum
Diectomis fastigiata	foldedleaf grass
Digitaria ciliaris	fingergrass
Diospyros oaxacana	a species in the family Ebenaceae
Dipodomys gravipes	San Quintín kangaroo rat
Dipodomys merriami	Merriam's kangaroo rat
Dipodomys merriami platycephalus	Merriam's kangaroo rat
Dipsas gaigeae	Gaige's thirst snail-eater
Dipsosaurus dorsalis	desert iguana
Distichlis palmeri	saltgrass
Distichlis spicata	saltgrass
Dodonaea	genus of about 70 species of flowering plants, often known as hop-bushes, in the family, Sapindaceae
Dodonaea viscosa	hop bush
Dolphina anchovy	short anchovy
Dolphinus delphis	short-beaked common dolphin
Dormitator latifrons	Pacific fat sleeper

(Continued)

Scientific Nomenclature	Common Descriptor(s) [Not Exhaustive]
Dosidicus gigas	squid species
Downingia spp.	downingia
Drymarchon corais	indigo snake
Dryocopus lineatus	lineated woodpecker
Dudleya anthonyi	Anthony's live forever
Echinaster luzonicus	asteroid species
Echinocactus platyacanthus	giant barrel cactus
Echinocereus leucanthus	clambering cactus
Echinocereus sciurus var. floresii	a species in the family Cactaceae
Echinothrix calamaris	echinoid species
Egretta caerulea	little blue heron
Egretta rufescens	reddish egret
Egretta thula	snowy egret
Egretta tricolor	tricolored heron
Eichhornia crassipes	common water hyacinth
Elaphe rosaliae	Baja California ratsnake
Eleocharis	a cosmopolitan genus of 250 or more species of flowering plants in the sedge family, Cyperaceae
Eleotridae	sleeper gobies
Eleotris sandwicensis	Sandwich Island sleeper, Hawaiian sleeper, or 'o'opu 'akupa
Eleutherodactylus modestus	species of frog in the family Eleutherodactylidae
Eleutherodactylus pallidus	species of frog in the family Eleutherodactylidae
Elgaria paucicarinata	San Lucan alligator lizard
Elgaria velazquezia	central peninsular alligator lizard
Elymus arenaria sedges	beach ryegrass
Elytraria imbricata	purple scalystem
Embiotocidae	surfperch
Empetrum nigrum	crowberry
Empidonax difficilis	Pacific-slope flycatcher
Empidonax difficilis difficilis	Pacific-slope flycatcher (difficilis)
Empidonax traillii extimus	southwestern willow flycatcher
Encelia farinosa	brittlebush
Engraulidae	anchovy family
Engraulis mordax	northern anchovy
Enhydra lutris	sea otters
Enhydra lutris nereis	southern sea otter
Enterolobium cyclocarpum	elephant-ear tree
Enteromorpha	genus of the family Ulvaceae (sea lettuce)
Enteromorpha plumosa	a species within the family Ulvaceae
Eptesicus fuscus	big brown bat
Eragrostis ciliaris	gophertail lovegrass
Eragrostis prolifera	Dominican lovegrass

(*Continued*)

Scientific Nomenclature	Common Descriptor(s) [Not Exhaustive]
Erethizon dorsatum	North American porcupine
Eretmochelys imbricata	hawksbill sea turtle
Eridiphas slevini	Baja California nocturnal culebra
Eriogonum	genus of the wild buckwheats
Eriogonum fastigiatum	San Antonio del Mar buckwheat
Eriophorum scheuchzen	cottongrass
Eryngium racemosum	delta button celery
Erythea brandegeei	San José hesper palm
Eschrichtius robustus	gray whale
Esenbeckia hartmanii	species of flowering plants in the family Rutaceae
Eucheuma uncinatum	species of red algae in the family Solieriaceae
Eucinostomus argenteus	spotfin mojarra
Eucinostomus currani	Pacific flagfin mojarra
Eucinostomus dowii	Dow's mojarra
Eucinostomus entomelas	dark-spot mojarra
Eucyclogobius newberryi	tidewater goby
Eudocimus albus	American white ibis
Eugerres axillaris	black-axillary mojarra
Eumeces lagunensis	cachora
Eumetopias jubatus	Steller sea lions
Euphorbia	refers to the family Euphorbiaceae, or spurge, which contains around 7,000 species in 218 genera, comprising herbaceous annuals and perennials, woody shrubs and trees, plus a few climbers
Euphorbia misera	milky grass
euphorbias of genus	croton
Eupomacentrus rectifraenum	Cortez damselfish
Eupsittula canicularis	orange-fronted parakeet
Euthynnus lineatus	black skipjack
Eysenhardtia adenostyles	a species of flowering plants in the family Fabaceae
Eysenhardtia polystachya	a species of flowering plants in the family Fabaceae
Falco femoralis	aplomado falcon
Falco mexicanus	prairie falcon
Falco peregrinus	peregrine falcon
Falco peregrinus anatum	American peregrine falcon
Falco peregrinus pealei	peregrine falcons
Falco sparverius	American kestrel
Farfantepenaeus brevirostris	crystal shrimp
Farfantepenaeus californiensis	yellowleg shrimp
Farfantepenaeus stylirostris	blue shrimp
Felis catus	domestic cat
Felis concolor	puma

(Continued)

Scientific Nomenclature	Common Descriptor(s) [Not Exhaustive]
Felis onca	jaguar
Felis yagouaroundi	jaguarundi
Feresa attenuata	pygmy killer whale
Ferocactus herrerae	twisted barrel cactus
Ferocactus townsendianus var. townsendianus	barrel cactus variety
Ficus	genus in the family Moraceae
Ficus cotinifolia	alamo tree
Ficus glabrata	common tropical tree in the fig genus of the family Moraceae
Ficus macrocarpa	Chinese banyan
Ficus mexicana	tree species in the family Moraceae
Ficus petiolaris	petiolate fig
Fimbristylis dichotoma	forked fimbry
Forestiera	genus of flowering plants in the olive family, Oleaceae
Forestiera rhamnifolia	species of flowering plant in the olive family, Oleaceae
Forpus cyanopygius	Mexican parrotlet
Fouquieria	genus of 11 species of desert plants, in the family Fouquieriaceae
Fouquieria columnaris	boojum tree
Fouquieria diguetii	palo Adán
Fouquieria diguetti	scrub
Fouquieria peninsularis	Baja ocotillo
Fouquieria splendens	ocotillo
Frankenia grandifolia	alkali heath
Frankenia palmeri	Palmer's seaheath
Frankenia salina	alkali heath
Frankenia spp.	sea-heath within the family Frankeniaceae
Franseria magdalenae	ragweed
Fregata magnificens	magnificent frigatebird
Fregata magnificens rothschildi	magnificent frigatebird
Fregata minor	great frigatebird
Fregata minor palmerstoni	'iwa or great frigatebird
Fuirena simplex	western umbrella-sedge
Fulica alai	alae ke'oke'o or Hawaiian coot
Fulica americana	American coot
Fulica americana americana	American coot
Fundulus lima	killifish
Fundulus parvipinnis	California killifish
Fungia curvata	species of disc coral in the family Fungiidae
Galeichthyes sp.	white barbel
Galeichthys caerulescens	catfish
Galeocerdo cuvier	blue shark
Gallinula chloropus	common moorhen

(Continued)

Scientific Nomenclature	Common Descriptor(s) [Not Exhaustive]
Gambelia copeii	Baja California leopard lizard
Gambelia wislizenii	horse-killing lizard
Gambelia wislizenii copeii	Cope's leopard lizard
Gambusia affinis	mosquito fish
Gaslerosleus aculearus	threespine stickleback
Gastrophryne olivacea	little toad
Gastrophryne olivacea;	narrow-mouthed toad
Gastrophryne usta	two-spaded narrow-mouthed toad
Gastrophyrne olivacea	western narrow-mouthed toad
Gavia immer	common loon
Gavia pacifica	Pacific loon
Gaviformes	genus of loons
Gecarcinus planatus	Socorro Island crab
Genyonemus lineatus	white croaker
Geococcyx californianus	greater roadrunner
Geococcyx velox	lesser roadrunner
Geomys bursarius	plains pocket gopher
Geothllypis beldingi goldmani	Belding's yellowthroat (goldmani)
Geothllypis trichas occidentalis	a subspecies of Belding's yellowthroat
Geothlypis beldingi	Belding's yellowthroat
Geothlypis trichas	common yellowthroat
Gerreidae	mojarra family
Gerres cinereus	yellow fin mojarra
Gerres cinereus	mojarra species
Gerrhonotus paucicarinatus	species of anguid lizards referred to as alligator lizards, in the family Anguidae
Gillichthys mirabilis	longjaw mudsucker
Gillichthys seta	fish
Ginglymostoma cirratum	nurse shark
Girella nigricans	opaleye
Girella simplicidens	gulf opaleye
Glaucidium minutessimum	East Brazilian pygmy owl
Glaucidium palmarum	colima pygmy owl
Gliricidia	genus of flowering plants in the legume family, Fabaceae
Gliricidia sepium	gliricidia
Globicephala macrorhynchus	long-finned pilot whale
Globicephala macrorhynha	pilot whale
Gobiesox juniperoserrai	clingfish
Gonyosoma oxycephalum	arboreal ratsnake
Gopherus agassizii	desert tortoise
Gracilaria	genus of red algae (Rhodophyta)
Gracilaria spinigera	species of red algae (Rhodophyta)

(Continued)

Scientific Nomenclature	Common Descriptor(s) [Not Exhaustive]
Grampus griseus	gray dolphin
Grus canadensis	sandhill crane
Guaiacum coulteri	guayacan
Guarea glabra	alligatorwood
Guavina guavina	guavina
Guavina micropus	Pacific guavina
Guayacum coulteri	shrub
Guazuma ulmifolia	west Indian elm
Guettarda insularis	species of tree in the family Rubiaceae
Guettarda sp.	trees such as
Gulo gulo	wolverine
Gygis alba	white tern
Hackelochloa granularis	pitscale grass
Haematopus palliatus	oystercatcher
Haematopus palliatus frazari	American oystercatcher
Haematoxylon brasiletto	Mexican logwood
Haematoxylum brasiletto	brazilwood
Haliaeetus leucocephalu	American bald eagle
Haliaeetus leucocephalus	American bald eagles
Halichoeres nicholsi	spinster wrasse
Haliotis sp.	an abalone species
Halodule wrightii	sea grass
Halymenia californica	species of red algae in the family Halymeniaceae
Halymenia templetonii	species of red algae in the family Halymeniaceae
Handroanthus chrysanthus	guayacán
Haplopappus venetus	coast goldenbush
Haplopappus venetus var. vernonioides	coast goldenbush
Hoplopagrus guentherii	Mexican barred snapper
Hazardia berberidis	barberry-leaf goldenbush
Helicteres guazumaefolia	a species in the family Malvaceae
Heliornis fulica	sungrebe
Heliotropium curassavicum	salt heliotrope
Heloderma horridum	Mexican beaded lizard
Heloderma suspectum	gila monster
Hemidactylus frenatus	common house gecko
Herpailurus yagouaroundi	jaguarundi
Hesperoleucus symmetricus	California roach
Heteranthera limosa	blue mud plantain
Heteroderma gibbsii	species of red algae in the family Corallinaceae
Heteromeles arbutifolia	toyon
Heteromys pictus	painted spiny pocket mouse

(Continued)

Scientific Nomenclature	Common Descriptor(s) [Not Exhaustive]
Heteroscelus incanus	wandering tattler
Hibiscus tiliaceus	variegated mahoe tree
Hibiscus tiliaceus	hau trees
Hilaria rigida	big galleta
Himantopus mexicanus	black-necked stilt
Himantopus mexicanus knudseni	aeʻo or Hawaiian stilt
Hippoglossus stenolepis	Pacific halibut
Hippomane mancinella	manchineel
Hirundo rustica	barn swallow
Hodomys	a genus of the rodents in the family Cricetidae
Holbrookia maculata	lesser earless lizard
Holothuria atra	Holothurian
Holothuria edulis	edible sea cucumber
Hura polyandra	a euphorb
Hydrocotyle mexicana	pennywort
Hydrolithon decipiens	a species that belongs to the family Corallinaceae
Hydrolithon farinosum	a species that belongs to the family Corallinaceae
Hydroprogne caspia	Caspian tern
Hyla regilla	Pacific tree frog
Hyla sartori	amphibian species
Hylocharis xantusi	hummingbird
Hylocharis xantusii	Xantus's hummingbird
Hymenaea courbaril	stinking toe
Hypnea cervicornis	species of red algae in the family Florideophyceae
Hypnea johnstonii	species of red algae in the family Florideophyceae
Hypomesus transpacificus	delta smelt
Hypo rhamphus	genus of halfbeaks, in the family Hemiramphidae
Hypsiglena torquata	cat's eye nocturnal snake
Hypsopsetta guttulate	diamond turbot
Hysterocarpus traskii traskii	tule perch
Ictalurus meridionalis	southern blue catfish
Icteria virens auricollis	yellow-breasted chat
Icterus cucullatus	hooded oriole
Icterus spurius	chestnut lark
Icterus wagleri	black-vented oriole
Ictinia mississippiensis	Mississippi kite
Iguana iguana	green iguana
Ilex brandegeana	a species in the holly family, Aquifoliaceae
Indigofera miniata	scarlet pea
Ipomea arborescens	morning glory tree
Ipomoea pes-caprae	bayhops

(Continued)

Scientific Nomenclature	Common Descriptor(s) [Not Exhaustive]
Ipomoea stolonifera	fiddle-leaf morning glory
Ischnura ramburii	damselfly
Isostichopus fuscus	brown sea cucumber
Istiophorus platypterus	Indo-Pacific sailfish
Ixobrychus exilis	least bittern
Ixobrychus exilis	western least bittern
Jacana spinosa	northern jacana
Jacaratia Mexicana	bonnet
Jacquinia aurantiaca	cudjoewood
Jania adhaerens	a species of red macroalgae in the family Corallinaceae
Jatropha	genus of flowering plants in the spurge family, Euphorbiaceae
Jatropha chamelensis	yellow paper
Jatropha cinerea	Arizona nettlespurge
Jatropha cuneata	matacora
Jatropha curcas	a euphorb
Jouvea pilosa	a species in the family Poaceae
Juncus	genus of monocotyledonous flowering plants, commonly known as rushes
Juncus acutus	spiny rush
Juncus acutus var. sphaerocarpus	spiny rush
Juncus mexicanus	Mexican rush
Karwinskia calderonii	huilihuiste
Katsuwonus pelamos	skipjack
Kinosternon integrum	Mexican mud turtle
Kinosternon oaxacae	Oaxaca mud turtle
Kinosternon scorpioides	scorpion mud turtle
Kinosternon scorpioides cruentatum	red-cheeked mud turtle
Kinosternum integrum	freshwater turtles
Krameria	genus in the family Krameriaceae, commonly known as rhatany
Kuhlia sandwicensis	aholehole
Kyphosus	genus of chubs
Lagomorpha	order that consists of two families, including the Leporidae (rabbits and hares)
Lagopus lagopus	willow ptarmigan
Laguncularia racemosa	white mangrove
Lampropeltis getula	common king snake
Lampropeltis getula nigritus	Mexican black kingsnake
Lampropeltis triangulum	milk snake
Lampropeltis triangulum sinaloae	Sinaloan milk snake
Lantana camara	common lantana
Larrea tridentata	creosote

(Continued)

Scientific Nomenclature	Common Descriptor(s) [Not Exhaustive]
Larus argentatus	European herring gull
Larus atricilla	seagulls terns
Larus californicus	California gull
Larus heermanni	Heermann's gull
Larus livens	yellow-footed gull
Larus occidentalis	western gull
Larus pipixcan	Franklin's gull
Larus sp.	gull species
Lasiurus borealis	eastern red bat
Lasiurus ega	southern yellow bat
Lasthenia burkei	Burke's goldfields
Lasthenia spp.	goldfields species
Laterallus jamaicensis	black rail
Laterallus jamaicensis coturniculus	California black rail
Layia spp.	tidy tips
Lemna aequinoctialis	lesser duckweed
Lemna gibba	gibbous duckweed
Leopardus pardalis	ocelot
Leopardus wiedii	endangered margay
Lepidochelys olivacea	olive ridley sea turtle
Lepidurus packardi	vernal pool tadpole shrimp
Leporidae	family of rabbits and hares
Leptodactylus melanotus	reddish-brown white-lipped frog
Leptodon cayanensis	gray-headed hawk
Leptonycteris curasoae	Curaçao long-nosed bat
Leptophis diplotropis	Pacific Coast parrot snake
Leptotila verreauxi	white tipped dove
Leptotyphlops humilis	western blind snake
Lepus alleni	antelope jackrabbit
Lepus callotis	white-sided jackrabbit
Leucaena leucocephala	river tamarind
Leucaena macrophylla macrophylla	subspecies of tree in the family Fabaceae
Leuresthes sardina	gulf grunion
Lichanura trivirgata	Mexican rose boa
Lile gracilis	graceful herring
Lile stolifera	sardine
Limnanthes vinculans	Sebastopol meadowfoam
Limnodromus griseus	short-billed dowitcher
Limnodromus scolopaceus	long-billed sandpiper
Limonium	sea lavender
Limosa fedoa	curlews

(Continued)

Scientific Nomenclature	Common Descriptor(s) [Not Exhaustive]
Limosa fedoa	marbled godwit
Linx rufus	bobcat
Lithophyllum imitans	species of thalloid red algae belonging to the family Corallinaceae
Lithophyllum margaritae	species of thalloid red algae belonging to the family Corallinaceae
Lithothamnion crassisuculum	species of thalloid red algae belonging to the family Corallinaceae
Litopenaeus	genus of prawns
Litopenaeus stylirostris	blue shrimp
Litopenaeus vannamei	whiteleg shrimp
Lobelia laxiflora	cardinal flower
Loligo opalescens	market squid
Lonchocarpus rugosus	species in the legume family, Fabaceae
Lontra canadensis	North American river otter
Lontra longicaudis	neotropical otter
Lophocereus schotii	senita cactus
Lotus scoparius	deerweed
Ludwigia octovalvis	kamole
Lutjanidae	family of the snappers
Lutjanus	genus of marine ray-finned fish, snappers belonging to the family Lutjanidae
Lutjanus aratus	mullet snapper
Lutjanus argentiventris	snapper
Lutjanus colorado	Colorado snapper
Lutjanus griseus	gray snapper
Lutjanus guttatus	spotted nose snapper
Lutjanus novemfasciatus	Pacific dog snapper
Lutra annectens	southern river otter
Lutra canadiensis	river otter
Lycium	genus of box-thorns
Lycium brevipes	Baja desert-thorn
Lycium spp.	species of flowering plants in the family Solanaceae
Lygodium venustum	monarch climbing fern
Lynx rufus	Lynx
Lyonsia gouldii	razor clam
Lyropecten subnodosus	lion's paw scallop
Lysiloma candidum	white stick
Machaerocereus eruca	creeping devil
Machaerocereus gummosus	flowering plant in the family Cactaceae
Macrobrachium	genus of freshwater prawns or shrimps in the family Palaemonidae

(Continued)

Scientific Nomenclature	Common Descriptor(s) [Not Exhaustive]
Macrobrachium acanthurus	cinnamon river shrimp
Macrobrachium americanum	cauque river prawn
Macrobrachium grandimanus	Hawaiian river shrimp
Macrobrachium hobbsi	genus of freshwater prawns or shrimps in the family Palaemonidae
Macrobrachium lar	Tahitian prawn
Macrobrachium michoacanus	long-arm shrimp
Macrobrachium olfersii	bristled river shrimp
Macrobrachium tenellum	longarm river prawn
Macrotus californicus	California leaf-nosed bat
Madracis pharensis	genus of cnidarians belonging to the family Pocilloporidae
Maiopsis panamensis	panamic spidercrab
Malva parviflora	mallow
Mammillaria dioica	strawberry cactus
Mammillaria insularis	cactus species
Mammillaria mazatlanensis	pincushion cactus
Mammillaria occidentalis	room cactus
Mammillaria peninsularis	pincushion cactus
Mammillaria thornberi	old man's head
Manfreda	genus of flowering plants in the family Asparagaceae
Mangifera indica	mango
Marmosa canescens	grayish mouse opossum
Mastichodendron capiri	tree species in the family Sapotaceae
Masticophis anthonyi	Clarion Island whip snake
Masticophis aurigulus	Baja California squeaky snake
Masticophis flagellum	coachwhip
Masticophis lateralis	California whipsnake
Matelea	genus of flowering plants in the family Apocynaceae
Maytenus phyllanthoides	sweet mangrove
Megapitaria	genus of bivalves of the family Veneridae
Megapitaria squalida	chocolate clam
Megaptera novaeangliae	humpback whale
Megasorex	genus of shrew from the subfamily Soricinae
Melaleuca quinquenervia	paperbark
Melanerpes aurifrons	woodpecker
Melanerpes chrysogenys	golden-cheeked woodpecker
Melanerpes uropygialis	Gila woodpecker
Melanitta fusca	velvet scoter
Melanitta nigra	black white-winged scoter
Melanitta perspicillata	surf scoter
Melanoides	genus of freshwater snail of the family Thiaridae.
Melanotis caerulescens	blue mockingbird
Melia azedarach	chinaberry

(Continued)

Scientific Nomenclature	Common Descriptor(s) [Not Exhaustive]
Melocactus delessertianus	Turk's cap cactus
Melospiza lincolnii lincolnii	Lincoln's sparrow
Menticirrhus sp.	berrugata
Mephitis macroura	hooded skunk
Mergus serrator	red-breasted merganser
Mesembryanthemum	genus of flowering plants in the family Aizoaceae
Mesembryanthemum crystallinum	common ice plant
Micorpterus dolomieui	smallmouth bass
Micrastur semitorquatus	collared forest falcon
Microtus californicus aequivocatus	San Quintín meadow rat
Micruroides euryxanthus	Arizona coral snake
Micrurus distans	west Mexican coral snake
Micrurus proximans	Nayarit coral snake
Mimodes graysoni	Socorro mockingbird
Mimosa albida	species in the family Fabaceae
Mimosa pigra	giant sensitive tree
Mimosa pudica var. unijuga	sensitive plant
Mimulus dentilobus	sharpwing monkeyflower
Mimulus floribundus	monkeyflower
Mimulus glabratus var. glabratus	round-leaf monkey-flower
Mimulus guttatus	seep monkey flower
Mimus polyglottos	northern mocking bird
Mimus polyglottos polyglottos	northern mocking bird
Mirounga angustirostris	elephant seals
Mirtyllocactus cochal	candelabra cactus
Mniotilta varia	black-and-white warbler
Mollossidae	Molossidae is the fourth-largest family of bats, containing about 110 species
Monachus schauinslandi	Hawaiian monk seal
Monanthochloe littoralis	shoregrass
Monantochloe littoralis	saline grass
Monopterus albus	Oriental rice eel
Montipora	genus of small polyp stony corals of the family Acroporidae
Mormoops megalophylla	wrinkled-bearded bat
Mugil cephalus	flathead grey mullet
Mugil curema	white mullet
Mugil hospes	hospe mullet
Mugilidae	mullet family
Muhlenbergia	muhly grass
Muntingia calabura	strawberry tree
Murex elenensis	Chinese snail
Muricanthus nigritus	black-and-white murex

(Continued)

Scientific Nomenclature	Common Descriptor(s) [Not Exhaustive]
Muricanthus princeps	genus of sea snails in the family Muricidae
Mus musculus	house mouse
Musa paradisiaca	banana
Musonycteris harrisoni	banana bat
Mustela frenata	long-tailed weasel
Mustela vison	mink
Myiarchus cinerascens	ash-throated flycatcher
Mycteria americana	Stork
Mycteroperca	grouper
Myioborus pictus pictus	painted redstart
Myliobatis californica	bat ray
Myotis californicus	California myotis
Myotis peninsularis	peninsular myotis
Myotis vivesi	fishing bat
Myotis vivesi	fisher-myotis-bat
Myozetetes similis	flycatchers
Myrichthys maculosus	tiger snake-eel
Najas marina	sea naiad
Nassarius luteostomus	a species of nassa mud snail
Nassau narica	Badger
Nasua	Coati
Nasua narica	white-nosed coati
Nasua nasua	South American coati
Natalus stramineus	Mexican funnel-eared bat
Neostapfia colusana	a colusa grass
Neotoma	genus of the woodrat
Neotoma lepida	desert woodrat
Neotoma phenax	Sonoran woodrat
Neotoma phexna	field rat
Nephrolepis hirsutula	sword fern
Nerita funiculata	funiculate nerite
Neritina sp.	genus of medium-sized to small sea snails of the family Neritidae
Nitzschia	genus of diatom in the family Bacillariaceae
Nitzschia punctata	species of diatom in the family Bacillariaceae
Noctilio leporinus	greater bulldog bat
Nodipecten suibnodosus	giant lion's paw
Nomonyx dominicus	masked duck
Notiosorex crawfordi	Crawford's gray shrew
Numenius americanus	long-billed curlew
Numenius phaeophus hudsonicus	whimbrel
Numenius phaeopus	picocurvo

(Continued)

Scientific Nomenclature	Common Descriptor(s) [Not Exhaustive]
Numenius tahitiensis	bristle-thighed curlew
Nyctanassa violacea	yellow-crowned night heron
Nyctanassa violacea gravirostris	yellow-crowned night-heron
Nyctibius griseus	common potoo
Nycticorax nictycorax	black-crowned night herron
Nycticorax nycticorax	black-crowned night heron
Nycticorax nycticorax hoactli	auku'u or black-crowned night heron
Nyctomys sumichrasti	Sumichrast's vesper rat
Nymphaea	genus of water lily
Nymphaea ampla	dotleaf waterlily
Oceanodroma melania	black storm petrel
Oceanodroma microsoma	least storm-petrel
Oceanodroma tethys tethys	wedge-rumped storm petrel
Odocoileus	genus of medium-sized deer in the family Cervidae
Odocoileus hemionus	mule deer
Odocoileus hemionus peninsulae	while mule deer
Odocoileus virginianus	white-tailed deer
Oenothera primiveris	yellow desert evening primrose
Oenothera wigginsii	evening primrose
Okenia hypogaea	beach peanut
Oligoplites altus	longjaw leatherjacket
Olneya tesota	iron stick
Olneya tesota	ironwood
Onagraceae	family of flowering plants known as the willowherb family or evening primrose family
Onchorhynchus tshawytscha	chinook salmon
Oncorhynchus gorbuscha	pink salmon
Oncorhynchus keta	chum
Oncorhynchus kisutch	coho salmon
Oncorhynchus mykiss	steelhead
Oncorhynchus nerka	sockeye
Oncorhynchus spp.	Pacific salmon
Oncorhynchus tshawytscha	chinook
Onychoprion fuscatus	sooty tern
Onychoteuthis banksii	common clubhook squid
Ophioglossum engelmannii	limestone adder's-tongue
Oporornis tolmiei tolmiei	MacGillivray's warbler
Opuntia	genus of about 90 species in the family Cactaceae
Opuntia bergeriana	red-flower prickly pear
Opuntia bigelovii	teddy-bear cholla
Opuntia cholla	chain link cholla
Opuntia clavellina	a species of the family Cactaceae

(*Continued*)

Scientific Nomenclature	Common Descriptor(s) [Not Exhaustive]
Opuntia comonduensis	comondu cactus
Opuntia excelsa	lofty prickly pear
Opuntia fulgida	jumping cholla
Opuntia puberula	puberula cactus
Opuntia rileyi	a species of the family Cactaceae
Opuntia spraguei	a species of the family Cactaceae
Opuntia velutina	velutina cactus
Opuntia wilcoxii	Wilcox cactus
Orbignya	genus of palm of the family Arecaceae
Orbignya cohune	cohune palm
Orbignya guacuyule	coquito palm
Orcinus orca	killer whale
Oreopanax	genus of shrubs and trees in the family Araliaceae
Ortalis leucogastra	white-bellied chachalaca
Ortalis poliocephala	west Mexican chachalaca
Oryzomys melanotis	rat
Oryzomys palustris	marsh rice rat
Oscillatoria	a genus of filamentous cyanobacteria
Osgoodomys banderanus	Michoacan deer mouse
Ostrea palmula	mangrove oysters
Otus kennicottii	western screech owl
Ovis canadensis	bighorn sheep
Ovis canadensis mexicana	desert bighorn sheep
Ovis canadensis weemsi	Weems' bighorn
Oxybelis aneus	Daudin's vine snake
Oxyura jamaicensis	ruddy duck
Pachira aquatica	Guiana chestnut
Pachycereus pecten	one of several species in genus Pachycereus of the family Cactaceae
Pachycereus pecten-aboriginum	hairbrush
Pachycereus pringlei	Mexican giant cactus
Pachymedusa dacnicolor	Mexican leaf frog
Padina	genus of brown macroalgae
Palafoxia rosea	rosy palafox
Pandion haliaetus	osprey
Panicum maximum	guinea grass
Panthera onca	jaguar
Panulirus sp.	lobster
Parabuteo unicinctus	Harris' hawk
Parabuteo unicinctus superior	Harris' hawk
Paralabrax clathratus	kelp bass
Paralabrax maculatofasciatus	a grouper

(Continued)

Scientific Nomenclature	Common Descriptor(s) [Not Exhaustive]
Paralichthys californicus	California flounder
Parmenteria aculeata	cucumber tree
Parophrys vetulus	English sole
Parula pitiayumi graysoni	Socorro parula
Paspalum conjugatum	Hilo grass
Paspalum convexum	Mexican paspalum
Paspalum paniculatum	Russel river grass
Paspalum vaginatum	seashore paspalum
Passer domesticus domesticus	house sparrow
Passerculus sandwichensis	savannah sparrow
Passerculus sandwichensis beldingi	Belding's savannah sparrow
Passerculus sandwichensis rostratus	large-billed savannah sparrow
Passerina amoena	lazuli bunting
Pavona clivosa	a species of coral of the family Agariciidae
Pavona gigantea	a species of coral of the family Agariciidae
Pecari tajacu	collared peccary
Pecten voqdesi	flying clam
Pectis arenaria	flowering species within the family Asteraceae
Pectis saturejoides	species of flowering plants in the family Asteraceae
Pediastrum	genus of green algae in the family Hydrodictyaceae
Pelamys platurus	sea snake
Pelecaniformes	order of the pelicans and herons
Pelecanus erythororchinchus	white pelican
Pelecanus erythrorhynchos	American white pelican
Pelecanus occidentalis	brown pelican
Pelecanus occidentalis californicus	brown pelican
Penaeus californiensis	brown shrimp
Penaeus stylirostris	blue shrimp
Peniocereus marianus	reina de la noche
Penneus	genus of prawn
Pennisetum ciliare	buffelgrass
Pennisetum setosum	fountain grass
Peponocephala electra	melon-headed whales
Pereskiopsis porteri	species of the family Cactaceae
Peromyscus eremicus	cactus mouse
Peromyscus eva	Eva's desert mouse
Peromyscus maniculatus	deer mouse
Peromyscus maniculatus coolidgei	mouse
Peromyscus perfulvus	mouse
Persea americana	avocado
Petrosaurus repens	short-nosed rock lizard

(Continued)

Scientific Nomenclature	Common Descriptor(s) [Not Exhaustive]
Petrosaurus thalassinus	Baja California stone lizard
Phaethon aethereus	red-billed tropicbird
Phaethon rubricauda rothschildi	red-tailed tropicbird
Phaeton rubricauda	red-tailed tropicbird
Phainopepla nitens	phainopepla
Phainopepla nitens lepida	phainopepla
Phalacrocorax	genus including the cormorants, within the family Phalacrocoracidae
Phalacrocorax auritus	cormorant
Phalacrocorax brasilianus	neotropic cormorant
Phalacrocorax olivaceus	neotropic cormorant
Phalacrocorax penicillatus	Brandt's cormorant
Phalaropus lobatus	red-necked phalarope
Phalaropus tricolor	Wilson's phalarope
Pheucticus melanocephalus	black-headed grosbeak
Phlebodium decumanum	calaguala fern
Phoca virulina richardsii	harbor seals
Phoca vitulina	common seal
Phocoena sinus	vaquita
Phoenix dactylifera	date palm
Phragmites australis	common reed
Phragmites communis	reeds
Phrynosoma coronatum	coast horned lizard
Phrynosoma mcallii;	flat-tailed horned lizard or flat-tailed false chameleon
Phyllanthus elsiae	ciruelillo
Phyllantus standleyi	species of plant in the family Phyllanthaceae
Phyllodactylus tinklei	leaf-toed gecko
Phyllodactylus unctus	San Lucan gecko
Phyllodactylus xanti	Xantus' leaf-toed gecko
Phyllorhynchus decurtatus	spotted leaf-nosed snake
Phyllospadix scouleri	Scouler's surfgrass
Phyllospadix torreyi	Torrey's surfgrass
Phyllostachys nigra	bamboo
Phyllostomidae	family of new world leaf-nosed bats
Phymatosorus scolopendria	lau'ae fern
Physalis minuta	groundcherry
Physeter macrocephalus	sperm whale
Picoides scalaris	ladder-backed woodpecker
Pinctada mazatlanica	Panama pearl oyster
Pinna rugosa	callus
Pipilo chlorurus	green-tailed towhee
Pipilo erythrophthalmus socorroensis	Socorro towhee

(Continued)

Scientific Nomenclature	Common Descriptor(s) [Not Exhaustive]
Pipistrellus hesperus	canyon bat
Piranga ludoviciana	western tanager
Piriqueta cistoides	morning buttercup
Pisonia	genus of the bird-catcher tree
Pisonia grandis	grand devil's-claws
Pistia stratiotes	water lettuce
Pitangus sulphuratus	great kiskadee
Pitar spp.	oysters clams
Pithecellobium	genus of the tree referred to as monkeypod
Pithecellobium dulce	Madras thorn
Pithecellobium lanceolatum	flowering plants in the family Fabaceae
Pituophis melanoleucus	pine snake
Pituophis vertebralis	cape gopher snake
Plagiobothrys spp.	popcorn flower
Plantago insularis	plantain
Platalea ajaja	roseate spoonbill
Platymiscium dimorphandrum	guapinol
Plicopurpura pansa	purple snail
Pluchea	genus of flowering plants in the family Asteraceae
Pluchea indica	Indian pluchea
Pluchea odorata	sweetscent
Pluchea sericea	cachanilla or pluchea
Pluvialis fulva	Pacific golden plover
Pluvialis squatarola	lapwing
Pluvialis squatarola	black-bellied plovers
Pocillopora	genus of stony corals in the family Pocilloporidae
Pocillopora capitata	a stony coral species in the family Pocilloporidae
Pocillopora damicornis	a stony coral species in the family Pocilloporidae
Pocillopora eydouxi	a stony coral species in the family Pocilloporidae
Pocillopora meandrina	a stony coral species in the family Pocilloporidae
Pocillopora verrucosa	a stony coral species in the family Pocilloporidae
Podiceps nigricollis	black-necked grebe
Podilymbus podiceps	pied-billed grebe
Poecilia	genus of guppies; in the family Poeciliidae
Poecilia reticulata	guppy
Polioptila caerulea obscura	blue-grey gnatcatcher
Polioptila californica	California gnatcatcher
Polioptila californica atwoodi	Northern Baja California perlet
Polyborus plancus	crested caracara
Polysiphonia pacifica	pretty polly
Polysiphonia simplex	a species of filamentous red algae

(*Continued*)

Scientific Nomenclature	Common Descriptor(s) [Not Exhaustive]
Polysticta stelleri	Steller's eider
Pomacea spp.	apple snails
Populus brandegeei var. glabra	gueribo
Populus fremontii	poplar trees
Populus monticola	mountain cottonwood
Populus trichocarpa	black cottonwood
Porites lobata	lobe coral
Porites panamensis	species of hard coral in the family Poritidae
Porolithon	a genus of red algae
Porphyrio martinica	American purple gallinule
Potamogetonaceae	the pondweed family
Potomegeton pectinatus	sago pondweed
Prionurus puncatus	coral
Procambarus clarkia	crayfish
Procyon insularis	Tres Marias raccoon
Procyon lotor	common raccoon
Procyon lotor grinelli	Baja California raccoon
Prosopis	genus of mesquite
Prosopis articulata	ironwood or mesquite species
Prosopis glandulosa	honey mesquite
Prosopis juliflora	ironwood or mesquite species
Prosopis palmeri	ironwood or blue mesquite
Prosopis sp.	ironwood or mesquite
Prunus serotina	black cherry
Psammocora brighami	a coral species in the family Siderastreidae
Psammocora stellata	a stony coral species
Psedoleptodeira uribei	cat-eyed snake
Pseudacris	genus of chorus frogs
Pseudacris regilla	Pacific tree frog
Pseudobalistes spp.	triggerfish
Pseudobalists spp.	cochi
Pseudorca crassidens	false killer whale
Psidium galapageium	galapgos guava
Psidium guajava	common guava
Psidium socorrense	guavaberry tree
Pteria sterna	mother-of-pearl shell
Pternia sterna	Pacific wing oyster
Pterocereus gaumeri	a columnar cactus species
Pteronotus davyi	lesser baldback bat
Pteronotus parnellii	Parnell's whiskered bat
Puffinus auricularis	Revillagigedo shearwater

(Continued)

Scientific Nomenclature	Common Descriptor(s) [Not Exhaustive]
Puffinus griseus	sooty shearwater
Puffinus opisthomelas	Mexican shearwater
Puffnus opisthomelas	black-vented shearwater
Pulvinaria urbicola	scale
Puma concolor	puma
Puma concolor improcera	cougar
Purpura pansa	species of sea snail in the family Muricidae
Purpura patula	widemouth rocksnail
Pyrocephalus rubinus flammeus	vermilion flycatcher
Quiscalus mexicanus	great-tailed grackle
Rallus limicola	Virginia rail
Rallus limicola limicola	Virginia rail
Rallus longirostris	mangrove rail
Rallus longirostris levipes	light-footed clapper rail
Rallus longirostris obsoletus	California clapper rail
Rallus longirostris yumanensis	Yuma Ridgeway's rail
Rana	genus of frogs commonly known as the Holarctic true frogs, pond frogs or brown frogs
Rana aurora draytonii	California red-legged frog
Rana boylii	foothill yellow-legged frog
Rana catesbeiana	Bullfrog
Rana forreri	Forrer's grass frog
Rana magnaocularis	northwest Mexico leopard frog
Randia armata	Indigoberry
Rangifer tarandus	barren ground caribou
Rathbunia alamosensis	octopus cactus
Rattus rattus	black rat
Rattus sp.	genus of rodents in the family Muridae
Recurvirostra americana	American avocet
Regulus calendula calendula	ruby-crowned kinglet
Reithrodontomys raviventris	salt marsh harvest mouse
Rhincodon typus	shark
Rhinoclemmys pulcherrima	painted wood turtle
Rhinobatos	a genus of fish in the Rhinobatidae family, commonly referred to as guitarfish
Rhizophora mangle	red mangrove
Rhizophora x harrisonii	hybrid mangrove
Rhizosolenia	diatom genera
Rhus integrifolia	lemonade berry
Rhus laurina	laurel sumac
Rhynchelytrum repens	natal grass
Rhynoclemmys rubida	Mexican spotted wood turtle

(Continued)

Scientific Nomenclature	Common Descriptor(s) [Not Exhaustive]
Ricinus	genus of the castor oil plant
Rissoina stricta	species of minute sea snails in the family Rissoinidae
Rodentia	the order of rodents
Roseodendron donnell-smithii	primavera
Rostrhamus sociabilis	snail kite
Ruellia californica	rama parda
Ruellia peninsularis	chamizo
Ruppia maritima	widgeon grass
Rynchops niger	black skimmer
Sabal mexicana	Mexican sabal
Saccarum officinarum	sugarcane
Saccostrea palmula	species of rock oyster in the family Ostreidae
Sagittaria lancifolia	lanceleaf arrowhead
Sagittaria latifolia	arrowhead
Salicaceae	the willow family
Salicornia	genus of glassworts
Salicornia bigelovii	dwarf glasswort
Salicornia europaea	glasswort
Salicornia pacifica	brine grass
Salicornia subterminalis	glasswort
Salicornia virginica	pickleweed
Salix bonplandiana var. bonplandiana	Bonpland willow
Salix chilensis	Chilean pencil willow
Salix gooddingii	willow trees
Salíx lasiolepis	arroyo willow
Salix sitchensis	sitka willow
Salix spp.	of the willow tree species
Salmo gairdneri	rainbow trout
Salpinctes obsoletus obsoletus	rock wren
Salvadora hexalepis	desert patch-nosed snake
Salvadora hexalepis virgultea	coast patch-nosed snake
Salvadora mexicana	Mexican patch-nosed snake
Salvelinus malma	Dolly Varden
Samanea saman	monkeypod
Sardinops caeruleus	Pacific sardine
Sardinops sagax caerulea	Pacific sardine
Sargassum	genus of large brown seaweed in the family Sargassaceae
Sargassum herporthizum	species of large brown seaweed in the family Sargassaceae
Sargassum johnstonii	species of brown algae in the family Phaeophyceae
Sargassum lapazeanum	species of brown algae in the family Phaeophyceae
Sargassum macdougalii	species of brown algae in the family Phaeophyceae

(Continued)

Scientific Nomenclature	Common Descriptor(s) [Not Exhaustive]
Sargassum sinicola	a brown algae
Sarotherodon	large genus in the family Cichlidae (cichlids). It belongs to the tribe Tilapiini in the subfamily Pseudocrenilabrinae
Sauromalus hispidus	Angel Island chuckwalla
Sauromalus obesus	chuckwalla
Sauromalus obesus townsendi	chuckwalla
Sayornis nigricans semiatra	black phoebe
Scaphiopus couchii	Couch's spadefoot toad
Sceloporus	genus of spiny lizards
Sceloporus clarkii	Clark's spiny lizard
Sceloporus horridus	horrible spiny lizard
Sceloporus hunsaker	Hunsaker's spiny lizard
Sceloporus licki	cape scaly lizard
Sceloporus magister	desert spiny lizard
Sceloporus nelsoni	Nelson's spiny lizard
Sceloporus orcutti	granite spiny lizard
Sceloporus siniferus	longtail spiny lizard
Sceloporus zosteromus	Baja California spiny lizard
Scheelea preussii	manaca palm
Schoenoplectus americanus	American bulrush
Schoenoplectus californicus	Bulrush
Schoenoplectus lacustris	neki or swamp fern
Sciades troschelii	chili sea catfish
Sciadodendrom excelsum	jobo de lagarto
Sciaenidae	family of ray-finned fishes belonging to the order Acanthuriformes, commonly called drums or croakers
Scirpus	genus of grass-like species in the sedge family Cyperaceae
Scirpus koilolepis	keeled bulrush
Scirpus maritimus	alkali bulrush
Scirpus maritimus var. paludosus	makai sedge
Scirpus spp.	bulrush species
Sciurus	genus of tree squirrels in the family Sciuridae
Sciurus aureogaster	Mexican gray squirrel or Mexican red-bellied squirrel
Sciurus colliaei	Collie's squirrel
Scomberomorus maculatus	sierra
Scomberomorus sierra	Pacific sierra
Scorpaenidae	rockfish
Scyllarides princeps	spiny clams
Sedum alamosanum	alamosa stonecrop
Seiurus noveboracensis	northern waterthrush

(Continued)

Scientific Nomenclature	Common Descriptor(s) [Not Exhaustive]
Senecio californicus var. ammophilus	flowering species within the family Asteraceae
Seriola lalandi	yellowtail amberjack
Sessuvium verrucosum	verrucose seapurslane
Sesuvium portulacastrum	sea purslane
Sesuvium verrucosum	beach purslane
Sideroxylon capiri	species of shrub in the family Sapotaceae
Sigmodon mascotensis	west Mexican cotton rat
Silvilagus sp.	cottontail rabbit
Simmondsia chinensia	jojoba
Skeletonema	genus of a cylindrical diatom of the family Skeletonemataceae
Smilisca baudinii	common Mexican tree frog
Solanum	genus of nightshade
Somateria mollissima	common king eider
Somateria spectabilis	king eider
Sonora semiannulata	western ground snake
Sorex ornatus	ornate shrew
Sorex ornatus sinuosis	Suisun shrew
Sorex vagrans haliocetes	salt marsh wandering shrew
Sorghastrum	genus of the family Poaceae
Sorghum bicolor	great millet
Spartina foliosa	cordgrass
Spatula clypeata	northern spatula
Spatula discors	blue-winged teal
Spermophilus atricapillus	Baja California rock squirrel
Speyeria zerene myrtleae	Myrtle's silverspot butterfly
Sphaeralcea hainesii	wild poppy
Sphoeroides annulatus	botete
Sphoeroides spp.	botetes
Sphyraena barracuda	barracuda
Sphyrapicus nuchalis	red-naped sapsucker
Sphyrna lewini	scalloped hammerhead
Sphyrna zygaena	smooth hammerhead
Spilogale gracilis	western spotted skunk
Spilogale gracilis leucoparia	spotted skunk
Spilogale pygmaea	pygmy skunk
Spionidae	tube worms
Spirodela polyrhiza	common duckweed
Spizella breweri breweri	Brewer's sparrow
Spizella pallida	clay-colored sparrow
Spizella passerine	chipping sparrow
Spondias purpurea	plum

(*Continued*)

Scientific Nomenclature	Common Descriptor(s) [Not Exhaustive]
Spondylus calcifer	species of the spiny oyster
Spondylus princeps unicolor	species of bivalve mollusc in the family Spondylidae
Sporobolus	cosmopolitan genus of plants in the family Poaceae
Sporobolus airoides	alkali sacaton
Sporobolus domingensis	coral dropseed
Sporobolus virginicus	seashore dropseed
Sporophila minuta	rusty-breasted seedeater
Spyridia filamentosa	red algae in the family Spyridiaceae
Stathmonotus sinucalifornici	California worm blenny
Staurotypus salvinii	giant musk turtle
Steganopus tricolor	Wilson's phalarope
Stelgidopterix serripennis	swallow species
Stemmadenia mollis	evergreen tree species
Stenella attenuate	dolphin
Stenella longirostris	spinner dolphin
Steno bredanensis	rough-toothed dolphin
Stenocereus	a genus of cactus
Stenocereus alamosensis	sina
Stenocereus gummosus	sour pitaya
Stenocereus standleyi	pitaya
Stenocereus thurberi	organ pipe cactus
Stenocionops ovata	velvet spidercrab
Stenogobius hawaiiensis	Naniha goby
Sterna	genus of the terns
Sterna anaethetus	bridled tern
Sterna caspia	Caspian tern
Sterna elegans	elegant tern
Sterna fuscata	sooty tern
Sterna hirundo	common tern
Sterna maxima	royal tern
Sternula antillarum	least tern
Sternula antillarum browni	California least tern
Stichopus chloronotus	greenfish
Strix occidentalis caurina	northern spotted owl
Strombus galeatus	eastern Pacific giant conch
Strongylocentrotus	genus of sea urchins in the family Strongylocentrotidae
Suaeda	sea-blite
Suaeda brevifolia	southern sea-blite
Suaeda californica	sea blite
Suaeda californica var. californica	wooly seablite
Suaeda esteroa	estuary seablite

(*Continued*)

Scientific Nomenclature	Common Descriptor(s) [Not Exhaustive]
Suaeda fruticosa	shrubby seablite
Suaeda nigra	romerito
Suaeda puertopenascoa	Puerto Penasco suaeda
Suaeda ramosissima	a species of seablite
Suaeda spp.	jauja or saldillos
Suaeda torreyana	Torrey sea-blite
Sueda fruitcosa	seep-weed
Sula dactylatra	masked booby
Sula dactylatra californica	masked booby
Sula leucogaster	brown boobies
Sula leucogaster brewsteri	brown booby
Sula nebouxii	blue-footed booby
Sula sula	red-footed booby
Sylvilagus audubonii	desert cottontail
Sylvilagus audubonii confinis	desert cottontail
Sylvilagus cunicularius	Mexican cottontail
Sylvilagus floridanus	eastern cottontail
Sylvilagus graysoni	Tres Marias cottontail
Syncaris pacifica	California freshwater shrimp
Syngnathus leptorhynchus	bay pipefish
Synthliboramphus craveri	Craveri's murrelet
Tadarida brasiliensis	Mexican free-tailed bat
Tabebuia chrysotricha	golden trumpet tree
Tabebuia donnell-smithii	primavera tree
Tabebuia palmeri	pink trumpet tree
Tabebuia rosea	pink poui
Tachybaptus dominicus	least grebe
Tachycineta albilinea	mangrove swallow
Tachycineta bicolor	tree swallow
Tamandua mexicana	northern tamandua
Tamarix	genus of salt cedar
Tamarix juniperina	tamarisk
Tamarix sp.	salt pine
Tantilla planiceps	western black-headed snake
Taxidea taxus	American badger
Tayassu tajacu	collared pecari
Tellina straminea	species in the widely distributed genus of marine bivalve molluscs in the family Tellinidae
Thais kiosquiformis	kiosque rock shell
Thalasseus elegans	elegant tern

(Continued)

Scientific Nomenclature	Common Descriptor(s) [Not Exhaustive]
Thalassia	genus commonly known as turtlegrass in the family Hydrocharitaceae
Thalassionema	genus of phytoplankton belonging to the family Thalassionemataceae
Thaleichthys pacificus	Eulachon
Thamnophis digueti	Diguet's two-striped garter snake
Thamnophis elegans	western terrestrial garter snake
Thamnophis gigas	giant garter snake
Thamnophis hammondii	two-striped garter snake
Thamnophis sirtalis tetrataenia	San Francisco garter snake
Thelypteris puberula var. sonorensis	Sonoran maiden fern
Thomomys bottae	Botta's pocket gopher
Thryophilus sinaloa	Sinaloa wren
Thula egretta	snowy egret
Thunnus alalunga	albacore
Thunnus albacares	yellowfin tuna
Thunnus thynnus	bluefin tuna
Tigrisoma mexicanum	bare-throated tiger heron
Tilapia mossambica	Mozambique tilapia
Tilapia spp.	invasive fish species
Tilapia zillii	redbelly tilapia
Tlalocohyla smithii	dwarf Mexican tree frog
Tomicodon boehlkei	Cortez clingfish
Totoaba macdonaldi	Totoaba
Toxopneustes roseus	roese flower urchin
Toxostoma cinereum cinereum	gray thrasher
Trachemys scripta	pond slider
Trachemys scripta grayi	Guatemalan slider
Trachemys scripta nebulosa	Baja California slider
Trachypenaeus similis	roughback shrimp
Triakis semifasciata	leopard shark
Tridacna maxima	giant clam
Trimorphodon biscutatus	western lyre snake
Trinectes fonsecensis	spotted-fin sole
Tringa flavipes	lesser yellowlegs
Tringa melanoleuca	greater yellowlegs
Tripsacum dactyloides	eastern gamagrass
Troglodytes aedon parkmanii	house wren
Troglodytes sissonii	Socorro wren
Tubastraea coccinea	orange cup coral
Turciops truncatus	common bottlenose dolphin
Typha angustifolia	narrowleaf cattail

(Continued)

Scientific Nomenclature	Common Descriptor(s) [Not Exhaustive]
Typha domingensis	southern cattail
Typha latifolia	broadleaf cattail
Typha spp.	cattail species
Tyrannus forficatus	scissor-tailed flycatcher
Tyto alba	barn owl
Uca	genus of the fiddler crabs
Ulva lactuca	sea lettuce
Uma notata	Colorado desert fringe-toed lizard
Urochloa mutica	para grass
Urocyon cinereoargenteus	gray fox
Urocyon cinereoargenteus californicus	California gray fox
Urocyon cinereoargenteus peninsularis	Baja California gray fox
Urosaurus auriculatus	Socorro Island tree lizard
Urosaurus bicarinatus	tropical tree lizard
Urosaurus clarionensis	lizard species
Urosaurus microscutatus	small-scaled lizard
Urosaurus nigricaudus	black-tailed tree lizard
Ursus arctos	Alaskan brown bear
Uta palmeri	San Pedro side-blotched lizard
Uta stansburiana	common side-blotched lizard
Uta thalassina	Baja blue rock lizard
Vaccinium vitis-idaea	mountain cranberry
Vaejovis mexicanus decipiens	scorpion
Vallesia glabra	pearlberry
Vermivora celata	orange-crowned warbler
Vireo atricapillus	black-capped vireo
Vireo bellii	Bell's vireo
Vireo bellii pusillus	least Bell's vireos
Vireo cassini cassini	Cassin's vireo
Vireo plumbeus plumbeus	plumbeous vireo
Vireo vicinior	gray vireo
Volvox	genus of chlorophyte green algae in the family Volvocaceae
Vulpes macrotis mutica	San Joaquin kit fox
Waltheria americana	sleepy morning
Washingtonia robusta	Mexican fan palm
Wedelia trilobata	wedelia
Wilsonia pusilla	Wilson's warbler
Wilsonia pusilla chryseola	Wilson's warbler
Xenomys nelsoni	Magdalena tree rat
Xenomys sp.	species of rodent in the family Cricetida
Xiphophorus helleri	swordtail

(Continued)

Scientific Nomenclature	Common Descriptor(s) [Not Exhaustive]
Xiphophorus maculatus	southern platyfish
Xylorhiza frutescens	a species of woody aster
Zalophus californianus californianus	California sea lion
Zalophus orca	sea lions
Zannichellia palustris	horned pondweed
Zea mays	corn
Zenaida asiatica	white-winged dove
Zenaida graysoni	Socorro dove
Zenaida macroura	mourning dove
Ziziphus sonorensis	nanche de la costa
Zonotrichia leucophrys gambelli	white-crowned sparrow (Gambel's)
Zonotrichia leucophrys oriantha	white-crowned sparrow
Zostera marina	eelgrass

Index

Note: *Italic* page numbers refer to figures.

Afro-Mestizo 255, 259, 266
agriculture 15, 20, 23, 28, 29, 51, 61, 75, 80, 83, 86, 91, 93,
    98, 108, 139, 143, 146, 154, 157, 158, 162, 171,
    172, 186, 189, 196, 199, 200, 202, 208, 210, 214,
    222, 224, 229, 233, 236, 240, 247, 255, 258,
    259, 266, 267, 270, 274, 278, 282, 285, 291,
    292, 297
agrochemicals 123, 161, 162, 189, 193, 196, 203, 206, 210,
    267, 285, 289, 292
'Āina 41
Alaska 2, 3, 6, *7–9*, 12, 28, 29
Aleuts 6
alien species 48, 248
amphidromous 44
Amuzgo 266
Areneros 120
Aztecs 289, 292

bird watchers 33, 34
birding 16, 48
BirdLife International 4, 296
Bolinas Lagoon 20, 23, *24*, *25*, 26, 49

Cahítas 143, 154
California 2, 3, 16, 19, 20, 21, 22, 23, 26, 28, 29, 30, 31, 32,
    33, 35, 37, 39, 40, 41, 48, 49, 54, 57, 59, 60, 61,
    64, 68, 97, 98, 109, 110, 115, 116, 126, 142, 147,
    172, 173, 175, 176, 260
California Department of Fish and Game 23, 29, 49
Canada 3, *7–9*, 12, 13, 15, 16, 23, 33, 44, 48, 111, 182,
    282, 303
Capacha 161
carbon 1, 2, 26, 36, 52
cattle 19, 20, 26, 41, 43, 71, 83, 94, 158, 177, 181, 196, 213,
    236, 247, 255, 259, 274, 282, 289, 292, 303
Chatino 266
Chichimecatlalli 123
chlorophytes 57
Chontal 274
climate 1, 2, 23, 33, 52, 54, 57, 60, 69, 87, 119, 145,
    175, 247
Coastal Highway 120
Cochimíes 64, 90, 97
Comcaác 111, 123, *124*, *125*, 126, 127, 131, 139, 172
CONABIO 54, 63, 78, 142, 157, 158, 220, 221, 296
conchales 138
Conference of the Contracting Parties 1, 3, 296
conservation 1, 2, 3, 5, 13, 16, 20, 28, 36, 44, 48, 54, 60,
    63, 68, 72, 75, 80, 86, 87, 90, 94, 98, 105, 109,
    115, 116, 132, 138, 139, 143, 146, 154, 158, 162,
    164, 167, 174, 181, 182, 189, 192, 196, 198, 199,
    203, 209, 210, 215, 218, 223, 229, 233, 236, 237,
    240, 244, 248, 251, 259, 267, 269, 275, 285, 293
contamination 65, 146, 206, 210, 233, 236, 259, 267, 274,
    285, 289

Convention on Wetlands 1, 2, 5
COP 1, 3, 4, 296
coral 2, 46, 47, 61, 90, 94, 96, 101, 103, 104, 123, 182, 185,
    191, 267, 270, 309, 314, 322, 326, 328, 329,
    334, 336
coral bleaching 2
Cucapá 53

deforestation 65, 168, 189, 193, 196, 199, 200, 203, 214,
    215, 236, 254, 278, 282, 285
duck hunters 33
ducks 6, 12, 23, 53, 69, 115, 143, 145, 146, 168, 181, 199,
    205, 214, 221, 236, 254, 289, 300, 311
dumps 289, 293

ecological balance 138
ecological goods and services 218, 221
ecological services 1, 26, 54, 266
economic importance 57, 97, 130, 165, 181
economy 33, 153, 214, 222, 255, 266
ecosystem functions 54, 105, 115
ecosystem processes 114
ecosystem services 33, 36, 52, 68, 111, 116, 138, 186
ecosystems 1, 2, 3, 5, 12, 13, 37, 62, 69, 72, 86, 90, 91, 94,
    96, 105, 108, 116, 120, 130, 147, 153, 161, 162,
    167, 171, 175, 244, 278, 281, 292, 293
eco-tourism 33, 51, 54, 72, 75, 108, 116, 120, 131, 158, 167,
    200, 206, 210, 218, 248, 275
education 16, 36, 40, 48, 51, 54, 72, 75, 80, 90, 116, 120,
    139, 146, 162, 164, 167, 182, 206, 215, 229, 233,
    236, 244, 248, 266, 271, 278, 282, 293
educational 33, 97, 98, 131, 135, 139, 164, 218, 233
ejido 210
environmental education 54, 293
estuary 13, 15, 20, 26, 29, 33, 34, 35, 36, 37, 40, 52, 54, 56,
    57, 78, 105, 107, 108, 109, 120, 123, 130, 131,
    137, 138, 139, 142, 149, 150, 153, 168, 193, 196,
    196, 199, 200, 202, 203, 206, 210, 212, 215,
    218, 236, 248, 254, 258, 334
exotics 57

First Nations 15
fish larvae 57
fishers 68, 75, 97, 101, 127, 135, 153, 154, 158, 168, 172,
    181, 185, 186, 192, 196, 218, 221, 236, 244, 247,
    266, 270, 285, 289
fishing 34, 48, 51, 54, 57, 60, 61, 65, 68, 69, 75, 87, 90, 98,
    101, 103, 105, 116, 119, 120, 123, 126, 127, 130,
    131, 135, 139, 142, 143, 146, 150, 152, 153, 154,
    156, 157, 158, 161, 162, 165, 168, 171, 172, 177,
    182, 185, 186, 189, 192, 193, 196, 198, 199, 200,
    202, 203, 208, 210, 214, 221, 222, 224, 229,
    232, 233, 236, 240, 244, 247, 248, 251, 255,
    258, 266, 270, 274, 278, 282, 285, 289, 291,
    293, 323

forestry 83, 86, 108, 186, 189, 236, 248
Fraser River Delta *14*, 15, *15*, 16
fresh water 52, 56, 57, 68, 77, 86, 87, 91, 118, 127, 158,
        220, 230, 259, 281

garbage 65, 186, 189, 218, 259, 274, 289, 293
geese 6, 12, 16, 23, 30, 143, 145, 146, 307
goats 71, 78, 90, 158, 232, 236, 259, 274
Guaímas 139
guano 46, 67, 68, 131, 134, 135
Guaycuras 90, 97, 101

habitat 1, 2, 36, 254
habitat destruction 181, 254
Halimeda sands 47
Hawaiian xi, 41, 43, 44, 47, 48, 49, 303, 312, 314, 317,
        321, 322
highway 19, 120, 138, 165, 214, 218, 222, 275
Hohokam 123, 174, 175
Holy Week 147, 150, 154, 202, 203, 210, 282
hunting 51, 54, 60, 65, 75, 91, 135, 139, 143, 146, 147, 158,
        162, 172, 210, 213, 258, 259, 267, 274, 282,
        285, 293
hydrologic 1, 19, 116, 130, 139, 162, 189

Important Bird Areas 16
indigenous people 57, 60, 68, 90, 97, 101, 131, 135, 172,
        189, 224, 232, 266
indigenous populations 138, 255
industrial development 19
International Union for the Conservation of Nature 4
International Water Management Institute 4
introduced species 37, 44, 108, 185, 233
invasive species 19, 44, 48, 57, 108, 146, 161, 229
IOP 4
isolated wetlands 114
IUCN 4, 6, 12, 19, 20, 40, 44, 47, 48, 116, 134, 137, 147,
        158, 193, 209, 221, 271, 275, 297
IUCN Red List 6, 12, 19, 20, 44, 47, 48
Izembek 6, *10*, *11*, 12
Izembek Lagoon National Wildlife Refuge 6, *10*, *11*

Jesuit 65, 72, 90, 94, 97, 108

Kingman Reef 46, 47
Kiribati 44, 47
Kumiai 54, 57

Laguna de Santa Rosa Wetland Complex *17*, *18*, 19
land conversion 19
landscape-level 114
laundry detergents 259, 274
livestock 54, 83, 91, 93, 94, 98, 139, 143, 146, 162, 186,
        189, 190, 200, 202, 206, 208, 214, 221, 222,
        233, 236, 255, 258, 267, 274, 278, 281, 282, 291
logging 26, 236, 278, 282, 289, 293
longshore flow 56
Low Pimas 138

Migratory Corridor of the Pacific 168
missionaries 61, 65, 90, 97
missions 87, 90
Miwok 23
Mixtecs 266, 270

monitoring 16, 36, 48, 68, 101, 105, 175, 222, 244, 248,
        267, 282, 293
mother of pearl 90, 97
motorized vehicles 90
Muwekma Ohlone 29

Nahuas 244, 266
Nahuatl 214, 244, 259
National Commission for the Knowledge and Use of
        Biodiversity 54, 63, 98, 157, 281
National Oceanographic and Atmospheric Administration
        23, 297
National Park Service 23, 297
National Wildlife Refuge 6, 39, *45*, *46*, 297
negative estuary 130, 131
NOAA 23, 297
NPS 23, 297

Oasis 78, 80, 91, 105, 108, 110, 295
Ohlone 29
Olmec 292
Ootam 123
outreach 54, 72, 80, 86, 101, 162, 164, 192, 215, 229, 233,
        236, 274, 278, 282, 293
overexploitation 68, 90, 97, 146, 181, 214, 258, 267, 285

Pacific Flyway 15, 20, 23, 37, 52, 111, 116, 120, 138
Pacific Migration Corridor 57, 158, 158
paleoecology 139
pasture 20, 206, 247
Pericúes 90, 97, 101, 105
pets 259, 274
phaeophytes 57
phanerogams 57
photography 40, 48
pigs 177, 229, 232, 236, 243, 247, 289, 292
Pinacateños 120
poaching 94, 161, 189, 236, 240, 275, 278, 281, 282, 289
pollution 16, 23, 33, 103, 105, 150, 189, 192, 203, 236, 278
Pomo 20

quarantine 48

Ramsar Convention 1, 2, 3, 4, 5
ranching 51, 91, 94, 105, 143, 177, 196, 208, 282, 292
recreation 13, 16, 28, 36, 40, 98, 167, 203
recreational activities 23, 90, 120, 123, 196, 215, 282
recreational fishing 23, 40
Red List 12, 19, 20, 47, 126, 135, 147, 193, 199, 200, 205,
        209, 215, 221
reefs 2, 3, 44, 46, 47, 61, 62, 68, 90, 94, 96, 101, 103, 104,
        105, 123, 126, 182, 185, 191, 267, 269, 270
research xi, 16, 28, 33, 34, 37, 40, 44, 48, 51, 68, 80, 90,
        94, 101, 109, 116, 130, 135, 139, 150, 154, 158,
        162, 172, 182, 185, 196, 199, 203, 206, 210, 218,
        222, 229, 233, 240, 244, 251, 259, 266, 271,
        274, 278, 282, 293
Rhodophytes 57
riparian 16, 19, 26, 30, 32, 33, 54, 86, 91, 94, 105, 107, 108,
        115, 171, 212, 270, 302
Rose Atoll 47

sailing 23, 48, 101, 158, 189, 192

salt 20, 29, 30, 33, 35, 36, 37, 48, 57, 61, 71, 75, 89, 90, 105, 116, 120, 123, 130, 143, 157, 161, 174, 177, 182, 189, 203, 206, 222, 224, 229, 309, 311, 316, 330, 333, 335
salt marsh 36, 130
salt water 36, 229
San Francisco Bay/Estuary 27, 28, 297
Seris 126, 131, 138, 172, 173, 174, 175
sewage 146, 193, 248, 255
SFBE 27, 28, 28, 29, 30, 297
sheep 23, 65, 71, 83, 86, 87, 137, 158, 229, 233, 236, 289, 325
shell dump 57
songbirds 15, 33, 60
sport fishing 57, 65, 135, 222, 270, 291
STRP 3, 4, 297

The Nature Conservancy 5, 44, 46, 297
Tohono O'odham 120
Tomales Bay 10, 20, 21, 21, 22, 22, 23
Totorames 171
tourism 28, 33, 54, 65, 75, 80, 86, 94, 98, 101, 105, 108, 109, 120, 123, 125, 127, 130, 135, 139, 143, 146, 154, 158, 164, 168, 181, 182, 186, 192, 193, 199, 203, 206, 214, 222, 233, 236, 244, 247, 248, 251, 255, 267, 270, 271, 274, 275, 278
tourist 57, 65, 72, 75, 90, 108, 120, 130, 150, 158, 162, 172, 181, 182, 192, 196, 209, 210, 215, 218, 222, 233, 236, 240, 244, 247, 255, 266, 270, 278, 282, 291, 293
traffic 20, 35, 103
trafficking 285, 289, 293
Trincheras 123
Tututepec 262, 267, 270

Unangan People 6
underemployment 214
unemployment 214
United Nations Educational Scientific and Cultural Organization 1, 5, 298
United Nations Environmental Program 5, 298
urban 2, 15, 16, 20, 29, 30, 37, 51, 54, 83, 139, 143, 150, 189, 192, 215, 218, 236, 240, 244, 271
urbanization 19, 29, 80, 162, 222, 244

volcanoes 41, 46, 61, 233, 292

wastewater 19, 20, 30, 51, 105, 146, 147, 161, 168, 172, 189, 293
waterfowl 1, 2, 6, 12, 13, 15, 16, 26, 33, 40, 44, 53, 60, 83, 105, 111, 115, 130, 131, 138, 143, 145, 146, 147, 154, 157, 181, 193, 199, 200, 203, 206, 208, 214, 218, 220, 247, 275
wetland degradation 1
Wetland Ecology xi, 5
Wetlands International 4
Wetlands of International Importance 1, 2, 3
Wildfowl and Wetlands Trust 4, 298
World Wetlands Day 5
World Wildlife Fund 4, 298

yachts 48, 57, 68, 90, 185, 221, 222
Yaqui Yoreme Nation 143
Yaquis 131, 138, 142, 143
Yokut 33
Yumans 57

Zapotec 266, 274